Organising the Farmers

Organising the Farmers

Cocoa Politics and National Development in Ghana

Björn Beckman

The Scandinavian Institute of African Studies
Uppsala 1976

ISBN 91-7106-101-0
Printed in Sweden by
Uppsala Offset Center AB, Uppsala 1976

Contents

Preface

This book is based largely on the records of the United Ghana Farmers' Council. It was commenced as part of a broad study of economic and political strategy in Ghana during the CPP (Nkrumah) regime. While looking for material on the cocoa sector—a vital one in Ghana's political economy—I was granted unrestricted access to the records of this organisation, which was the farmers' wing of the ruling party. I decided to grasp this unique opportunity to penetrate more deeply into the economic and political organisation of the CPP state. The material was far too rich to be compressed into one or two chapters of a general study and I decided to use it for a separate and more specialised one. Nevertheless, the key role of cocoa and the strategic importance of rural political organisation seem to justify this close examination also from a broader perspective.

I studied in Ghana from 1967 to 1971. The last two years were devoted to the Farmers' Council study. I was supported by a travelgrant from the Scandinavian Institute of African Studies in Uppsala, a scholarship from the University of Stockholm, and, at a later stage, a grant from the Swedish Council for Social Science Research. The University of Ghana at Legon generously provided accommodation and research facilities, including association with the Institute of African Studies. Other institutes and departments (Economics, Agricultural Economics, Institute of Statistical, Social and Economic Research) provided valuable opportunities to discuss my research proposals and preliminary findings.

The Farmers' Council records had been taken over in 1966 by the Purchasing Department (later the Produce Buying Agency) of the Cocoa Marketing Board. The staff of this organisation—also partly taken over from the Council—assisted me greatly during the year-long investigations. Other Ghanaians, officials, cocoa farmers and traders, in Accra or in the villages, helped to document and explain the facts about the economics and politics of cocoa. Cadbury officials in Accra and Birmingham contributed their comments.

I am grateful to all friends and collegues in Ghana, England and Sweden who have read the whole or parts of my drafts and manuscripts. Nobody has done so more carefully and critically than Jitendra Mohan, formerly with the University of Ghana. Dennis Austin, Nils Elvander, Sverker Gustavsson, Leif Lewin, Colin Leys, Richard Rathbone, Roger Southall and Daniel Tarschys are others who have offered valuable comments and encouragement. Most of the writing-up has been done since I became attached in 1972 to the Department of Political Science at the University of Uppsala. Here, largely due to Lars Rudebeck, favourable conditions have been created for the comparative study of third world strategies for national liberation and development. A viable and expanding research group has been established. I am particularly grateful to Lars Rudebeck and the members of that group for their close and friendly discussions of my work. Financial support for my attachment to the group has been provided by the Swedish Council for Social Science Research. Access to the excellent facilities of the Scandinavian Institute of African Studies is another important factor in creating these favourable conditions in Uppsala.

A few points of practical information should be added:

Notes are used in this book for documentary purpose. I have avoided the inclusion of additional information or discussion. The notes have therefore been placed at the end of the book in order not to distract the reader. A comment on the system of references used will be found at the beginning of the notes section.

A list of *abbreviations* and a *glossary of terms* covering institutions, technical terms, etc. will be found at the very end of the book.

Appendix B contains a written statement by Martin Appiah Danquah, the former General Secretary of the UGFC. It gives the official Council version of much of what is treated in this book from a rather different perspective.

Currency: After independence the Ghana currency was the Ghana Pound (£G) with an official exchange rate equal to Pound Sterling. In 1965 a decimal currency was introduced, the CEDI, equal to 8 shillings 4 pence. Throughout the text £ will be used to designate the colonial as well as the post-colonial Pound.

Täby and Uppsala, April 1976

Björn Beckman

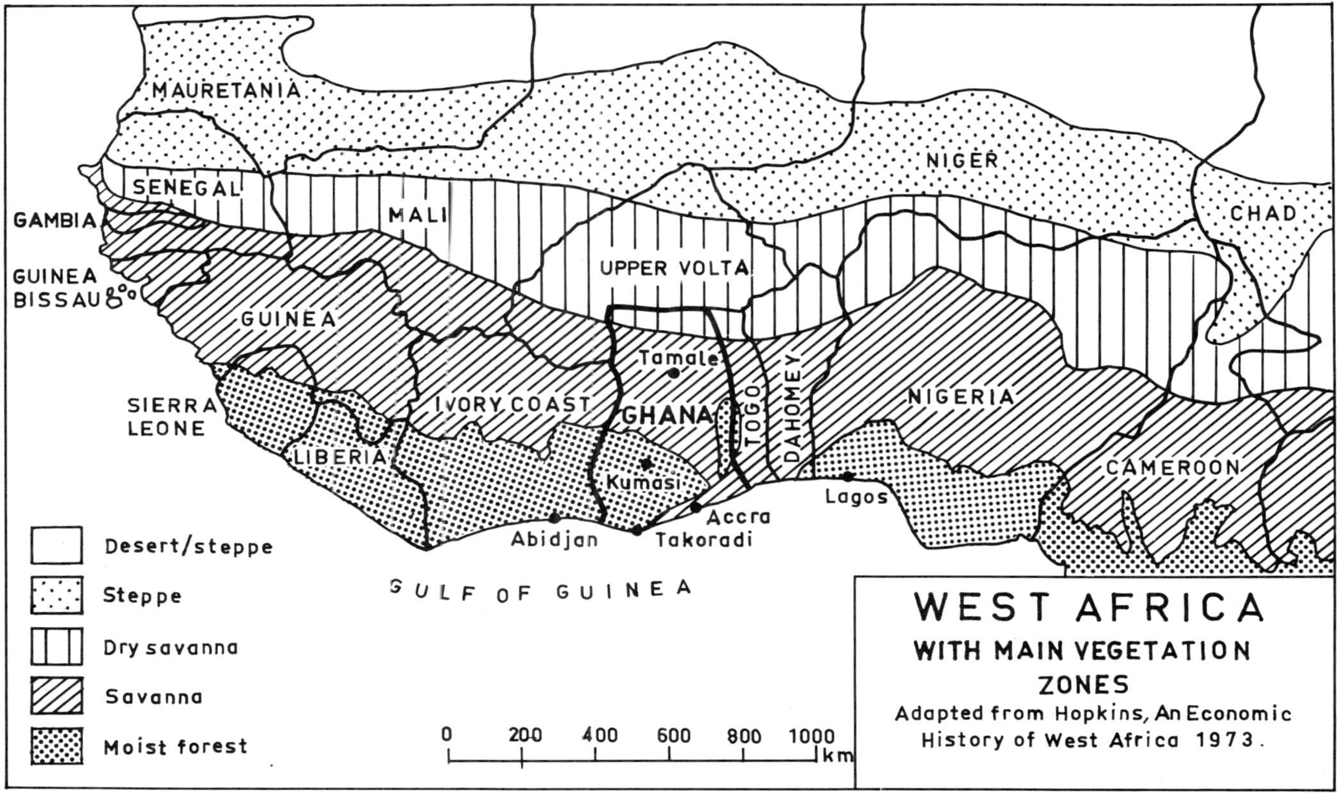

WEST AFRICA
WITH MAIN VEGETATION ZONES
Adapted from Hopkins, An Economic History of West Africa 1973.

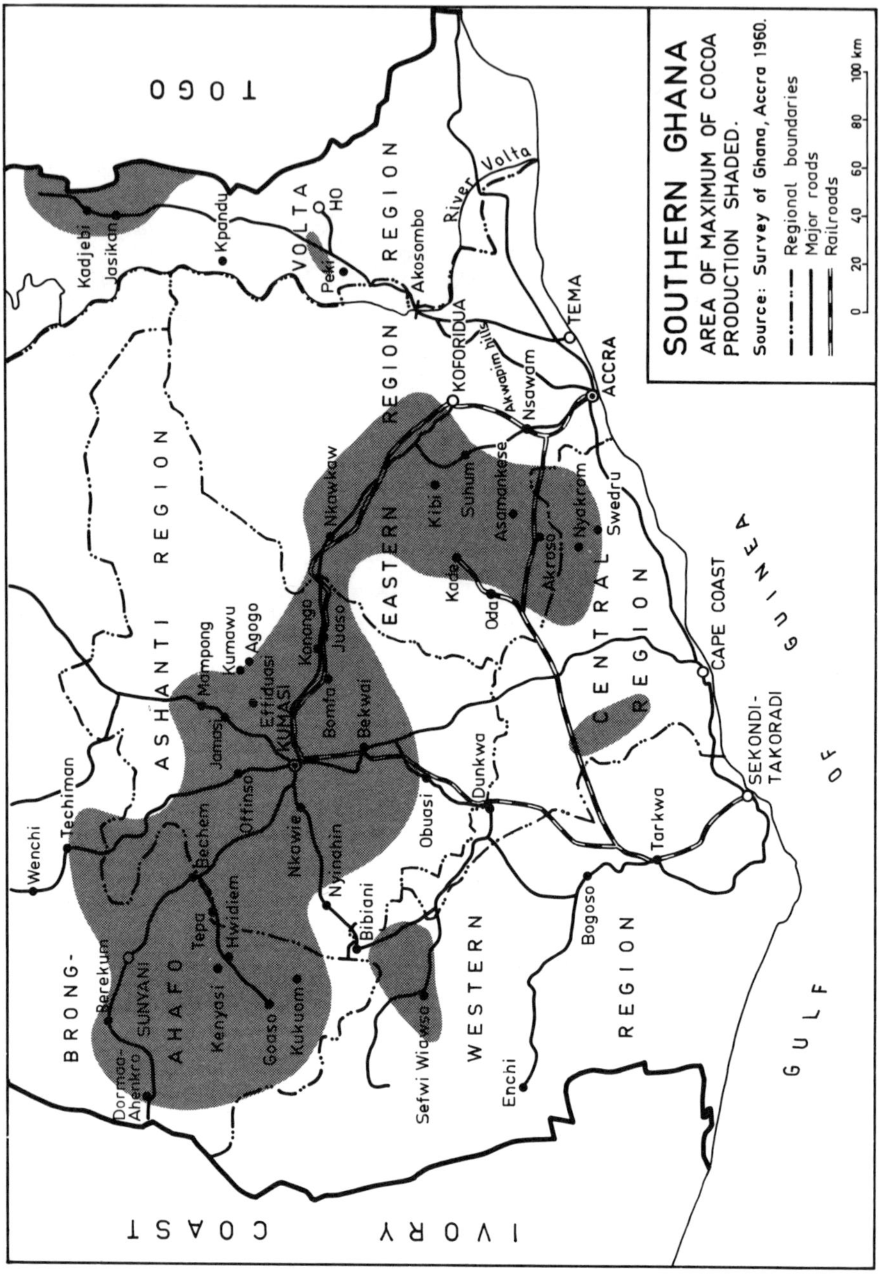
SOUTHERN GHANA
AREA OF MAXIMUM OF COCOA PRODUCTION SHADED.
Source: Survey of Ghana, Accra 1960.
Regional boundaries
Major roads
Railroads
0 20 40 60 80 100 km
TOGO
IVORY COAST
GULF OF GUINEA
VOLTA REGION
EASTERN REGION
ASHANTI REGION
BRONG-AHAFO
CENTRAL REGION
WESTERN REGION
River Volta
Akwapim hills
Kadjebi
Jasikan
Kpandu
HO
Peki
Akosombo
TEMA
ACCRA
KOFORIDUA
Nsawam
Nkawkaw
Kibi
Suhum
Asamankese
Akroso
Nyakrom
Swedru
Kade
Oda
CAPE COAST
SEKONDI-TAKORADI
Tarkwa
Bogoso
Enchi
Sefwi Wiawso
Bibiani
Nyinahin
Obuasi
Dunkwa
Bekwai
Bomfa
Juaso
Konongo
KUMASI
Effiduasi
Agogo
Kumawu
Mampong
Jamasi
Offinso
Nkawie
Bechem
Techiman
Wenchi
Tepa
Hwidiem
Kenyasi
Goaso
Kukuom
SUNYANI
Berekum
Dormaa-Ahenkro

I. Ghana and the CPP: an Introduction

The United Ghana Farmers' Council was the farmers' wing of the Convention People's Party, which, led by Kwame Nkrumah, gained independence for Ghana in 1957. The Farmers' Council was dissolved and banned in 1966, together with other CPP organisations, after a military coup had overthrown the Nkrumah government. This book is primarily about the Farmers' Council and its functions in the political and economic system established by the CPP. It was founded in 1953 as a general farmers' organisation with political objectives, as part of the CPP effort to organise the cocoa farmers, the economically most important section of the Ghanaian peasantry. Ghana's economy was and continues to be heavily dependent on cocoa, which contributes some 60 per cent of the export earnings and provides the backbone of the public economy. The cocoa trade was used by the CPP as a platform for its political penetration of the cocoa areas. The Council was simultaneously an instrument for organising the agricultural population in general in support of the party and the government. After independence, it was recognised by the CPP government as the only organisation entitled to represent the country's farmers. The proclamation of this political monopoly was followed by a campaign to establish control over the local cocoa trade. As a result the foreign firms which bought cocoa locally were expelled from the market. In 1961, the Farmers' Council was made the sole buyer of Ghana's cocoa after a brief struggle with rival interests. A combined political and commercial monopoly was thus established. For five seasons, from 1961 to 1966, the Farmers' Council handled the entire cocoa crop (at this time averaging well over 400,000 tons—36 per cent of world production), while organising the peasantry for political purposes.

This book is concerned with exploring and documenting this combined political and commercial monopoly. It is intended as a contribution to the political history of the CPP period and the analysis of decolonisation and national development in Africa. The CPP state, as well as its successors, was highly dependent on the surplus extracted from the cocoa economy. This surplus formed the basis for a rapidly expanding public sector and an ambitious national development programme. I examine the role of the Council in securing a *political* basis for the cocoa appropriations of the state. How were the producers organised for this purpose? What methods of political mobilisation and control were applied? I argue that the way in which the CPP organised the cocoa farmers highlights the party's approach to rural political organisation.

The struggle over the local cocoa trade was primarily a struggle over the distribution of cocoa income between the producers and the state. But it was simultaneously a struggle between various social groups claiming a share in the commercial inheritance left by the foreign firms, which hitherto had dominated the trade. The monopoly of the Farmers' Council meant the elimination of the mass of local "middlemen", often farmers, who had been doing much of the actual buying for the firms. Farmer cooperatives, which had prospered under colonial protection, were forcibly "merged" with the Council and in practice eliminated. This book discusses the social forces engaged in this struggle and the significance of the Farmers' Council victory in relation to the development and organisation of these forces.

In my attempt to understand the class character of the Council's commercial and political monopoly I am particularly concerned with exploring the interaction between the "national" aspirations of the party and the government on one hand, and the "sectional" aspirations of their cadres on the other. The Farmers' Council monopoly is studied here both as a vehicle for mobilising resources for national development and as an instrument of class appropriation, serving the self-advancement of a new commercial-bureaucratic class. I discuss the way in which such national and sectional interests reinforced as well as contradicted each other and how the former were compromised by the latter. The principal theoretical interest of the study lies probably in the documentation of this interaction. I believe that the exploration of the nature of this relationship is important for an understanding of the class character of the CPP state. I hope that it will throw some light on the political organisation of the economy in post-colonial societies in general.

The study is organised as follows. This present chapter continues with a brief outline of Ghana's experience of decolonisation and national development in the Nkrumah period. Most of the chapter is a presentation of the way in which this experience has been discussed and analysed by others. The wide international reputation of Kwame Nkrumah's government, its avant-garde role in the decolonisation of Africa and in the Pan-African movement, its massive development programme, are all features which have attracted wide attention and ensured an unusually rich literature of analysis and discussion. I shall concentrate on those aspects of these writings which most directly concern the class character of the regime and its strategy for national development. The arguments may perhaps at times seem somewhat abstract, but I hope that the very concrete account which constitutes the major part of this book will help to give flesh and blood to these general propositions. Some of the more important and problematic terms used are discussed in a concluding note.

Chapter 2 introduces the Ghanaian cocoa economy. It indicates the role of cocoa in the economy and examines the way in which production is organised and how it has developed over time. The Cocoa Marketing Board, the state export monopoly, and its background in the colonial economy are presented. The organisation of the local cocoa trade—the most immediate arena for the events discussed in this study—is outlined. In the following two chapters (3 and 4) I discuss the early history of the Farmers' Council and the struggle which preceded its take-over of the cocoa trade in 1961. Chapter 3 traces the Council's roots in the tradition of farmer-trader organisations. I look at the early relations between the Council, the CPP and the CPP-supported Cocoa Purchasing Company. The development of the Council in the pre-independence period was greatly affected by the political struggle which followed on the government's decision in 1954 to freeze producer prices and raise the level of taxation.

Chapter 4 (The Struggle for Monopoly) gives an account of the campaign against the foreign firms, the local middlemen, and the "old" cooperatives. I discuss the significance of the foreign withdrawal from this sector of the economy and the position of the local capitalist class. Opposition was most vocal against the liquidation of the old cooperatives. I discuss the nature of the parliamentary and government support for the latter and why the monopoly solution offered by the Council finally triumphed.

The trading monopoly of the Farmers' Council was a drastic intervention in the affairs of the cocoa economy. It brought the new party-state right on to the door-steps of the farmer-producers. The quality of its economic and political organisation was subjected to severe testing. The internal records of the Council and the mass of petitions submitted by the farmers before as well as after the military coup offer a rich source of information on this day-to-day confrontation at the village level. The examination of the performance of the Council makes it possible to discuss more closely what social forces manifested themselves in the monopoly. What did the methods used and the difficulties encountered reveal in this respect? In whose interests was the monopoly administered? These are the main issues raised in Chapter 5, where the impact of the monopoly on the cocoa trade is discussed. I examine in particular the emergence of a new class of middlemen, and their official and unofficial methods of appropriating a growing share of the cocoa surplus.

The political organisation of the Council, as examined in Chapter 6, reveals clearly the contradictions between its two main roles: on one hand, its role as a farmers' organisation, officially governed by the elected representatives of the farmers, charged with the administration of the cocoa trade; on the other, its function as an instrument of fiscal and political control over the farmers on behalf of the state. Not only farmers, but also the government had difficulties in making the organisation serve its interests. The government made several attempts to reform the Council for its own purposes but with little effect. These failures underline the unsolved nature of the contradictions and the room it left for the new middlemen to operate not only between the producers and the market but also between the producers and the state.

In Chapter 7 (Cocoa and the Public Economy) the focus is shifted from the coccoa trade to the cocoa appropriations of the state. The chapter examines the producer price and tax policies of the CPP government in the context of changing market conditions. I discuss the importance of cocoa income for the development of the CPP state and the role of the Farmers' Council in securing the political basis for these appropriations. I examine tentatively the consequences of the redistribution of cocoa income for national development in general.

The last chapter summarises the main argument and suggests certain general conclusions about the political and economic organisation of the CPP state, which can be drawn from the Farmers' Council experience.

Decolonisation: the Ghanian experience

Following the outbreak of popular violence in 1948, Nkrumah and the CPP assumed the leadership of the nationalist movement in a brief and successful campaign for "self-government now". The party was formed in 1949 as a breakaway from a more moderate nationalist grouping. Independence, until then only a distant and rather obscure prospect, suddenly came within reach. The party won a sweeping victory in the elections of 1951 (the first of any substance in Ghana) and Nkrumah was brought from prison to lead a transitional government under colonial control. A period of growing political rivalry between national and regional groupings followed. The CPP was twice confirmed in power in general elections, although with a much reduced majority. Independence was proclaimed in March 1957.

The decade which followed—until the Nkrumah government was overthrown in a military coup in February 1966—is one of the most closely studied and intensely discussed periods in contemporary African history.[1] We may distinguish at least three major themes. A first one concerns the outstanding role played by Nkrumah as a Pan-Africanist leader and writer, seeking continental unity as a necessary means for the liberation and emancipation of all African people.[2] At midnight on Independence Eve he proclaimed that "our independence is meaningless unless it is linked up with the total liberation of the African continent".[3] The following year, in 1958, the All-African People's Conference was held in Accra and gave a powerful boost to the independence movement elsewhere in Africa. Ghana was engaged in the Congo crisis in support of Lumumba and broke with Britain over Rhodesia. The Organisation of African Unity, although very different from Nkrumah's vision, was designed in response to the pressures of the Pan-African movement where Nkrumah was a leading force.[4] Growing awareness of the limitations imposed on national development and economic independence by existing territorial divisions makes Nkrumah's vision not less relevant to-day.[5]

A second theme in the discussion of the Nkrumah-CPP experience lies closer to the present study. It concerns the process of political decolonisation, the dismantling of the parliamentary type of government introduced in the transition period, and the rise of the one-party state. A number of important issues have been raised in that controversy, including problems of competitive politics, political liberties, violence and repression. Closely related are issues of national unity and the articulation of regional and ethnical interests. Confrontation between government and opposition took increasingly non-parliamentary forms. Opposition groups refused to cooperate with political leaders which they did not regard as legitimate and whose intentions they suspected. The government in its turn sought to enforce its hegemony by authoritarian methods. The Preventive Detention Act of 1958 was an important step in this direction. The once substantial parliamentary opposition disintegrated. A presidential constitution introduced in 1960 strengthened Nkrumah's personal power. A one-party referendum confirmed the one-party state in 1964, after a series of real and suspected plots and assassination attempts.[6]

The political and commercial monopoly of the United Ghana Farmers' Council was part of the consolidation of the one-party system. A study of the political processes which led to the establishment of this monopoly and the internal political organisation of the Council is clearly also a contribution to the understanding of this system. The emphasis given the present study differs, however, from that in studies which have focused on the evolution of the "political system" as such. Problems of politicial organisation are approached here in the context of national economic development. It links up with the third main theme of the "Ghana debate", which concerns the achievements and failures of the CPP government in the attempt to build a national economy. What methods were used and what difficulties encountered? What is the wider relevance of the CPP experience? Let us first recall some of the principal features of this performance.[7]

Few African leaders have emphasised as strongly as Nkrumah the limitations of political independence and the necessity to struggle for economic independence if one kind of colonialism is not merely to be substituted for another.[8] What did it in practice mean in the Ghanaian case? The large resources available to the government, the high ambitions of its leaders, and the popular expectations stirred by the nationalist movement gave an unusual force and breadth to the national development effort. The CPP was brought into government at a time of booming cocoa prices. It was decided to implement the colonial ten-year plan of 1951 in half the time. The revised colonial plan was followed by more ambitious "national" ones. The economic and social infrastructure was expanded at a remarkable pace. A massive educational programme and the giant Volta hydro-electric scheme were major manifestations. An extensive network of new or improved roads and the new harbour and industrial town of Tema were others. Economic independence was increasingly perceived as depending on the growth of the manufacturing industry. The second five-year plan (1959) announced that "specially high priority will be given within the next five years to promoting the establishment of not less than 600 factories of varying size producing a range of over 100 different products".[9] Attempts to encourage foreign private investment in this field had only limited success. Emphasis was placed instead on the leading role of the public sector in the development of the advanced sectors of the economy. The state itself became a major industrial entrepreneur. The post-coup regime inherited a variety of completed and uncompleted projects, including plants for cocoa processing, steel-works, aluminium products, electronics, glass, meat products, textiles, sugar, etc.

Stagnating food production and growing demand from the rapidly expanding urban economy brought large scale state intervention in agriculture, as well, with state farms and mechanisation on a cooperative basis (partly managed by the United Ghana Farmers' Council). Important national state-owned enterprises were established in commerce and services, including a national shipping company, airways, banking, insurance, construction and trade in general, in order to break the dominance of foreign capital in these sectors. Public export monopolies, previously mainly in cocoa, were extended to other major export products, such as timber and diamonds. Five out of seven goldmining companies were bought by the state when threatened with a close down by their foreign owners.

Redirection of foreign trade and other external sources of finance and assistance was an important element in the strategy to free Ghana from the restrictions imposed by the colonial economy. A number of agreements on trade and economic cooperation were signed with the socialist countries of Eastern Europe and China. The British share of Ghana's foreign trade declined sharply. By 1965, it was surpassed by that of the socialist countries both on the export and import side.

1961 is usually regarded as a major turning point. Up to this point, the government had pursued fairly "liberal" economic policies. Public investments were primarily in the service sector. Foreign and national private enterprise was expected to do the lion's share of actual production. The Ghana Pound was freely exchangeable for Sterling and there were few restrictions on imports. A high rate of growth was achieved, reflecting primarily the export boom of the first post-war decade and the impact of the accumulated savings which had thus been placed at the disposal of the independent government.

By 1960–61, however, the conditions for this "liberal" high-growth strategy had fundamentally changed. Export earnings had stagnated and terms of trade were turning strongly against Ghana (as well as most other exporters of "soft" primary products). Imports, however, continued to rise at a high rate to feed the growing advanced sectors of the economy with capital and consumer goods. A series of foreign-payment deficits rapidly ate into the accumulated reserves of the boom years. In 1961 alone these reserves were halved. At this point the government decided to "close" the economy. Foreign-exchange controls were introduced on a comprehensive basis. Imports were subjected to licensing. An "austerity" budget sought to raise declining public revenue at the same time as it sought to contain the inflationary pressure of the new situation.

At this point public intervention in production and trade accelerated. The 1959 plan was discontinued and work began on a more advanced, "comprehensive" exercise, the *Seven-year Development Plan,* launched in

1964. At the political level it had been preceded by a new party programme in 1962, *For Work and Happiness,* which confirmed the leading role of the state in the economy. The development programme was greatly expanded. It pushed ahead, in spite of the rapidly deteriorating financial position of the country and of the government.

The turning-point of 1961 is the immediate historical context of the nationalisation of the local cocoa trade under the United Ghana Farmers' Council. It was a step in the accelerated effort to establish national control in the foreign dominated sectors. But more importantly, it was an attempt to reinforce government political control over the cocoa economy and the cocoa producers at a time when the financial position was deteriorating and "sacrifices" were demanded from the Ghanaian people to sustain the intensified national development effort.

The cost of the overall effort was high. The resources of the CPP state in terms of finance, administrative capacity, and political support were greatly overextended. The programme was financed, as summarised by one student, by large and growing budget deficits and heavy foreign borrowing; a growing share of output was mobilised by the state through taxation and by compression of private consumption through deficit financing; future export earnings were heavily mortgaged.[10] Public expenditure on social and welfare services and general administration (defence, police, and diplomacy, in particular) continued to rise, absorbing all current revenue and much of the savings of the government. The bulk of the economic and industrial programme was therefore financed through domestic and foreign borrowing. Much of the foreign credits were short- or medium-term (suppliers' credits). Few if any of the projects were able to pay for their own capital costs on such terms. The pressure on regular revenue grew sharply as these loans fell due for payment. Interest payments alone placed a heavy burden on the public economy.

Many projects were poorly conceived and met with great difficulties in implementation. Most state enterprises continued to require subsidies instead of contributing revenue. While the rate of capital investment in the economy was kept at a very high level, the production generated was low. Public investment not only failed to break the stagnation caused by adverse export conjunctures; it had an aggravating impact on growth. The administration of import licences and foreign exchange met with major difficulties and contributed to stagnation. Shortages and irregular supplies added to the rising inflation, caused mainly by the methods used by the government to finance its domestic borrowing. Falling real wages and producer prices in combination with such shortages caused widespread popular discontent. Corruption and private profiteering among politicians and public officers added to the growing disenchantment.

It was the foreign-payments situation, however, which brought about the acute crisis in the Ghanaian economy by 1965. It was aggravated by a further sharp decline in cocoa prices which the international Cocoa Producers' Alliance failed to contain by their decision to withhold cocoa from the market in late 1964. The hold-up itself added to the payment crisis. Appeals for assistance from the International Monetary Fund and the World Bank met with demands for major changes in Ghana's economic policies, including anti-inflationary measures, deflation of local demand, a halt to new projects financed by suppliers' credits, a halt to subsidies to state enterprises, more liberal policies towards foreign private investment, and revision of the bilateral agreements with the socialist countries, "with a view of reducing their harmful impact on the Ghana economy".[11]

In its main outline this was to become the economic programme of the military regime which took over the following year, although the latter added more drastic steps to reestablish the "health" and international respectability of the Ghanaian economy. Personnel from the socialist countries were expelled, a number of industrial and agricultural projects were abandoned, and state enterprises were offered for sale. The attempt to liberalise the economy continued under the elected government which succeeded the military interim regime in 1969. Most importantly, an attempt was made to dismantle the system of import and foreign-exchange controls. After some good cocoa years, however, this policy was again pursued in the face of falling export earnings. The liberal (IMF) recipe was a drastic devaluation in order to prevent another collapse in the foreign-payments situation. A new military coup (1972) followed shortly after this devaluation and its accompanying inflationary upsurge.

The failure of the economic policies of the Nkrumah regime may be explained in terms of the lack of clear priorities, an attempt to do too much at one time. There was also an apparent lack of realism and foresight in the assessment of the resources available and the difficulties involved in this massive development effort. Inexperience, incompetence, corruption, and private profiteering may all have contributed significantly to the distortion of policy and implementation. Opinions differ about the relative importance of these various factors and about the soundness of particular projects and solutions. Opinions also differ about the scope for corrective measures. All are issues which require close study in order that similar mistakes may be avoided elsewhere and at other times. These are not, however, the issues to be discussed here. There are other issues which concern the nature of the strategy itself and its relevance for the overall objectives of economic decolonisation and national development. In what relation did this strategy stand to the principal social forces of the Ghanaian political economy? In what way do the failures and achievements of the CPP reflect the balance of social forces and the way in which these were brought to bear on the political organisation of the Ghanaian state?

These are the broad issues which underlie the present inquiry into the political and commercial monopoly of the United Ghana Farmers' Council. These are also the issues which are raised in the more general discussion on the Ghanaian experience of economic decolonisation and national development to which I now turn.

The political economy of the CPP: the wider discussion

This brief survey does not pretend to cover the whole spectrum of strategic issues involved. Arguments, for example, related primarily to a liberal foreign-investment strategy based on further integration in an international market economy are not included. Those arguments doubtless require serious attention and discussion. In the present context, however, I have limited the selection to such contributions as seem to share certain basic strategic assumptions with the nationalist leaders themselves about the need to create a national economy by way of breaking the prevailing colonial or neo-colonial pattern dominated by foreign private capital. This position is summarised by Reginald Green, who says that:

> A state whose economy is characterised by concentrated external dependence can hope to neither a significant degree of control over the rate of growth and nature of allocation of domestic resources nor a high level of external credibility from which to bargain.[12]

It is therefore necessary to create a national political economy, which means, in Green's words, a strategy which "consists in both directly productive and infrastructural investment policies being led by a sector which is basically closed, national government-directed and, in Africa, at least substantially publicly owned".[13]

Within such common ground, however, widely different views are held about the nature of the CPP achievement. At one end we find those like Fitch and Oppenheimer, who see the CPP strategy as basically misconceived, a product of a dependent, petty-bourgeois regime unable to produce any substantial results in the way of decolonisation and national development.[14] At the other end we find those like Roger Genoud to whom the strategy was "reasonably correct" and reasonably well implemented and more or less the only possible one in view of the social and economic structure of the economy.[15] An intermediary position is suggested by Roger Murray, who emphasises the contradictions in CPP strategy between neo-colonial and genuinely nationalist (anti-imperialist) forces.[16] Other important contributions to be discussed here have been made by Jitendra Mohan och Judith Marshall.[17] Relevant material is also found in the writings on "non-capitalist development", where special attention has been paid to the Ghanaian case.[18] These references by no means exhaust the list of important contributions. I hope, however, that this selective survey will indicate at least some of the more important elements in this debate.

Bob Fitch and Mary Oppenheimer's *Ghana: End of an Illusion* is a forceful interpretation of the "radical break" type which has also had wide circulation beyond those with a special interest in Ghana. Although clearly rash in some of its conclusions and not always accurate in its details, it had the great advantage of having been produced soon after the 1966 coup which took many by surprise. The coup seemed to call for a reexamination of the Ghanaian case with its well-established "progressive" reputation. Following by now well-known writers such as André Gunder Frank, Samir Amin, and Charles Bettelheim, they claimed that only a radical break with the colonial economy could lay the basis for real national development and real growth. Old relationships had to be severed, old structures smashed.[19] This was never done in Ghana. In the early period (before 1961) there was "no effort whatsoever to restructure Ghanaian society". The CPP leaders seemed to believe, according to the authors, that in order for Ghana to achieve rapid economic growth, "the institutions of colonialism needed only to be administered by Africans". In the later period they find an attempt to break with the colonial pattern but it failed because it was not made "soon enough, and when it was made, it was not complete enough".[20]

CPP "socialism", from this point of view, built on three basic erroneous assumptions: firstly, state ownership by itself leads to socialism (and national development); secondly, the state sector can overcome the private sector without engaging in a "life-and-death struggle"; and finally, foreign private capital will let itself be used to build socialism. The failures in the public-sector economy are seen by these writers as essentially caused by this attempted policy of "coexistence". State enterprises can only compete in a capitalist system when backed by a unified system of state control.[21]

Jitendra Mohan, with a greater first-hand knowledge of the CPP system, gives a similar picture of the pre-1961 period. The petty-bourgeois socialism of the CPP "simply maintained and lubricated a fundamentally private enterprise economy dominated by colonial interests". While more cautious when evaluating the economic achievements of the "radical" phase, he nevertheless concludes that "petty-bourgeois nationalism which was enough to win the country its independence was not enough to transform it".[22]

Judith Marshall, with the help of indicators of economic dependence suggested by Reginald Green, has taken a closer look at the radicalised strategy of the 1960's. She sees a mixture of state capitalism and private capitalism, with state intervention "purely in capitalist terms". This was a genuine attempt to create a national economy but it failed to "extricate Ghana from its structures of dependency towards a self-realiant, integrated and selfsustaining political economy".[23] The strategy of import-substituting industrialisation did not reduce dependence, but led rather to increased commitments of imports and foreign exchange.[24] With other critics of import-substitution, Marshall also stresses increasing technological dependence.[25] She gives a good summary of the criticism raised against the Volta river hydro-electric scheme as geared essentially to the requirements of an extended neo-colonial enclave economy.[26]

Few critics have failed to point out the contradiction between the professed strategy of economic independence and the extreme reliance on foreign borrowing to finance state investments. This is a point also stressed by those who take a more positive view of the strategy as such. Reginald Green, for example, writes that an economy "bereft of external reserves and lumbered with short-term debts is an economy delivered into the hands of its creditors whatever the strategy which led to the impasse".[27]

While emphasising the failures and misconceptions of the CPP strategy, many radical critics have seen important achievements in certain directions. These include the redirection of foreign trade and the curtailment of foreign private interests in a number of sectors. The structural changes brought about by state intervention were real, according to Marshall. It was also inevitable that they involved short-term dislocations of a serious kind. Part of the failure to achieve much progress, in spite of the increase in national control, was seen as attributable to an adverse international economic situation. Under such conditions the very attempt to accelerate growth was bound to cause major difficulties.[28]

The recognition of the genuine nature of the national development effort and the concrete achievements points to an alternative interpretation, focusing on the complexity of the tasks facing the CPP, the limited resources at its diposal, and the basic restrictions imposed by the social, economic and political realities of the Ghanaian situation. From that perspective, the substantial results, rather than the failures, come to the fore in the analysis. The "illusions" are attributed, not to the Ghanaian leaders and their supporters, but rather to those who hold exaggerated views about the scope for a radically different and more revolutionary strategy.[29] Reginald Green's careful "reflections on economic strategy, structure, implementation, and necessity" is such a "balanced" account, stressing in particular the achievements in terms of a strategic reorientation of economic policy, while pointing at the same time to inconsistencies and failures in implementation.[30]

Roger Genoud has challenged more directly some of the major assumptions of the radical interpretation of the CPP. His book, *Nationalism and Economic Development in Ghana* (1969), is in large part a polemic against Fitch and Oppenheimer. He has stressed the logical or even inevitable nature of the CPP response to the social conditions of Ghana. CPP strategy in his view was a middle-way between "neo-colonial laissez-faire" and a radical break-away strategy as determined largely by the balance of social forces and the objective situation of the Ghanaian economy. The first alternative had rapidly been undermined by the deteriorating export conditions. The second alternative was impossible because of the small size of the economy and the lack of an experienced and disciplined political party capable of leading such a break-away move. These restraints, however, did not prevent major achievements. Heavy infrastructural investments created conditions for a radical reorientation and major structural change.[31] Despite modest economic growth and continued financial dependence, a basis was laid for the modernisation and industrialisation of the country. In important respects, Ghana succeeded in establishing national control over the economy, although the instruments of control were not coherently applied.[32]

Favourable cocoa prices permitted, according to Genoud, at first substantial results within the framework of "non-controversial" economic policies, that is, without confronting the institutions of the colonial economy. As the cocoa situation deteriorated, the "closing" of the economy and massive state intervention in production became necessary in order to keep up the national development effort already begun.[33] Genoud draws at this point on a much-quoted "model" formulated by Dudley Seers, delineating the stages in the development of a primary produce exporting economy.[34] In this light, Genoud sees the CPP programme as a consistently "nationalist" one and sees little scope for a more "socialist" alternative. Nor was there much scope, according to him, for a "liberal" strategy in this situation. He points at the limited differences between the CPP and the "right" opposition in terms of general economic strategy. The longterm consequences of the military coup and CPP's "failure" are played down. He regards coups as part of the natural perils of the situation. The nationalist strategy would sooner or later have to be resumed "for the simple reason that the program is fundamentally the only one and it was not completed by the former regime".[35]

Fitch and Oppenheimer's "neo-colonial" interpretation and Genoud's "nationalist" one are based on different understandings of the way in which the CPP strategy was determined by the existing social situation and how the CPP itself was rooted in society. The neo-colonial model stresses the narrow national class basis of the regime and how, as a consequence, it was unable to overcome the overriding neo-colonial determination of the situation. The narrow base was reinforced, as we shall discuss below, by the process of "bureaucratisation". The road was blocked for the participation of working class and broader mass forces capable of sustaining the vital break with the colonial economy. Genoud, on the other hand, rejects this narrow class determination of the CPP regime. Its petty-bourgeois cadres are seen as belonging to a broad middle stratum which in the absence of either a bourgeoisie or a proletariat of any significance was the natural recruiting base. The party itself had a broad mass following. The strategy was determined, according to Genoud, by the need of the leadership to balance all groups in society in the fluid class structure of the post-colonial situation.[36]

An alternative both to the narrow social determination of the Fitch and Oppenheimer "model" and the broad one argued by Genoud is suggested by Roger Murray. While in agreement with much of Fitch and Oppenheimer's analysis he rejects what he calls their "pseudo-Marxist determinism", demonstrated in particular by their attempt to explain the industrial policy after 1961 in terms of the class interests of a bureaucratised petty-bourgeoisie seeking new sources of surplus when old ones were running dry.[37] Murray criticises the tendency, as he sees it, in much radical analysis to apply "naive and undialectical dichotomies" (roughly: to see things in either black or white), when characterising new states and leaders as either revolutionary or neo-colonial. He emphasises the need to take account of the presence of contradictory tendencies. The post-colonial state "has simultaneously to be perceived as the actual instrument of mediation and negotiation with external capitalism and as a possible instrument of a continuing anti-imperialist and socialist revolution".[38]

Murray and Genoud agree about the fluid class structure of the post-colonial situation but while the latter sees it as necessitating a "policy of equilibrium", Murray puts emphasis on the contradictions within the governing stratum and the scope for widely different strategies. In the Ghana case, the anti-imperialist line within the CPP was not sufficiently strong to ensure a consistent and successful strategy of its own. This did not mean, however, that the strategy of the CPP could be reduced to the self-advancement of the petty-bourgeoisie or of the party and state bureaucrats.

A position close to Murray's has been developed by John Saul in an analysis of the Tanzanian situation as part of an attempt to explore a general theory of the post-colonial state, focusing on the "relative autonomy" of the state apparatus vis-à-vis local class forces.[39] The relatively "open" class situation has created conditions, according to Saul, for a struggle within the "unformed" governing "class" between those who seek to consolidate the neocolonial structure and those who challenge it.[40] Murray similarly speaks of the need to understand "the contradictions inherent in the accession to state power of unformed classes".[41]

The undecided class character and the radical potential of petty-bourgeois leadership in post-colonial societies is a basic assumption in the discussion of a "non-capitalist road" as developed above all by Soviet and Eastern European marxists.[42] Nkrumah's Ghana was regarded by writers within this tradition as a case of such non-capitalist development during the first half of the 1960's. The growing difficulties and final collapse of the CPP regime contributed to a reappraisal and further development of the theory.[43] By the non-capitalist road is usually meant a strategy which leads to rapid growth of the productive forces while at the same time restricting the growth of private capitalism. It makes use of foreign as well as local private capital, while the state simultaneously plays a leading role in the advanced sectors of production and in the overall organisation of the economy. Its purpose is to prepare for a stage where material and political conditions will permit a transition to socialist development, based on an advanced industrial structure and an organised working class capable of providing leadership. Such a non-capitalist road is possible because of the limited development of capitalism in the third world and the contradiction between national liberation and capitalism as a result of colonialism. The ascendancy of socialist states on the international scene is seen as another important condition for this desengagement of former colonies from the road of capitalist development.[44]

The CPP made no distinction between non-capitalist and socialist development. The former term was never used. Its strategy, however, may be seen as a "non-capitalist" one in the sense developed by this school of thought. As understood in this perspective, the CPP was a national-liberation movement which had been brought under the leadership of a radical section of the petty-bourgeoisie. These "national democrats" were gradually adopting the position of the working class and of socialism.[45]

A Soviet student, writing before the military coup, emphasised the advanced influence exercised by the state over the Ghanaian economy, including elements of "properly socialist reforms". The deliberate restrictions imposed on the national capitalist class and the selective use of foreign capital were commended, although there existed a problem of how to find and use "such measures of control and regulation as will prevent the private sector from moving beyond the framework of tasks placed before it and from turning the country on to a path of capitalist development".[46]

The contradictions inherent in the situation have been more fully developed in the analysis which followed the coup. Ghana was not the only failure of countries once thought to be on a non-capitalist road and the reassessment was mostly given a general formulation. The potential of the radical petty-bourgeois leaders and the post-colonial state apparatus in general to lead and sustain non-capitalist development was critically discussed. Experience had shown that often a "clearly unsupportable burden is hoisted on to the shoulders of the young state apparatus". Greater attention was paid by the students to the social basis of the post-colonial regimes. Individual leadership by "outstanding national democrats" like Nkrumah did not make up for deficiencies in organisation and material conditions. Ideology, state expansionism, and leadership were running far ahead of the actual potentialities created by the level of social forces and organisation.[47]

Two problems of special relevance for the present study have been given particular attention in this discussion of the CPP experience of national development. They both concern the political organisation of the state: one, the process of bureaucratisation and how it affected the class character of the regime and its policies, the other, the nature of the organised political base provided by the party. The two issues are often linked to each other as, for example, in the common suggestion that the party was absorbed by the state, that it was bureaucratised. The United Ghana Farmers' Council provides good illustration of these problems. Let us look at some of the main propositions involved.

Writers of the non-capitalist school have pointed to the "degeneration" of the leadership groups and their "gradual transformation into bureaucratic elites divorced from the people, who put non-capitalist development in cold storage, identify the egoistic interests of their own groups with the interest of the state and nation and exploit in their name the working masses".[48] "Bureaucratic state capitalism" is a term often used by these writers to characterise the political-economic system emerging out of such a process.[49] An analysis of a similar type has been suggested by Claude Meillassoux when discussing the "bureaucratic process' in Mali, a country with a radical experience of the Ghanaian type. The political-bureaucratic elite, in his view, served first as an instrument of the colonial power, later temporarily as the mouthpiece of the peasantry, only to end up as a class of its own, exploiting the peasantry through the economic control it was able to exercise through the state apparatus.[50]

The bureaucratic theme has been developed forcefully in the Ghanaian case by Jitendra Mohan. Out of the petty-bourgeois mass, he suggests, emerged a "substantial bureaucratic bourgeoisie entrenched in the state apparatus at all levels". The massive intervention of the state in the economy, given this essentially bureaucratic political base, simply multiplied, according to him, the opportunities for self-enrichment of those manning the double state and party apparatus.[51] Genoud, on the other hand, sees the rise of bureacratic government in less ominous terms. He suggests that "a national bureaucracy" must be the natural leading force in the absence of other significant class forces.[52] John Saul emphasises the confluence of economic and political power in the hands of the political-bureaucratic elites and the degree of "autonomy" which it gives them vis-à-vis other groups in society.[53]

The problem of the independent power and class character of the political-bureaucratic elite is part of the wider problem of the organised political base of the Nkrumah regime. The nature of the CPP itself and its social roots come into focus. One proposition which has been widely accepted is that the party declined as an independent force and was "absorbed" by the state.[54] We may remember Frantz Fanon's well-known account of the disintegration of nationalist parties once independence had been achieved. They no longer helped to formulate popular demands and increase the awareness of the people of their real needs, but instead became the instruments of the leaders, who used them to pass on their decisions to the people. Party cadres were absorbed into administrative posts and the parties themselves became part of the state apparatus.[55]

The CPP was long regarded as one of the genuine mass parties emerging out of the anti-colonial struggle in Africa.[56] Its "disintegration" therefore attracted particular attention. The apparent absence of any significant resistance to the change of regime in 1966 was by many taken as the ultimate evidence that the popular strength of the CPP was largely a myth.[57]

In line with Fanon's argument, a distinction is made by many between the CPP of the pre-independence struggle and the bureaucratised party of later years. Genoud, for example, speaks of the rapid decline of the party after it had first served as an effective mobilising mass force.[58] Others, like Mohan and Fitch and Oppenheimer, take a more critical view of the role of the party in the early period, as well. They find more manipulation of regional division, traditional loyalties, and economic patronage than real mass mobilisation.[59] Apter speaks of "payoffs and petty manipulations" as the typical method used by local party organisers to gain support.[60] The view of the party as essentially based on economic patronage is pursued most persistently by Maxwell Owusu in his study of politics in a rural small town. The strength of the CPP, according to him, was its ability, as a government party, to provide economic opportunities, jobs and a sense of power for all classes of people ("particularly the disadvantaged") on the almost sole basis of party membership. "When the basis of patronage suffered, the regime lost its friends".[61]

Most observers agree that the strains imposed by the economic conditions of the early 60's and the attempts to "close" the economy and to industrialise changed the demands made on the party as a mobilising force.[62] The failure of the CPP to function in this situation is seen by Genoud as a more or less natural consequence of the need to pursue a policy of "equilibrium", that is, to balance the interests of all major groups. In the absence of a dominant internal contradiction which could form the basis of the economic struggle, the party withered away and was replaced by the bureaucracy as the leader of the national development effort.[63]

Certain attempts were made to reorganise the party at this time, including closer control over the party "wings" like the Ghana TUC and the Farmers' Council. Attempts were also made to improve on the ideological training of the cadres at all levels. Such exercises, according to Mohan, proved mostly mechanical and superficial. Political control was simultaneously reinforced at the expense of democratisation. Fresh barriers were raised against popular participation and mobilisation.[64] The integration of the party wings stifled rather than galvanised popular initiative, according to another student.[65] Marshall speaks of party control over trade unions und farmers' organisations as an instrument of the "de-politisation" of the workers and the peasants.[66] We may compare this to Saul's discussion of demobilisation and bureaucratic consolidation in the Tanzanian case.[67] Apter sees the CPP in 1965, the year before the coup, as consisting of "vast networks of committees which did not meet, organisations which failed to function, and personal manipulations which aroused mutual suspicion, mistrust and recriminations". What grew, he concludes, were the number of party posts and positions whose main function was to provide "opportunities for political and economic entrepreneurship".[68]

Bureaucratisation, the failure to broaden the political base, the inadequate ability of the party to provide mass support for the radicalised policies of the 60's—these are some of the common formulations of the political failure of the CPP. Waste, self-enrichment, corruption, poor priorities, and misconceived policies are in various degrees seen as direct consequences of this basic failure of political organisation. To Fitch and Oppenheimer, the failure of the CPP to organise the "most oppressed strata of the working class" and make their demands "the moral imperative of the Party organisation" was the root cause of its bankruptcy and the proof of its petty-bourgeois and neo-colonial character.[69] Marshall takes a more positive view of the intentions of the leadership in the radical phase, which she sees as "generally aimed at the interests of the workers and peasant". But the party never succeeded, according to her, in linking itself in an organised manner to these classes. It failed to build a "base of power actually rooted in the forces of production". Original sources of support, such as the commercially-oriented section of the petty-bourgeoisie, were at the same time alienated by the increasingly "statist" strategy. Marshall observes that while the struggle for political independence had brought about a convergence of various local interests, the struggle for economic independence led to the fragmentation of political support. Groups which had participated fully in the former struggle were unwilling to give up short-term interests for the long-term goal of building an "integrated, self-sustaining national economy".[70] Ruth First similarly notes that "the regime alienated those who hoped to improve their personal and political fortunes by independence; but it did not disarm them, or displace their influence". She points in particular at the way in which the reorientation of the trade to the East, import licensing and state trading contributed to restrict opportunities for "a tough but frustrated propertied and trading class" impatient with the barriers to its growth.[71] The conflict is illustrated concretely at the local level in Owusu's Swedru study.[72]

Many writers stress the destructive impact which the gap between the private and public morality of many of its leaders had on the party's wider support. "For the people to accept sacrifices and hardships for the country's development", according to Mohan, "complete elimination was required of those leaders and advisers who were themselves anything but self-sacrificing". Nor was this enough. There was also the necessily of creating democratic political structures capable of promoting an understanding of what was at stake:

> This could only be possible through active and continuous association of the people with political and economic decisions involved in reconstruction and development, especially at the grass-roots level . . .[73]

The CPP regime failed to transcend the decisive limitations on its own capacity caused by its narrow political base. This is a view taken also by those with a more positive view of the strategy as such. Genoud, for example, although anxious to dedramatise the failure of the CPP, sees the major "internal" cause of its failure in the inability of the party to find forms of political control and support which would match the rising costs of development after 1960.[74] Reginald Green similarly asks what political pattern would have been consistent with this strategy of economic decolonisation. He makes a distinction between three alternative systems, an "aristocratic" (elitist), a coercive, and a mass-support system. Neither of the first two could be effectual in the Ghanaian situation, the first being unable to secure public acceptance, and the second failing because of the absence of the necessary instruments for such coercion. Ghana, according to Green, sought to approximate to a mass-support system, but mainly rhetorically. "The reality was an increasingly less effective, less honest and less acceptable elite system".[75]

Summarising this discussion, it should first be noted that many of the differences of interpretation are linked to different assumptions about possible and desirable strategies. Looking at the factual account alone, the picture is less contradictory. The dominant picture presented by these writers shows a dependent, neocolonial regime which attempted to develop a national economy but which was restrained both by limitations in its own political capacity and by deteriorating economic conditions outside its control. To some writers the efforts were either too late or too ineffective to achieve substantial results; others see more substance in the achievements. The system which evolved was more nationalist than socialist, although some writers (especially in the pre-coup period) saw elements of socialist reforms. The state played an important role in this national development effort, adding an increasingly "state-capitalist" or, according to some, a "non-capitalist" flavour to the new political economy. Those using the latter term stress the potential for a development in a socialist direction. In explaining the state-capitalist features some writers emphasise the weakness of the national private capitalist forces in providing leadership while others stress the self-advancement of the bureaucracy: the growth of a "bureaucratic bourgeoisie". The two perspectives can of course be combined.

Those emphasising the limited achievements attribute particular importance to the narrow class basis of the regime and its failure to root its strategy in the social organisation of production. It is seen both as a question of how the capacity of the regime was determined by its social and political base, and of its failure to surpass these limitations and extend (or maintain) its organised following. The petty-bourgeois background of the leaders and the cadres is underlined, as opposed to a broader working class or peasant one. Others define the social base of the party more broadly, although still mainly in terms of the "middle" strata of the Ghanaian society.

The failure of the party to provide an organised political base for the regime, particularly in the post-1961 period, is partly attributed to the process of bureaucratisation, arising both from the expansion of the state bureaucracy and the weakness of the democratic process within the party itself. Bureaucratisation is seen as the "degeneration" of leadership and as the consequence of the weakness of other class forces capable of taking the lead in national development. The political failure of the CPP is explained also in relation to the changing basis of political support. Important groups of traders and small businessmen, for example, which originally supported the regime were alienated by the growing emphasis on state enterprise and state economic control. Others were alienated by the corrupt and oppressive performance of many of the new leaders. The party failed at the same time to extend its organised support to new social groups.

The picture is far from conclusive. Many important and difficult theoretical issues underlie the differences of emphasis and interpretation. Some of the difficulties are conceptual or terminological. The present study cannot aspire to provide a more conclusive overall interpretation of the CPP experience as a whole, nor will it suggest a more comprehensive theory or terminology. Its contribution to the wider discussion lies in the exploration of the empirical relevance and meaning of some of the general propositions in the concrete context of the Farmers' Council experience. Let me list some of the major points where I believe that this experience may help to illustrate and clarify general issues.

This can be done, first of all, with regard to questions about the organised political base of the CPP regime. The Farmers' Council records allow us to study closely the way in which the CPP approached the problem of organising the agricultural population and the cocoa farmers in particular. The roots of the party in the social organisation of production, the bureaucratisation of the party, and the problems of internal democracy and participation can discussed and documented in considerable detail. Many of the specific issues raised in the general discussion can be illuminated. This applies, for example, to the relationship between political mobilisation and *de*-mobilisation and the impact of "state capitalist" or "non-capitalist" schemes on the composition of the party's following.

The takeover of the local cocoa trade by the Farmers' Council was a step towards the establishment of national economic control. The expulsion of the foreign cocoa buyers was hardly a strategic decision, in view of the fact that the state already controlled the export trade through the Cocoa Marketing Board. Nevertheless, the commercial monopoly of the Council permits a discussion of the nature of the CPP strategy and the character of the "state capitalist" institutions which were established. Was the monopoly a dependent, petty-bourgeois, neo-colonial solution along the lines suggested by, for example, Fitch and Oppenheimer? Or was it the natural, national solution—a reflection of the balance of social forces—in accordance with the interpretation ventured by Roger Genoud? To what extent does the monopoly illustrate the contradictions between neo-colonial and anti-imperialist forces within the post-colonial situation for which Murray and Saul suggest that we should look? How substantial were the achievements? How real were the structural changes involved? What was the impact on foreign capital? What restrictions were imposed on the growth of a national capitalist class? Was an "unsupportable burden" imposed on the young party organisation as a result of its wide commercial responsibilities? What did it mean in terms of bureaucratisation and ability of the organisation to perform more basic political functions? In which way does the monopoly clarify and illustrate propositions about an emerging "bureaucratic-bourgeoisie"? These are all issues raised in the general discussion of the CPP experience which can be explored in relation to the Council's takeover and administration of the cocoa trade.

The wider significance of this takeover, however, lay in the supportive role of the commercial and political monopoly for the cocoa appropriations of the CPP state. The political implications extend therefore beyond the cocoa trade as such. They concern the financial and political basis of the CPP state and its strategies for national development. The Farmers' Council experience helps to trace these financial and political roots in the cocoa economy. Cocoa was at the centre of the country's international dependence. But cocoa was also given a strategic role in the attempt to overcome this dependence. By investigating the class character of the Farmers' Council monopoly, its social roots and political organisation, it is possible to highlight the political dynamics of this process of national and class emancipation.

A note on terms and concepts

A few words must be added on some of the terms and concepts which appear both in the wider discussion above and in my own presentation. These terms and concepts include important general ones like "class", and those referring to specific social categories (petty-bourgoisie, bureaucracy) or political-economic systems and processes (state-capitalism, non-capitalist development, etc. There are important theoretical and conceptual problems associated with most of these terms which cannot be discussed here. The purpose of this brief note is merely to indicate the meaning which they have been given in the present context. I hope it is sufficient to allow the reader to follow the general argument.

A basic concept is that of *class*. It is used here for particular social groups and in combinations such as class character and class basis in my discussion of the nature of political institutions and policies. Class as a concept referring to the differentiation of society into large groups of people with opposing or different stakes in production has been developed primarily within the Marxist tradition, but it has also become part of a broader framework of social and political analysis.[76] As used here it involves the ownership of the means of production, the organisation of labour, and the mechanisms for producing and sharing social wealth.[77] By the class character of institutions and policies is usually meant the way in which class interests or the development of class forces are expressed at the political level. The class basis of policies and institutions is similarly taken to mean the way in which they are rooted in the social organisation of production. Such expressions do not necessarily imply the presence of full-fledged classes or groups, consciously and collectively pursuing their own interests.[78] Class analysis, in this sense, refers more generally to the way in which politics influences and is influenced by the social organisation of production. An underlying assumption, of course, is that the way production is organised is of overriding importance in determining the nature and quality of society. It decides largely the way in which social wealth is used and consumed. Classes or, more abstractly, social forces engaged in maintaining or changing this organisation are consequently seen as principal actors in the development of societies, including their political institutions. Political analysis, starting from such assumptions, is therefore less concerned with important individuals (such as Kwame Nkrumah or Martin Appiah Danquah, the forceful General Secretary of the Farmers' Council) than with the way such collective actors—abstract forces or concrete groups—stamp their mark on developments and institutions.

Some class categories have been used widely in the debate and require comment. *Petty-bourgeois* is a term commonly used to characterise the class orientation and class basis of the CPP. The separate class identity of the petty-bourgeoisie is not altogether clear. In its original Marxian usage it seems to refer to small commodity producers (with their own means of production) in a society dominated by capitalist relations of production. The meaning becomes more obscure when we deal with societies like Ghana, where the bulk of the population is still engaged in pre-capitalists forms of production. If the entire peasantry, for example, were to be included, the concept would lose much of its usefulness. On the other hand, it is often used—as can be seen from the discussion above—to include groups which are not primarily commodity producers but belong to the "service" sector, or in the Marxist vocabulary, the "intermediate strata". This category includes traders, officials, professionals and other groups. Mohan, for example, lists a broad range of groups when specifying the petty-bourgeois base of the CPP: elementary-school leavers, teachers, clerks, messengers, petty traders, store-keepers, artisans, drivers, small contractors and businessmen, urban wage-earners, and ex-servicemen.[79] The importance of most of these categories in the composition of the CPP cadres is supported by many observers, although the overall designation varies.[80] The inclusion of wage-earners in the list raises questions about the lines to be drawn between "working class" and "petty-bourgeois" elements. No comprehensive defintion of the petty-bourgeoisie will be attempted here. It seems, however, as if the use of the term in most of the writings on Ghana focuses on the urban and rural "entrepreneurial" strata, those engaged in commodity production as well as in commerce, transport, housing, and other commercial services. These are categories which draw most of their earnings and livelihood from the market; they own property and capital, and occasionally or regularly hire labour.[81] This emphasis on the market-oriented entrepreneurial groups will be preferred when the term is used in the present study. It leaves out the mass of agricultural producers who depend primarily on production for consumption within the household. It also leaves out wage-earners as such. On the other hand, this study is much concerned with the petty-bourgeois, or in this sense, the entrepreneurial aspects of wage employment, especially in the case of public officials who use their office as a platform for private enterprise.

Bureaucracy is another term freely used in the discussion above as a category of class, although more rarely to denote a distinct class of its own. It is most commonly treated as an "intermediary stratum" and, as we have seen, it is sometimes taken to overlap both with the working class and the petty-bourgeoisie. Some writers speak of a "bureaucratic bourgeoisie" when referring to the incumbents of a system with extensive state production, but without effective working class or popular control. Others use this term to characterise the bureaucratic "wing" of the bourgeoisie (the capitalist class) in a society where the state is closely involved in the administration of an essentially private capitalist system. We have seen how these usages are reflected in the Ghanaian debate. With Murray and Saul I hold it to be important to stress the mediating role of the bureaucracy in relation to various class forces in society rather than its separate class identify. The capacity of the bureaucracy in a particular situation to disengage itself from the influence of the social groups (which it is supposed to serve) is nevertheless a most important issue also in the present study.

Capitalism, socialism, state capitalism and non-capitalist development are terms used to characterise various models of social organisation or aspects of such models. Only the latter two may require some comment here. *State capitalism* usually refers to the extensive participation of the state in production either directly as owner or as provider of essential supporting functions for the private sector. This is a concept which has above all been applied to advanced capitalist societies (for example, in the combination "state monopoly capitalism") but it has also been used by critics of certain socialist states in order to suggest that the bureaucracy rather than the working class is in control. The distinction between socialism and state capitalism in this context is therefore a matter of how the class basis of the system is defined. These different meanings have also been extended to third-world countries, either to stress the dependent role of state interventionism in relation to foreign or local private capital, or to underline the self-advancement of a new bureaucratic ruling class through the state apparatus.

Some writers have emphasised the special role of state capitalism in third-world countries as a consequence of the limited development of productive forces. Tamás Szentes and Ignacy Sachs, for example, underline the distinction between the regulative and supplementary role of state capitalism in advanced capitalist societies and its more strategic role in the development of productive forces in the third world.[82] They suggest that state capitalism "substitutes for the non-existing capitalists, performing actually the function of 'collective capitalism'".[83] Sachs distinguishes between two main lines of state capitalist development in the third world; in one, the public sector acquires a permanent and dominant role. Strategic branches of production are reserved for the state and the development of private capital is restricted. According to the second model, the state acts as the caretaker for the interests of a capitalist class which is not yet strong enough to take on all the necessary tasks of production. The public sector finances private capitalist development and public enterprises are privatised when possible. Sachs sees these models as two extremes; although they are not found in a pure form anywhere, individual cases tend to approximate to one or the other while containing elements of both.[84] This "open" conception of state capitalism in the third world, that is, as containing elements pointing in widely different directions, fits well with my understanding of the Ghanaian situation. The present use of the term will therefore not necessarily imply either a capitalist or a "state bureaucratic" type of development. The class character of the Farmers' Council monopoly should be discussed in relation to both perspectives.

The first line of state-capitalist development indicated by Sachs comes close to the concept of *non-capitalist development* as discussed above (p. 23), but without suggesting a long-term socialist direction. I think that the latter concept may serve as a useful complement in characterising certain aspects of state involvement in third-world economies. It seems to have relevance in the Ghanaian case during the latter part of the CPP period. The leading role given to the public sector, the extension of national control over sectors previously dominated by foreign capital and the deliberate restrictions imposed on local private capital may be usefully discussed both in terms of "state capitalism" and "non-capitalist development". The distinction is by no means clear and requires further discussion. I use the terms here with hesitation and with the hope that they jointly will convey something of the special flavour of the CPP system.

National development, as used in this book, should be understood in the context of the postcolonial situation and in contrast to such concepts as *colonial, neo-colonial or dependent development.* National development is taken as largely synonymous at this stage with continued decolonisation, especially in terms of establishing national control over the economy. Important theoretical and analytical problems are involved. Where are lines to be drawn, for example, between dependence and national control? The difficulties become particularly apparent when more advanced forms of dependence and "international divisions of labour" are considered.[85] National development in the Ghanaian case involved in the first place an attempt to break with the classical pattern of a *colonial economy.* That pattern implied, among other things, extreme dependence on production of raw materials for the world market and dependence on the advanced industrial countries for most of the supply of manufactured goods. It implied also large enclaves of extractive export industries (mining and timber) with little linkage to the national economy and dependence on one or a few major trading partners. The picture of the colonial economy includes a lopsided agricultural sector developed mainly in response to external markets and with difficulties in feeding the population. The public sector is similarly underdeveloped and geared primarily to the requirements of the export economy. Commerce and banking at more advanced levels are controlled by foreign capital.[86]

It is against the background of this classical pattern of colonial dependence that it is appropriate to speak of problems of policy in Ghana during the CPP period in terms of national development. CPP policies represented a massive attempt to achieve decolonisation in the economic field. Growing foreign indebtedness, unsuccessful bargains with foreign capital and other failures of strategy and implementation do not alter the basic national, anti-colonial character of the development effort. This study is not the appropriate forum for a closer discussion of the overall achievement. It is important, however, that the Farmers' Council and the contradictions between national and class forces which it exposes be understood in the context of such a general perspective.

It is in this context that I use the term *state accumulation,* in order to describe the appropriation of cocoa income by the CPP state, a central process examined in this study. Accumulation is a concept commonly used as synonymous to capital formation. There is an everyday meaning to the term which points at the fund-raising rather than the expenditure side of the process, which makes it particularly relevant here. I use it in order to emphasise the role of cocoa in financing the "state-capitalist" programme of the CPP. It may be argued that most of the cocoa income of the government was absorbed in current expenditure rather than in capital formation. It seems less useful, however, to discuss what money was used for what purpose. Cocoa contributed much of the real resources of the government. Even foreign loans would ultimately have to be repaid out of that source. Directly or indirectly, the financial and administrative basis of the programme was largely provided by cocoa. The rate of capital formation in the public economy was high.[87] It included, of course, considerable expenditure on buildings and equipment of doubtful value to the development effort. But there was also heavy investment in the economic infrastructure and in the industrial programme led by the state sector. State capital formation played a crucial role in the CPP strategy of national development and decolonisation. The combined political and commercial monopoly of the Farmers' Council is examined in relation to this process of state accumulation.

II. The Cocoa Economy

The general structure of the colonial economy as it existed in Ghana at the time of independence was formed largely in the closing years of the last and the early decades of the present century.[1] Robert Szereszewski has summarised this development:

> Historically the present economic structure of Ghana started to evolve when the natural resources of the forest belt began to be exploited and exported by Ghanaian and foreign enterprise. Capital started to accumulate around the cocoa, mining and timber industries, and strong migratory movements of labour, both internal and external, followed. Urbanization, the development of new services and the provision of social and economic infrastructure were largely restricted to the areas of favourable natural resources and to the coastal centres which were the seats of administration and points of contact with the outer world.[2]

The process brought a degree of "prosperity" which was high by third-world standards. It is estimated that Ghana's per capita Gross National Product by 1960 was over twice that of Nigeria's and almost three times that of India's.[3] The dominant role of peasant production brought about a relatively wide distribution of this prosperity, unlike colonial economies based mainly on foreign-owned mines and plantations. Commercial development and urbanisation were stimulated even outside the main colonial centres. This was of course not an entirely new process. The cocoa economy could build on a long tradition of commercial agriculture and long distance trade.[4] The advanced character of the Ghanaian economy is seen, for example, in the low proportion of agriculture in total employment (62 per cent in 1960) and the high proportion of commerce (15 per cent).[5] It is also seen in the low rate of "subsistence" consumption in total consumption (20 per cent) and the relatively low share of food in total household expenditure. The capital stock of the economy, primarily represented by buildings and cocoa plantations, was also unusually high.[6]

These are measures which refer to the economy as a whole. The picture looks different when we add the regional dimension.[7] Wide gaps exist between different parts of the country. This is primarily a difference between the poor northern savannah regions, of little interest to the colonial economy except as labour reserves, and the "rich" forest regions to the south, with their cocoa, timber, gold and diamonds. But this is also a difference between the urban and commercial coastal areas, with their long tradition of European penetration and trade, and the less penetrated "interior". Politically the coastal regions ("the Gold Coast") were effectively incorporated into the British colonial empire by the mid-19th century, after centuries of partial occupation based on forts and trading posts. The Ashanti kingdom in the central forest belt was not fully pacified until the turn of the century, after a series of wars. This is also the time when the northern "protectorate" was established. The three parts continued to be administered in important respects as separate units up to the Second World War.[8] The strong regional differences in colonial economic penetration had important consequences for social differentiation of labour and wealth, not only between regions but within society as a whole.

Cocoa production[9]

The first exports of cocoa from the Gold Coast were recorded in 1885, about ten years after the consolidation of territorial control in the coastal regions (including the destruction of the Ashanti capital, Kumasi, by British troops). It was only by the turn of the century that exports reached substantial proportions (1,000 tons).[10] Roads and railroads built in this period facilitated the rapid expansion which now followed. By the outbreak of the First World War, cocoa exports exceeded 50,000 tons, and ten years later they reached the staggering figure of 200,000. After this extremely rapid expansion, growth slowed. The 300,000 level was reached by the mid-1930's. From this point a decline set in due to the impact of world economic depression, the new war and the spread of a dangerous crop disease, the swollen shoot. Not until the late 50's did production recover pre-war levels, now followed by a new remarkable fit of expansion taking output over the half million ton mark in 1964/65. From this climax a new period of decline and stagnation began. Over the last decades Ghana's share in total world production has usually varied between 30 and 40 percent.[11]

The importance of cocoa to the Ghanaian economy can be demonstrated in various ways. The share of total export earnings is of course a particularly striking measure. Looking at the post-independence period alone, we find that up to 1966, cocoa contributed on the average 62 per cent of the value of exports, followed by mining and timber with 23 and 12 per cent respectively.[12] Variations were not great: there was a bottom in 1957 of 56 per cent and a peak in 1965 of 67. The tendency in this period was for cocoa's share to rise, which had less to do with cocoa itself than with the decline in export earnings from other sources. The share of cocoa in the total value of production (Gross Domestic Product) was estimated at 15 per cent for the 1959–61 period. It equaled one-third of total agricultural production and about three times the shares of the mining and timber industries taken each by themselves.[13] According to the 1960 population census, one-fifth of the "employed" population was engaged in cocoa production, or about half a million people.[14] The number of people directly or indirectly dependent on cocoa for their livelihood was of course much greater but at the same time more difficult to estimate. Cocoa dominates the economies of the most populous regions, containing over half of the seven million recorded as the population of Ghana in 1960. It has been vital for the growth of towns, including the cocoa capital, Kumasi, and it has extended its impact not only into northern Ghana but deep into neighbouring countries by its heavy reliance on migratory labour. Cocoa has been equally vital for the growth of modern government, which continues to depend on the surplus of the export-import trade in financing its activities. The cocoa price-boom of the first post-war decade laid the basis for the ambitious development programme of the CPP. Falling cocoa prices in the post-independence period destroyed the political conditions for this massive exercise in state accumulation. We shall examine more closely this critical aspect of Ghana's dependence on cocoa at a later point in this study (Chapter 7).

The Ghanaian cocoa industry was the creation of local farmers, responding to the opportunities of a new cash crop. Foreign owned plantations or foreign settlers played no role in this development. As shown most clearly by Polly Hill, the driving force was above all a stratum of rural entrepreneurs investing in land and labour, moving to areas where soils were most suitable and where land could be acquired through purchase or leases. They reinvested the surplus in new land and new farms, but also in transport equipment and other complementary services.[15] In particular, she has documented the pioneering role of the farmers from the Akwapim Hills, where cocoa was first grown, and their migration to the forests of the Eastern Region. (See map on p. 10) A later wave of migrant farmers moved from central Ashanti westwards into the Ahafo forests.[16] But cocoa was also incorporated in the normal pattern of production throughout the already populated areas wherever suitable land was available. This meant that widely differing types of holdings were established, from small family farms near existing villages to large plantations with numerous labourers deeper into the forest.[17] Dramatic changes in the social organisation of production were involved. Not only were new relations of dependence and conflict established at the international level between primary producers and manufacturers. New relations were also created over land and labour replacing a system dominated by family production for the individual household within a framework of communal land tenure. Private or semi-private property rights in land were extended throughout the cocoa areas. Land was bought and sold. Farms were mortgaged and auctioned. Farm labourers were hired even on quite small holdings, most commonly as sharecroppers on mature farms, receiving one-third of the crop, the so-called *abusa*. Wage labour was also used extensively, especially when establishing new farms. A growing number of farmers became tenants outside the areas where they could claim a share in communal lands. Rents were sometimes merely minor tributes but could rise to one-third of the crop or more, paid to local chiefs as the representatives of the communal owners or to private landlords. The trade in cocoa and commodities sold to the producers, the growing demand for credit—backed by expected crops and private property—and the need for transport and other services increased social differentiation still further. Important sub-strata of farmer-traders, money-lenders, transport-owners, etc. developed within the farming community itself.

Two factors in particular seem important when explaining the dynamics of this process of social differentiation and class formation. Firstly, the nature of tree crop production and its impact on property rights and capital formation; secondly, the role of regional inequalities in creating conditions for the growth of a labour market. Cocoa is a tree crop. Once established it continues to yield over several decades.[18] Different varieties had different yield cycles. The one most common in Ghana in this period was long in maturing, about five to ten years, and it could take longer still before a substantial crop was obtained. The economic life of the tree was longer than the life expectancy of the planter. Communal land rights were linked to land rotation and annual crops. Private rights in a particular piece of land for sustained periods of time were in that context less essential. With the introduction of tree crop production this changed. The land was for all practical purposes occupied by individuals for an indefinite time. Rights of inheritance were revised. The mature cocoa farm represented a substantial investment of unrewarded labour. The most difficult work was the clearing of the heavy forest, followed by planting and maintenance of the young farm, including pruning, spacing, weeding and the arrangement of temporary shade. This capital, whether accumulated through the farmer's own labour or by hiring that of others, was property, distinct from the land itself. It was a platform from which bargains could be struck in the new labour market. Most commonly, share-crop labourers were employed on the mature farms. An annual flow of income was ensured, often with only limited managerial responsibilities undertaken by the owners. The farm-owner could move on to a new area and establish new farms, returning for occasional inspections and supervision, particularly at harvest time. Others would remain in the village to do the cocoa work themselves or to draw a part-rent, part-profit income while attending to other work (or retiring). This was capitalist farming, especially on the larger holdings, but with a strong peasant element, both in the important role of family labour as distinct from cash investment in capital formation, and in the landlord-tenant aspect of share-crop labour. "Peasant capitalism" may be a relevant name for this system of production.[19]

Inequality in the access to land was a condition for the rapid growth of a labour market and of capitalist relations of production. Cocoa cultivation is regionally restricted by factors of climate and soils. Only the tropical forest provides the necessary natural conditions. The value of land in the forest rose greatly as compared with that of the savannah. People outside the cocoa regions found that they could earn more as labourers on cocoa farms than from cultivating their own land, despite the large share taken by the farm-owners. The flow of cheap labour from the savannah was crucial for the rapid expansion of production. The bulk of Ghana's cocoa was soon produced on holdings using hired labour on a more or less permanent basis. The 1960 population census counted some 100,000 farm-labourers in the half million total employed in cocoa production. It is generally thought to be a gross under-count, partly due to the fact that it was taken in an off-season month when many migrant farm labourers were away, but also because of a confusion of share-crop labourers and farmers. Polly Hill suggests a figure closer to 200,000.[20] The farm labourers were employed mainly on the medium-size and larger holdings. As these generated the bulk of total output, hired labour was strategic for the cocoa economy as a whole.

Regional inequalities facilitated the growth of a labour hiring economy, affecting at the same time the nature of the contradictions between capital and labour. Migration delayed the consolidation of a rural working class. The migrants retained a hold in the land and peasant economies of their home areas. The flow of labour from outside also retarded the process of proletarianisation within the cocoa communities themselves. The remaining communal land rights gave a certain protection to members of the land owning group, but large-scale private occupation and the alienation of land to "strangers" reduced the opportunities for new generations of producers. There were also great differences in the access to land within the original communal framework. Nor was all land equally useful for cocoa. The process of differentiation accelerated with the accumulation of capital.

Various attempts have been made to chart the actual distribution of land, income and wealth in the cocoa economy. The pattern can only be roughly indicated. Using a sample based on the 1963/64 purchasing records of the Farmers' Council I have estimated the overall distribution of sales. Important problems of method are involved in translating these figures into a comprehensive picture of income distribution.[21] The general picture which emerges, however, points at a highly skew pattern of distribution, with a narrow stratum of big farmers controlling much of total output and a mass of small producers who together control very little. About half the sales were made by some ten per cent of the farmers. Three-quarters of the farmers sold less than one-third of the crop. The lower 50 per cent sold only 12 per cent of the crop. These figures confirm at a national level the picture given in local studies by Beckett, Hill and others of sharp differences in the size of holdings.[22]

The level of cocoa income varied with prices and the size of crops from one year to another. Average sales as recorded in the sample were 57 loads of 60 lbs. With a producer price of 54 shillings in 1963/64, this equalled a gross value of about £150. Is it possible to distinguish strata among the farmers with different class characteristics? Where may lines be drawn between "big" and "small" farmers and what do such lines mean in terms of the social organisation of production? These issues are discussed at some length in my study of social differentiation and class formation in the cocoa economy.[23] Theoretical problems are involved in deciding the purpose and basis for such distinctions. But there are also practical problems caused, for example, by the close integration of cocoa in a general pattern of production, including the growing of local food crops for the market or for domestic consumption. The use of sharecrop labour even on quite small holdings is another factor which makes clear distinctions difficult.[24] For the purposes of the present study, however, we may at least distinguish an upper layer of big farmers, set apart from the majority both in terms of their heavy reliance on hired labour and the dominant role of cash in the household economy. It is mainly within this stratum that we find the farmer-traders who have played such an important role in providing organisation and leadership in the cocoa community. We meet them in the pre-war farmers' associations, in the cooperative movement as well as in the Farmers' Council. As big cocoa farmers I have chosen to designate primarily those with 200 loads of cocoa or more. These farmers would normally employ several labourers. Their gross income by 1963/64 prices would be over £500. By the terms of the purchasing records, they may have constituted about five per cent of the farmers, controlling almost one-third of Ghana's cocoa. Going below 200 loads we still find holdings employing several labourers but we soon approach a level where farm owners themselves, together with their families, may supply a large portion of the labour necessary. Then we reach a level where hired labour may still be used but where the size of the holding no longer requires the labour of more than one adult person. That line may perhaps be put at 40 loads. Below such a line we may speak of small-holders, representing 60 per cent of the farmers in the 1963/64 sample, accounting for less than one-fifth of the sales. The stratum between the upper and lower ones, the "middle farmers", would then comprise one-third of the farmers, controlling half of the sales. These are clearly floating lines. What is important is not their precise location but the rough logic of the division. Above all it is important that we remember this highly differentiated pattern behind the facade of uniformity when we examine the methods used by the CPP (the Farmers' Council) in organising the agricultural producers.

The Cocoa Marketing Board and the local buying system

Cocoa production in Ghana has from the very beginning been in the hands of local farmers and their employees. The trade, on the other hand, was up to the Second World War controlled by foreign firms both on the exporting and local buying side. Occasional attempts by farmers to organise themselves to by-pass the firms in order to secure better prices and trading profits had mostly failed. The firms managed the local trade through a network of agents, big farmers and other "brokers". During the war a public export monopoly was established. It was permanented after the war as the Cocoa Marketing Board. The firms accepted the role as its "Licensed Buying Agents". This Board became a principal instrument of state accumulation in the CPP period. It also provided the institutional framework for the struggle over the local cocoa trade. We must therefore examine the structure and colonial origin of the Cocoa Marketing Board and the local trading system on which it was based.

In 1937, the foreign firms buying cocoa in the Gold Coast and in Nigeria entered into an agreement with the purpose of reducing "harmful" competition. Confidential market-sharing and price-fixing agreements were normal features of West African colonial trade.[25] The Gold Coast cocoa farmers responded with a hold-up of sales in combination with a boycott of imported merchandise. Similar actions had been undertaken before on a smaller scale.[26] This time, however, the cocoa-farming population was effectively mobilised. The hold-up was maintained successfully throughout the main crop season and was only called off when a commission appointed by the British government was able to arrange a "truce". The hold-up of 1937–38 is a major event in the political history of Ghana. The report of the commission is an important source of information, both on the cocoa trade and on the growth of farmers' organisations.[27] Although its recommendations were largely ignored it provided part of the rationale for the marketing board system.

Apart from the general aims of market-sharing the buying agreement of the firms seems to have been designed to "clip the wings" of the numerous intermediaries which did the actual buying on their behalf.[28] The commission accepted the view of the firms that the activity of these middlemen was a major reason why competition was excessive and marketing costs unnecessarily high, to the disadvantage of farmers as well as firms.[29] But the report was also critical of the methods used by the latter to introduce greater "order". Instead, it advocated an entirely new method whereby firms and farmers should be able to deal more directly with each other, dispensing of the services of the middlemen. It proposed a government-controlled organisation, a "Cocoa Farmers' Association", which should sell cocoa on behalf of the farmers.[30] The proposals were unacceptable to the firms, but they were never brought up for serious discussion. The war intervened, which permitted the government to prolong the "truce". The disruption of Atlantic shipping prompted the metropolitan government "to undertake the risk involved in purchasing the entire crops".[31] It became the monopoly buyer of West African cocoa. The actual trade was shared out among the foreign firms already engaged in accordance with previous performance. This meant in fact that the restrictions on competition sought by the firms were now given government backing.[32] The cocoa board set a "minimum price" to be paid to the producers.

The war-time scheme, rather than that proposed by the 1938 commission became the model for post-war marketing. The earlier proposals sought to strengthen the position of the farmers within a competitive market. The elimination of price competition through statutory control of producer prices was the central feature of the new scheme. The undesirable effects of price fluctuations formed a principal argument for not returning to the pre-war situation.[33] Cocoa Marketing Boards were established in the Gold Coast and Nigeria with a joint marketing company in London to do the actual selling. "Licensed Buying Agents" were appointed to handle local purchases. A producer price was to be fixed at the beginning of each season and buying agents were paid a fixed allowance per ton, including a profit margin varying with the price. Unlike the war-time scheme, the Board was free to license any buyer capable of handling a certain minimum tonnage.[34]

It was declared that the Board would under no circumstances make profits at the expense of the producers. The surplus would be used primarily to maintain stable producer prices. But it could also be used for other purposes "of general benefit to the cocoa producers and the industry", including research, control of crop diseases, credits, cooperatives and "the provision of other amenities and facilities to the producers".[35] However, the superior importance attributed to price stabilisation when justifying the scheme can be seen from the conclusion that "on the average of a period of years" the buying and selling transactions would approximately balance.[36] The ambiguous character, to say the least, of this declaration, became evident as the accumulated surplus of the new Board grew from the £14 million inherited from the war-time scheme to £76 million in 1950/51 season, well above the value of the entire crop. Except for one year the Board continued to obtain a surplus throughout the 1950's.[37] The character of the Board as an instrument of public accumulation was manifest well before the CPP began to steer these funds into its own development programme.

The immediate beneficiary was the British metropolitan economy. The massive surplus of the Board was invested in British government securities.[38] Fitch and Oppenheimer in particular have exposed this flow of resources from Ghana to Britain, which they regard as the real purpose of the colonial marketing boards.[39] Britain was anxious to control the hard currency earnings of the colonial producers for the reconstruction of the metropolitan economy after the war. This may also have involved postponing the demand from the colonies for industrial goods to a time when it better suited British industry. In justifying this outflow of funds it has been argued that it was necessary to withhold money from the producers in order to prevent inflation. But the surplus could have been used to build up productive capacity in Ghana instead of in Britain. The inflationary consequences would have been reduced even with higher producer prices, if the hard-currency earnings had been freely available to Ghana. After 1951, the CPP leaders shared the responsibility for allowing this continued outflow of resources. It may be argued that at this point they were not a in position to initiate any major local investments. Investing overseas may in this view have been the only way to save the funds for a later time when such local political capacity had been achieved. Some local expenditure from Marketing Board funds were undertaken in the early years, in particular for the control of the swollen shoot disease, including payments of compensation to farmers whose trees were cut out. In terms of the total resources available, however, these and other local payments were still quite small.[40] After 1951, an increasing portion of the overall surplus was channelled directly into the coffers of the local government by means of radically raised export duties. I return to these developments in Chapter 7.

What did the marketing-board system mean to the foreign-capital interests in the cocoa economy? At first glance there seemed little to commend itself in the new scheme from that perspective. The market sharing element in the pre-war and wartime agreements was no longer included. Farmers' cooperatives were officially encouraged. Statutory pricing and the export monopoly restricted the freedom of the trade. There was a risk that the West African marketing boards would be used for purposes which were not necessarily in the interest of the cocoa manufacturers. Their fears were hardly allayed by official declarations that the boards aimed "neither at the creation of state monopolies nor at the exercise by the West African cocoa producing territories of a dominating influence in the world market".[41] Opposition came primarily from American cocoa traders and manufacturers who had the additional fear that British trade interests would be given preferential treatment.[42] There was also some opposition at the British end, which contributed to the temporary postponement of the new scheme and the revision of the 1944 proposals. We need not go into the details of these objections. It is more important to focus on the factors which may explain the cooperation of the dominant British trade interests. This cooperation was a precondition for the functioning of the new system.

The position taken by Cadbury, the leading British cocoa manufacturer with its own buying organisation in the field, was crucial. At a manufacturers' conference in 1946 the company declared its general acceptance of the marketing boards. It supported the official argument that the trade cycle had to be evened out by statutory pricing in order to give African producers greater security and thus ensure a steady growth of supplies.[43] The company was alarmed by the critical state of cocoa production caused by a long period of reduced planting, poor maintenance and the rapid spread of swollen shoot. The boards were seen as a device to rehabilitate the West African cocoa economies. The powerful position of the company in the metropolitan market and its long-established and close connections with the local colonial government contributed to a justified confidence that its interests would not be compromised. The company was also directly represented on the Gold Coast Marketing Board. As a manufacturer, Cadbury was not primarily concerned with local trading profits. The main function of its local buying company was to ensure a steady, good-quality supply of cocoa at prices which at this stage would ensure a rapid recovery of production. The financial viability of its local operations would at the same time be ensured by the buying allowance agreed with the Board.

Although Cadbury's position was crucial, the largest local buyer was the United Africa Company group. They controlled almost half of the Gold Coast cocoa trade in the 1930's.[44] They did not represent cocoa manufacturers but general trading interests. Export-produce buying was closely integrated in their case with the trade in imported European manufactured goods, so-called merchandise. The same up-country trading organisation could be used for both purposes.[45] Overhead costs were shared. Why did a powerful firm like the United Africa Company, with its dominant position in the British

West African trade in general, agree to cooperate with a state monopolistic trading scheme which blocked the return to a "free" market after the war? It could be argued that the firm had little choice: either it would cooperate with a scheme supported by wider imperial interests and the principal cocoa manufacturers or find its share of the cocoa trade passing into the hands of others. On the other hand, it may also be argued that its position in the West African economy was sufficiently strong to influence the implementation of the scheme in its own favour. A "fair" return could be expected on the capital already engaged. As a temporary device state protection of profits may even have some attraction, especially against the background of the difficulties of the 1930's, as revealed by the 1938 commission. In a more long-term perspective, however, the position taken by the United Africa Company must be understood in terms of changes in the pattern of foreign commercial penetration and control.[46] There was a move towards a new "division of labour", with local entrepreneurs taking over "field" operations and with foreign capital specialising in more advanced forms of management and finance. The difficulties of the 1930's reflected a growing competition from below. Official support for cooperative marketing similarly reflected a recognition of these pressures and the attempt to create "constitutional" outlets for such local commercial aspirations and the desire to avoid confrontations of the 1937–38 type. The growing sophistication of local consumption patterns created at the same time profitable avenues for specialisation by foreign commercial capital. It could be left to local entrepreneurs to push the trading frontiers in the main lines of mass consumption goods already introduced on the market. It was less important to control cocoa buying in order to support the merchandise trade as far as foreign capital was concerned. I shall return to these changes in the structure of foreign capital interests when discussing the final "expulsion" of the firms from the local cocoa trade. (Chapt. 4).

When judging the reaction of British capital to the marketing-board system, it should be remembered that this system was introduced at a time when colonial political control had not yet been seriously challenged anywhere in British tropical Africa, nor in the metropole. The new colonial policies urged by the Labour government aimed at increased metropolitan responsibility for "colonial welfare and development", rather than at political disengagement.[47] It was certainly not realised that, in the Gold Coast, imperial control would be relinquished only a decade after the war. Nor was it realised that the operations of the Cocoa Marketing Board would be affected by strong nationalist pressures only a few years after its inauguration. The Board was gradually transformed into a vehicle of national aspirations. By 1950/51 the membership was reconstituted. Representatives of the cocoa manufacturers and the foreign cocoa traders were replaced by Ghanaians "with commercial experience".[48] After 1951, ministerial and thus increasingly CPP control over the Board grew. We shall return later to these changes. In the immediate post-war situation, however, it was the weakness rather than the strength of local political pressures which should be emphasised when explaining the confidence of the colonial government and firms alike when embarking on this major state capitalist venture.

Farmers, local traders and the export monopoly

The war denied farmers and local cocoa traders the fruits of their victory over the foreign firms in the 1937–38 cocoa hold-up. The temporary restrictions on competitive marketing accepted by their leaders in the truce of 1938 were continued in a more rigid form under the war-time marketing scheme. Representatives of the firms had even before the outbreak of the war suggested a prolongation of the quota-system of the truce for a substantial period (ten years or so), in order to eliminate "overlapping buying stations, as well as reducing supervisory and other costs".[49] This was rejected by the commission of inquiry with reference to the expected negative reactions of the producers. The truce itself had aroused strong dissatisfaction, which could be expected further to be strengthened if market-sharing were to be maintained on a long-term basis.[50] The views of the commission were ignored, at least for the duration of the war. Market-sharing was maintained, although in no way necessary for the function of either the state export monopoly or statutory pricing.[51] The war resulted in a freezing of the conflict at the point reached in the 1938 truce. This was highly unsatisfactory to the farmers and traders, who had successfully carried through the holdup. But the collapse of the international cocoa market drastically reduced their bargaining strength. As the end of the war and a recovery of the market was in sight, agitation revived. The immediate target was the low prices offered by the authorities, but the proposals for post-war marketing were also attacked. In September 1945, a delegation of cocoa farmers went to London to register their protests.[52] Not surprisingly, it was suspected that the foreign firms were sponsoring the new scheme. In October that same year, representatives of various farmers' organisations met in Suhum to decide whether a new holdup was to be organised or whether the announced producer price was to be accepted. After intervention from the Standing Committee of the Joint Provincial Council of Chiefs the latter line won.[53]

The market situation had turned strongly in favour of the producers. In 1937/38, prices had been falling. The resistance to the buying agreement of the firms was particularly strong because the agreement was believed to contribute to this decline. Now prices were rising sharply. Under such circumstances farmers' open resistance to continued export monopoly and price control subsided. Their resentment, as that of other cash earners in Ghana at the time, was instead chiefly directed against shortages and rapidly increasing prices in the consumers' market. Inflation and shortages played a central role in the dynamic political situation of the post-war years.[54] The rapid spread and devastating effects of the swollen shoot disease, the intense controversy over the appropriate methods of controlling it, including open resistance to the large-scale cutting of healthy trees for preventive purposes, contributed greatly to redirect the interest of the farmers away from the new marketing scheme, which at least for the time being seemed to offer higher prices.

With the intensification of the political struggle, farmers grew increasingly critical of the rising surplus of the Board. But the abolition of the Board was not widely demanded. It was difficult for the farmers to obtain a clear picture of the long-term implications of its price-stabilising function in the very erratic and unstable market situation of the immediate post-war years. As long as prices were rising, the removal of short-term fluctuations must have been seen as a definite advantage. But price fluctuations as such had not been part of the grievances which on several occasions in the past had mobilised farmers against the firms. Resentment had rather been directed against the way in which prices were thought to be manipulated by the latter. Both farmers and local traders wished to obtain a share of the profits of the firms. Statuatory pricing by the marketing board did not remove the threat of price manipulations, and experiences from the war had showed that the firms and the colonial government cooperated closely.

The rapidly rising surplus of the marketing board, however, redirected producers' interests away from the traditional conflict with the firms to what was happening to the money officially accumulated on their behalf. The all-important question of prices had at the same time been disengaged from its market context. The major confrontation was no longer between farmers and the foreign firms (still controlling the local trade), but between farmers and the government. I shall not attempt to discuss the nature of this confrontation in the pure colonial context of the pre-1951 period. Our concern is with the subsequent period, when elected leaders actively participated in shaping the policies of the Cocoa Marketing Board. As we shall see in the next chapter, this confrontation became acute soon after the CPP had entered the government and it even threatened to disrupt the party's plans for independence.

Something, finally, must be said about the organisation of the local trading system and how it was affected by the new public marketing scheme. The best source of information on the pre-war pattern is still the report of the 1938 commission of inquiry, although obviously based on rather hastily gathered facts. The report suggests that the total number of brokers and sub-brokers operating under the firms may have been as much as 40,000.[55] The brokers, including the very large ones, were often cocoa farm-owners as well. The figure implies that perhaps as much as every tenth farmer may have been engaged in cocoa trading. Average purchases per trader in such case would have been less than ten tons. As evidence, the report quotes information from one firm which engaged 20 brokers, buying a varying amount from 50 to 1,000 tons each or 4,000 altogether. The number of sub-brokers in this case was estimated at 500. A broker handling as much as 1,000 tons of cocoa was a substantial businessman in his own right. The distinction between brokers and sub-brokers was not a matter of two layers only. In many cases cocoa passed through a chain of intermediaries before it reached the firm, including different types of traders, from farmers selling a neighbour's crop to big "professional" traders handling the output of thousands of farmers. The dependence of individual brokers or sub-brokers on their respective foreign principals also varied. Some had their own buying equipment and storage facilities and employed their own current capital in the business. Others depended exclusively on their "employers" in both respects. Even when they had independent resources brokers were normally bound to deliver to a specific firm either by a simple agreement or by accepting a semi-employed status with a small salary or a retaining fee in addition to their normal trade commissions. The rate of such commissions was usually "progressive", that is, the more cocoa that was delivered, the higher was the rate per unit.

The growing power of the brokers vis-à-vis their principals was a major factor behind the 1937 market-sharing agreement concluded by the foreign firms. The agreement sought to protect declining profits against the "excessive" claims of the brokers. A contributing factor to this shift in the balance of power was probably tendencies of concentration among the brokers themselves. The report noted that larger brokers had become more important in recent years.[56] The firms were faced with a smaller number of big brokers, each exercising a high degree of independent control over the necessary supply channels of the market. Not all firms relied on brokers alone. Cadbury bought largely through its own directly employed staff, either at fixed buying stations or by touring "scale-buyers". Even this system permitted brokerage at a lower level, but it was less dependent on the big brokers.

The complex multi-layer system was held together by the flow of credit. Firms advanced trading capital to brokers, brokers to sub-brokers, and these in turn to their agents or to farmers, all against promises of future deliveries. The scope for commercial profits was to a large degree dependent on the skilful management of these flows. At each level, brokers in addition usually employed their own capital. The system depended ultimately on the farmers' demand for credits, which in turn was closely linked to the seasonal nature of production. Pledges and sales of future crops had created a wide-spread pattern of short-term indebtedness. The line between short-term trading advances and general moneylending (where crops were pledged as security) was floating. Many of the local brokers would perhaps be more correctly described as money-lenders than as cocoa traders. Certain non-Ghanaian groups, especially the so-called "Lagosians", a designation referring to Yorubas or Southern Nigerians in general, were prominent in this field of combined money-lending and cocoa trading. The typical creditor-broker, on the other hand, seems to have been the big or middle cocoa farmer, who in this way could find a natural outlet for his surplus capital, especially as investment opportunities in the rural areas, outside the cocoa economy itself, were limited.

How much of the pre-war trading structure survived the introduction of the export monopoly and statutory pricing? Bauer has noted that the new scheme did not seek to regulate the brokerage system, although dissatisfaction on that point expressed by firms as well as by the 1938 commission had been a major argument for reform.[57] But the seasonal price decisions of the Marketing Board had important implications for local trading. The scope for speculative profits based on short-term price changes was eliminated. Nor could profits be made on the basis of deficiencies in the transmission of price information at the local level. The importance of distance and transport costs in calculating the effective producer price was also reduced when the same official price became applicable at major railheads, irrespective of the distance to the port. Such factors may have cooperated in narrowing the trade margin available for the intermediaries. The result must have been a certain "rationalisation" of the local trading structure, reducing the number of stages through which cocoa passed on its way from the producer to the Licensed Buying Agent. On the whole, brokers were unlikely to be happy about the new marketing system and the restraints which it imposed. The sharp increase in the competition between buyers in the late 1940's and early 50's made it possible at the same time for brokers to demand higher commissions.[58] It is not clear how much of this had to be passed on to the farmers.

The new system, however, created opportunities for private buyers and cooperatives to trade directly with the Board without having to pass via the foreign firms. This was an important step in meeting some of the frustrated commercial aspirations of farmers as well as local traders. Although they were barred from establishing themselves as exporters in their own right, they found that the prospects of increasing their share of the local cocoa trade improved. This was particularly so in the case of the cooperatives which were given active support by the colonial government. Cooperative marketing was seen as a way to reduce antagonism between the farmers and the foreign traders and to limit the dependence of the industry on the "middlemen" who until then had occupied such a critical position between the firms and the farmers. The successful hold-up and boycott of 1937–38 was not only a threat to the short-term economic interests of the foreign firms. It also posed a more general threat to the peaceful evolution of the colonial system. It was essential to create institutional outlets for the "legitimate" commercial aspirations of the farmers in order to prevent the cocoa trade from becoming a platform for anti-colonial agitation. The 1938 commission had little confidence in the existing cooperatives, which at the time bought less than three per cent of the crop. It suggested instead an entirely new type of centrally controlled organisation which was to sell cocoa on behalf of the farmers.[59] The scheme had in fact much in common with the monopoly organisation to be established by the United Ghana Farmers' Council. At this time, however, the scheme received no official support. The 1944 and 1946 white papers declared it to be the intention of the colonial government to encourage the cocoa farmers to assume increasing responsibility in marketing the crop through cooperatives.[60]

The stage was set for a multicornered struggle over the colonial inheritance in the cocoa trade. Local commercial interests pursued their aspirations inside as well as outside the officially supported organisations. The licensing system ensured a strategic role for the Cocoa Marketing Board and the government in determining the pace and direction of the struggle. The Board supplied much of the current capital of the trade through a system of advances. The foreign buyers continued to dominate but important inroads were made by the cooperatives and later, by the CPP-supported Cocoa Purchasing Company. The decisive break with foreign control over local cocoa trading occurred after independence, when the CPP government, ensured of political control, backed the monopoly claims of its own farmers' organisation. Let us turn to the origins of this institution and its road to commercial and political supremacy.

III. The Origin of the Farmers' Council

Early on Saturday morning April 8, 1961, Kwame Nkrumah delivered a radio speech, the *Dawn Broadcast,* "in accordance with the cherished customs of our fathers, whereby advice is sought or given at early dawn."[1] The speech was largely an attempt at public self-criticism of a regime which at this time had been in full political control for four years. It has been widely cited as a major expression of the radicalisation which the regime underwent in the early 1960's.[2] Members of the ruling Convention People's Party who used their positions for private gains were attacked. CPP members of the National Assembly were asked to resign if they wished to pursue private business. Extortion, corruption and nepotism in the party were condemned. Criticism was also directed against the inefficiency and waste of the Civil Service. The *Dawn Broadcast* greatly stirred public discussion in Ghana, although its code of conduct was never seriously enforced. One part of the speech, however, had immediate, concrete and far-reaching consequences. It was the announcement of the government's final decision on the nationalisation of the cocoa trade. Directions had been issued that the United Ghana Farmers' Council should be given the sole responsibility for organising the purchase of all cocoa produced in Ghana on behalf of the Cocoa Marketing Board.[3] The Board had been instructed to terminate the licences of foreign firms, cooperatives and local private traders. Three years earlier the Farmers' Council had bought only seven per cent of a much smaller crop. For five years, from 1961 to the military coup in 1966 when it was dissolved and banned, the Farmers' Council handled the entire cocoa crop. The establishment of the local cocoa-buying monopoly was a step in the direction of a new political-economic order, where the party and the state increasingly merged and greatly extended their control over the economy. The *Dawn Broadcast* announced the integration of the Farmers' Council with the party. The Council, as well as the Trade Union Congress and other, "integral wings" of the party, would no longer have a separate membership.[4] The cocoa farmers were collectively enrolled in the party through the purchasing monopoly of the Council.

The present chapter is concerned with the development of the United Ghana Farmers' Council as a political and commercial organisation in the pre-monopoly period. The records for the early period at my disposal are fragmentary and the picture which can be drawn will be tentative and partly inconclusive. A systematic study of the history of farmers' and traders' organisations in the pre-independence period has not been attempted. The origin of the Farmers' Council in the tradition of farmer-trader organisations is emphasised here. It will be shown how the conflict between the sectional aspirations of these groups and the wider political objectives of the CPP isolated the Council from this tradition. In the late 1940's it had been possible to enroll cocoa farmers in an alliance of anti-colonial forces. But attempts to establish party control over existing organisations met with difficulties. The Farmers' Council was established as a CPP-sponsored organisation in 1953. It depended closely on the financial and administrative support of the Cocoa Purchasing Company, a CPP controlled subsidiary of the Cocoa Marketing Board. I will discuss how the price and tax policies of the CPP government caused a head-on clash with the cocoa farmers. The frail organisation established by the Council was disrupted. Its identification with fiscal policy reduced its support. But regional cleavages prevented the cocoa farmers from uniting in opposition to the party. The collapse of the Cocoa Purchasing Company as a result of financial irregularities and malpractices caused a new crisis in the development of the Council, but its fortunes were reversed as CPP power was consolidated after independence. With the financial support of the government, the Council took over the cocoa business of the defunct Company. It was awarded official recognition as a national organisation with exclusive rights to represent the country's farmers. The political monopoly of the Council preceded and was more comprehensive than the commercial monopoly. Cocoa trading provided the essential financial basis of the wider political activity of the Council. The close connection with the cocoa trade indicated at the same time its narrow base in the agricultural economy as a whole.

The cocoa farmers and the nationalist movement[5]

The early nationalist movement had mostly a narrow urban middle-class base. Leading members were lawyers and businessmen. The United Gold Coast Convention which was formed in 1947 was in this respect only one in a series of similar groupings which had emerged in the coastal urban societies of colonial West Africa. The numerical strength, the aspirations and frustration of these groups had no doubt grown over time. The expansion of the exportmarket economy had strengthened resentment against the continuing grip of foreign capital over the import-export trade. The colonial government and the big firms offered few openings for educated Africans, except at the lowest level. The government continued to play off the political aspirations of the urban groups against the more cooperative chiefs, the "natural rulers", which were acclaimed as the true representatives of the silent rural majority. The post-war situation made it possible to link the grievances of the urban middle-class to a broader anti-colonial movement stimulated by the war and the unsettled social and economic situation which followed. Shortages, inflation and unemployment, especially among the demobilised troops, were part of the hardships most frequently mentioned. The military and economic decline of the old colonial powers and the rise of anti-colonial forces on a global scale had radically changed the international conditions for continued colonial occupation. A broader section of the population was drawn into the nationalist movement, including businessmen, workers, clerks, teachers, ex-servicemen, unemployed school-levers and others, still largely from the urban sector, although many with close contacts in village society. Some of these groups had already established local and regional associations capable of mobilising their members for wider political purposes.

It was to these groups, especially the so called Youth Associations, that Kwame Nkrumah directed his party-building efforts when engaged as the full-time secretary of the United Gold Coast Convention in 1947. These groups were also to provide the main basis of the Convention People's Party, established under Nkrumah's leadership in 1949, after he had broken with the more conservative leadership of the Convention. The radicalisation of the movement was accelerated by dramatic events in 1948, when foreign firms were boycotted because of the increasing prices of imported goods. The confrontation culminated in a clash between a procession of ex-servicemen and the police in Accra and in riots in the major towns. Foreign firms were looted and burnt.

Although certainly stirred by these events and by the intensified political activities of the Youth Associations of the small towns, the rural masses—the farmers, the farm-labourers, the small-village traders and craftsmen—were only marginally engaged in the organised nationalist movement. The cocoa farmers certainly had grievances of their own. Initial opposition to the public export monopoly had receded, but the price policies and the surplus of the Cocoa Marketing Board were widely criticised. More acute was opposition to the compulsory cutting-out of cocoa trees. The government attempted to control the swollen shoot disease which already had destroyed production in some of the most prosperous cocoa districts and was rapidly spreading to new areas. The authorities considered it necessary to cut out not only trees already affected but also the surrounding, healthy ones. To the farmers it meant destruction of capital which might have been able to yield substantial returns for many more years to come. Hostility to the government among the affected farmers was intense and there were cases of open clashes.[6] Cocoa farmers' opposition to compulsory cutting-out contributed to the strength of the anti-colonial movement. The political leaders supported the demands of the farmers and attempts were made to bring the various existing farmers' organisations together into an anti-colonial front. The 1937/38 cocoa hold-up had demonstrated the organising potential of the cocoa farmers and their ability to threaten colonial interests. (See chapter 2) Nkrumah had been successful in using existing youth organisations, with their predominantly literate, urban membership, as a platform for the party. But the attempt to incorporate or establish party control over the farmers' organisations was fraught with much greater difficulties.

The information available to me on the political organisation of the farmers in the 1948–1953 period contains many unclear points. The official chronicle produced by the Farmers' Council for its ten-year anniversary suggests that the first move towards a united farmers' front was made in 1950, when representatives of all farmers' organisations in the country met in Accra on the invitation of Nkrumah.[7] This meeting was followed by two others, one in Accra and one in Nkawkaw. A committee chaired by E. A. Haizel was set up to draft a constitution for a united farmers' organisation. This was adopted at the founding meeting of the Council at Prempeh Hall, Kumasi, on the 26th September, 1953.

The Jibowu Commission, investigating the affairs of the Cocoa Purchasing Company referred in 1956, to evidence from one farmers' leader, Ashie Nikoi.[8] He maintained that he and Nkrumah discussed the question of organising the farmers when they were both political prisoners in James Fort Prison. The purpose of the organisation, according to Nikoi, was to enable the farmers to bargain collectively for the sale of their cocoa. The Report continued that "it is clear from the evidence of Dr. Nkrumah that when they both regained liberty Ashie Nikoi was entrusted with the duty of organising the farmers, which he did and the organisation was known as the Ghana Farmers Congress". The commissioners were satisfied that owing to a difference between Nikoi and Nkrumah over the amount of control which the CPP was to exercise, the Congress split, one section following Nikoi, the other being organised by A. R. Dennis and others on the instructions of Nkrumah, leading to the establishment of the Farmers' Council. According to Austin, Ashie Nikoi and John Ayew founded the Farmers' Committee of British West Africa in 1939.[9] It was one of many organisations of cocoa farmers and cocoa traders established during the inter-war period, with the main purpose of trading in cocoa. Nikoi and Ayew joined the leadership of the United Gold Coast Convention. Nikoi even became a member of the Central Committee of the CPP. In December 1949, still according to Austin, they formed the Ghana Farmers' Congress out of their own Farmers' Committee and two other farmer-trader organisations from the late 1930's, the Asante Farmers' Union and the Gold Coast Farmers Association (Sika Mpoano Akuafo Fekuw Ltd). Nkrumah addressed the inaugural meeting. Two leading members were arrested during the intensified political struggle in 1950 ("Postive Action"). In October 1950, the Congress met again in Nkawkaw (the date is substantiated by references to the minutes of the meeting) and was again addressed by CPP leaders. The meeting strongly condemned the cutting-out policy and the Cocoa Marketing Board. The first meeting had petitioned the colonial government and demanded direct representation of the farmers in the National Assembly but the second one agreed to sponsor candidates through the CPP.[10]

One who claims to have been among those who planned the Ghana Farmers' Congress has left his own account. This is Kwabena Koranteng Ameyaw who petitioned the National Liberation Council after the 1966 coup in the name of the Eastern Division of the National Farmers Union. He claimed that the Congress was not formed by the CPP but had been planned mainly by Nikoi and one Albert Owusu Buruku (later chairman of the Akwamu Farmers' Union). It was inaugurated in Nsawam (Accra, according to Austin) by "the veteran John Kwame Ayew, who started the Farmers' Liberation Movement in 1924". The Nkawkaw Conference had been organised and financed by the Congress itself and not by the CPP. But it was divided. Some of its leaders resented the attempts by the CPP to establish political control. Ameyaw noted that the CPP succeeded partially in splitting "the united front".[11] Apart from some confusion over dates and localities the various accounts are mutually compatible. Two aspects of particular importance for the present analysis should be underlined. Firstly, it seems as if the CPP did not establish a farmers' organisation of its own but sought to operate through existing organisations. This was during the most militant phase of the struggle, that is, up to the 1951 elections, when the victorious CPP leaders were given ministerial responsibilities. It also seems as if these farmers' organisations, unlike the important youth associations, were not closely integrated with the party. The second point relates to the social basis of these organisations, which more or less enthusiastically cooperated with the CPP. They had been established primarily in response to an interest in the marketing of cocoa. This formed a narrow social base in two ways. Firstly, and most importantly, these organisations did not involve the mass of non-cocoa farmers, not to mention the farm-labourers. Secondly, in the context of the cocoa-farming community itself, they were controlled by an upper stratum of farmer-traders, with commercial experience and/or aspirations.

It is not clear to what extent the Ghana Farmers' Congress succeeded in enrolling other farmers' organisations apart from those listed among the founders at the December 1949 meeting. It is particularly noteworthy that the cooperatives which by the end of the 1940's were by far the strongest local cocoa-buying organisations were not mentioned in connection with the Congress. Official colonial support and the successful expansion of their trading may have made them a less likely candidate for the anti-colonial front which the Congress sought to establish. The impression is thus gained that the mid- and late 1940's was a period when the cocoa-farming communities were engaged in a number of attempts to establish local and regional organisations. Several factors contributed to this ferment, including the impact of the export-monopoly, the unstable market situation, the swollen shoot, and the influence of nationalism. In the background was the important experience and self-confidence generated by the successful cocoa hold-up and trade boycott of 1937/38.

An important aspect of these organising activities was the need to identify local leaders capable of representing the farmers either for commercial or wider purposes. In 1937/38, local chiefs partly acted in this role. But the Nowell Report emphasised the importance of non-traditional leaders in the organisation of the hold-up. The commissioners were told in evidence that it had long been the practice, at least in certain districts, for Head Farmers to be chosen by the farmers as their advisers and as arbitrators in cases of dispute.

This had been done particularly in connection with efforts to sell cocoa directly to overseas markets. In the late 1930's, such Head Farmers were brought together in regional unions.[12] They were also called Chief Farmers. This was the term later to be adopted by the Farmers' Council for its local representatives.

The origin of the institution of Chief or Head Farmers is not altogether clear. A Farmers' Council document suggests that before the year 1948, "the year in which the country became politically conscious", there existed various farmers' organisations which "made room for the recognition of nominated 'Chief Farmers'". The document further suggests that these Chief Farmers were nominated by the Chief or Headman of the village without consulting the farmers and that it was the duty of the farmers to recognise the person so appointed. This person served as "the bridge (the liaison) between the Chief or Headman and the farmers in the locality".[13] Why was there a need for a special liaison officer between the chiefs and the farmers? Different explanations are possible. This type of Chief Farmer may have had special functions in relation to the export crop producing farmers, especially if the latter operated in outlaying camps where they were not permanently settled and therefore had no separate local authority. Considering the important role of migration across traditional boundaries of jurisdiction, such Chief Farmers may have been particularly adapted to the need to regulate the relations between local chiefs and stranger farmers. It is also suggested that Chief Farmers served as a liaison between farmers and the colonial government. A Memorandum by the Managing Director of the Cocoa Marketing Board in 1966 mentions that before the Second World War Chief Farmers had been appointed by the government.[14] This could be compatible with the proposition already noted that some Chief Farmers were nominated or appointed by local chiefs. Some of these may in this respect merely have acted on instruction to meet requirements of the government for access, especially to the cash crop producing farmers. Such nominated Chief Farmers could, for example, be encouraged to take the lead in introducing improved practices or otherwise communicate relevant information to the wider farming community. The connection between the Chief Farmers and the promotion of improved practices and cooperative marketing is stressed in a Memorandum from a farmers' group in Suhum to the deGraft-Johnson Committee in 1966. It stated that farmers had formed themselves into groups with Chief Farmers and elders as supervisors and that they were taught to improve quality and to adopt the "group marketing system".[15]

Nominated semi-official Chief Farmers thus existed alongside or perhaps prior to those elected by the farmers themselves. The tenth-anniversary chronicle of the Farmers' Council even claims that the election of Chief Farmers was a result of post-war developments. The 1948 political awakening in the country "brought with it the need for the election of officers thus putting an end to the former unprogressive step of nominating people who did not enjoy the full support of the farmers".[16] While waiting for a more systematic investigation of the origin of the Chief Farmer institution, we may conclude that by the 1940's a tradition of farmers' representatives was well established. Some may have been nominated by chiefs and colonial authorities, others elected by the farmers themselves. Some had been engaged in commercial and political actions, directed in particular against the foreign cocoa-buying firms, but also more lately, against the state export monopoly and the compulsory cutting-out of cocoa trees under the swollen-shoot control programme. They had organised themselves, or been organised, in regional associations, either for direct commercial ends or for general representative purposes. Efforts were now made by political leaders to unite these groupings and mobilise them on the side of the nationalist movement.

What role did chieftaincy play in the attempt to enroll the farmers in the anti-colonial movement? Chiefs cannot be treated as a monolithic category with uniform interests in this respect. The structure of chieftaincy varied strongly between different parts of the country, as did the personal background and outlook of individual chiefs. The radical turn of the political struggle, however, intensified in many places the conflict between chiefs and "commoners", as organised, for example in the "youth" associations. The conflict was influenced by the support awarded chieftaincy by the colonial government. Chiefs had long been acclaimed the "natural" spokesmen of the people in official attempts to undermine the rival claims of new nationalist leaders. But chiefs had been closely involved in support of the cocoa hold-up in 1937/38 and many supported similarly the 1948 boycott of European goods. In some areas, however, the conservative stand taken by leading chiefs on the national issues led to open confrontation. In Ashanti, the Asantehene attacked the "agitators". He deplored the intention of the youth to import a foreign system of government: "You people claim to be nationally conscious, and therefore undertook to fight against rising prices and black marketing. You claim to be waving the torch of emancipation, yet you have proved yourself to be after your own personal interests. What do you want us to make of your clamour for Self-government?"[17] The difficulties in some areas to by-pass the chiefs in the attempts to reach out to the farmers were considerable. This may be illustrated by the subsequent proclamation by the Asantehene that if Krobo Edusei, the leader accused of organising the 1948 unrest in Ashanti, "engineers another boycott or travels through Ashanti for funds, he should be arrested". The Asantehene instructed his chiefs to beat the gong-gong and make it clear to the people that "the recent reductions in prices have not been due to the efforts of Krobo Edusei".[18] In the less traditionalist southern provinces, the effective powers of the chiefs to obstruct the nationalist movement were not as great as in Ashanti and the North.

The suspicious and sometimes hostile attitude of certain chiefs was, however, only one of many difficulties which obstructed an effective direct appeal by the nationalist leaders to the rural masses. The problem of political mobilisation in this fragmented, essentially non-literate peasant society was great also without such particular barriers. The basic means of communication were poor and the leaders had extremely limited financial resources at their disposal. In this situation, it was the more necessary to seek to utilise as far as possible existing organisations. The farmer-trader groups of the cocoa areas were perhaps the most advanced organisations which so far had emerged in the rural economy. They were the natural target of the organising effort of the nationalists.

The alliance of farmers' groups within the Ghana Farmers' Congress did not survive the CPP election victory of 1951 for very long. Ashie Nikoi, the main architect, was expelled from the Central Committee and the party in early 1952. Two other leading party men, Dzenkle Dzewu and Mate Kole, who were said to have been engaged in the reorganisation of the Congress after the split over the relationship with CPP, were expelled at about the same time.[19] The material indicates an important difference between Nikoi and the CPP leadership over the local cocoa trade. It was an issue of central importance to the organisations which had joined the Farmers' Congress. The Congress demanded that a company sponsored by them should be licensed to buy cocoa. They also demanded that Cocoa Marketing Board funds should be used for loans to the cocoa farmers, and more specifically, to relieve indebtedness. Both schemes could be turned into powerful instruments for mobilising the farmers politically. The assumption of ministerial responsibilities by the CPP leaders after the 1951 elections opened ways for the party to make use of the Cocoa Marketing Board for such purposes. But it was no longer necessary to rely on the shaky alliance achieved in the Farmers' Congress. Resources were now available for the party to intervene more directly.

The new governmental powers of the CPP led to a revision of its attitude to the Cocoa Marketing Board. Before 1951, it had supported the strongly anti-CMB position taken by many of the farmers' groups. The Ghana Farmers' Congress declared in 1950 that the export monopoly meant that "the entire Cocoa Industry once owned by peasantry had as it were been nationalised and converted into State property with no compensation whatever".[20] It is perhaps unlikely that the party leadership would have underwritten this statement but it promised in its 1951 election manifesto that the "farmers will be given control of the vast funds with the Cocoa Marketing Board to use for the benefit of the farmers primarily, and the country in general".[21] The last words did of course hold the doors wide-open for a way of using the funds which was opposed by the farmers' groups. Once more closely associated with the government, the CPP leaders showed no desire to discontinue the policy of using Marketing Board funds for general development expenditure. On the contrary, the "conversion of cocoa farmers' reserve funds into public funds" which had been opposed by the farmers proceeded at an accelerating pace. Many farmers clearly saw this as contrary to the election promises.[22] Disagreement on this point contributed to the disruption of the Farmers' Congress, and to its ineffectiveness as a platform for continued political work among the cocoa farmers.

Leaders of the Farmers' Congress were as anxious as the CPP government to seek to employ the tremendous regulative and financial resources of the Cocoa Marketing Board to further their own plans. In their case such plans were primarily linked to the desire to control the local cocoa trade. In 1951, Nikoi was appointed a member of the Cocoa Marketing Board. It has been suggested that he later approached the Board for a loan in order to set up a Cocoa Purchasing Company to be the Board's sole agent for the purchase of cocoa in the country.[23] The demand for a monopoly of this type was clearly unacceptable to the government which was still largely under direct colonial control. The colonial government had recent commitments, not only to the foreign firms but also to the cooperative movement which it had helped to build. The idea that the Board should sponsor a local cocoa-buying company alongside other Licensed Agents, however, was actively pushed by the CPP leaders.

The Cocoa Purchasing Company established in 1952 was very different from that envisaged by Nikoi and other Congress leaders. It was wholly owned by the Cocoa Marketing Board and by-passed entirely the Congress leadership and existing farmer-trader organisations. A.Y.K. Djin, a businessman and the chairman of the CPP Finance Committee soon became Managing Director. Nikoi, according to deGraft-Johnson, had objected to CMB control over the Company and resigned from the Marketing Board when it decided to go ahead with its own (and CPP-controlled) project.[24] Ameyaw, one of the early Congress leaders with a non-CPP background in the farmer-trader movement, recalled in 1966 that a company, the Ghana Producers' Association, was established in 1951 by members of the Congress group. It is quite possible that this was the company for which Nikoi sought CMB finance and recognition. Ameyaw states that Nkrumah told the Executive Committee of the Company "that he was not going to give loans to the Association unless it was brought under his political control". At this juncture, opinions in the committee were divided. But they "refused to enter into any joint venture by allowing the Firm to be brought under the political control of the Cocoa Marketing Board, a statutory body . . . controlled by the Executive (Cabinet)". Nkrumah replied, still according to Ameyaw, that if they insisted on having an independent organisation, they would have to find their own capital.[25] Not willing to join hands with the CMB- and CPP-controlled Cocoa Purchasing Company, the farmer-trader groups supporting the Producers' Association decided to go it alone and to seek to finance the company from share-subscriptions by the members "without further demands being made on Kwame Nkrumah and his regime". Ameyaw says that he himself led the Colony Branch (the southern provinces) and J. K. Opoku the Ashanti Branch of the Association.[26] The extent of the rift in the relations between CPP and some of the important farmers' groups of the previously CPP-supported Farmers' Congress was indicated by the subsequent refusal of the CMB to grant a licence to the Association.[27] The CPP government, by establishing the Cocoa Purchasing Company, chose a way to deal with the cocoa trade and the question of how to organise the farmers which left little room for existing organisations. The Purchasing Company became a principal instrument for the promotion of a new famers' organisation, this time closely affiliated with the party, the United Ghana Farmers' Council.

The Farmers' Council and the Cocoa Purchasing Company

In a memorandum from the mid-1950's, the General-Secretary claimed that the Farmers' Council became known "only when the Cocoa Purchasing Company was created for its members to sell their cocoa".[28] In fact, the formal establishment of the Company preceded that of the Council by a year. The new CPP-affiliated farmers' organisation took shape within the framework set by the trading and credit operations of the Company. The latter provided the financial basis for the particular strategy chosen for organising the farmers. In the words of the General-Secretary, the Cocoa Purchasing Company was the "Holy Ghost" of a Trinity where the CPP was the Father and the Farmers' Council the Son which had been sent by the Father "to be born of poor parentage to rub shoulders with the common people and win them to Him".[29] With a rather different type of symbolism, the Cocoa Purchasing Company was described as the "atomic bomb of the CPP" by a leading member of the party.[30] The CPP leaders attached high hopes to the Company as a potentially powerful instrument in the local political struggle.

The Company was incorporated in June 1952, with a share capital of £2 million, wholly owned by the Cocoa Marketing Board. Its main function was to buy cocoa in competition with the other licensed agents of the Board.[31] In the 1952/53 season, the Company bought six per cent of the cocoa crop.[32] In 1953/54, its share had increased to 18 per cent, a level which was maintained for two additional years. In 1956/57 it declined and in 1957 the Company was finally liquidated as a consequence of large-scale financial irregularities, investigated by a commission of enquiry. This was the Jibowu Commission, named after its chairman. The success and failure of the Cocoa Purchasing Company as a commercial-political front organisation was related not only to its cocoa buying operations but also to its role as manager of a Loans Agency, established for the purpose of relieving indebtedness. This was done by giving farmers credit to redeem farms which had been pledged (mortgaged) to money-lenders. Again, the Marketing Board provided the capital. That CMB funds could be used for such ends had previously been suggested by the colonial government.[33] The Ghana Farmers' Congress had demanded action in this direction. The first funds for the Loans Agency were released by the Marketing Board in September 1953, the same month the Farmers' Council was founded in Kumasi. One year later, almost £2 million had been received. The last payment, in September 1955, brought the total to £2.8 million.[34] Irregularities in the administration of the loans scheme and the inability to recover outstanding loans were important contributions to the collapse of the Company.

In the General-Secretary's account of the origin of the Council, Djin, the managing director of the Cocoa Purchasing Company, was mentioned as one of the most important persons in conceiving the Farmers' Council.[35] Djin did not hold any position in the Council itself. It is probable that the organisers of the Company did much of the ground work in 1952/53, preparing the way for the final establishment of the Council later in 1953. The disintegration of the Farmers' Congress had created the need to establish a new network of representatives who could claim to constitute a national farmers' organisation. The limited trading activities of the Purchasing Company in its first year suggest that the Company itself had not come very far in its own organising work. The build-up of the Company and the Council may therefore be seen more as a parallel process. But the Company was first in the field and certainly behaved as the senior partner.

A "Joint Regional Executive Conference", in May 1953, seems to have taken the initiative in convening the constitutent meeting of the Council at Prempeh Hall in Kumasi, September 1953. The minutes of the meeting indicate that some kind of regional structure had already been established. The first and last President of the Council, the National Chief Farmer, was before his election referred to as the Ashanti Chief Farmer.[36] The character of the Farmers' Council as primarily a cocoa-farmers' organisation was clearly revealed on this occasion, not only in the complete absence of representatives from the regions to the north of Ashanti but also in the way in which regional representation under the new constitution was allocated according to the level of cocoa production. It was argued that this overrepresentation of the cocoa regions was justified because the organisation was to be financed from cocoa money. The Council's financial dependence on the Purchasing Company was almost complete. The Company had agreed that for every purchased load of cocoa it would deduct three-pence from the commission contained in the buying allowance. The deduction equalled only one-third of one per cent of the current price paid to the producers. In 1954, the Farmers' Council estimated that it would obtain over 80 per cent of its revenue from this source.[37] As other estimated revenue was very uncertain, it is probable that the cocoa deductions accounted for an even larger share of actual income. At then-current levels of purchases, expected revenue from the deductions amounted to some £25,000. The Farmers' Council pressed for additional direct grants-in-aid from the Company and for an increase in its share of the cocoa-buying commission. By 1955, its entitlements seem to have been approximately £35,000.[38] It is not clear how much actually was received. For all practical purposes, the Farmers' Council was financed from the commercial operations of the Cocoa Purchasing Company.

The activities of the Purchasing Company and the Farmers' Council were closely integrated. The Company appointed district officers (supervisors), whose task was to establish contacts at village level with farmers prepared to sell cocoa to the Company. A "Farmer-Receiver" did the actual buying, with funds obtained from the Company. Particularly important was the distribution and collection of advances. The Farmer-Receivers were not necessarily either local men or farmers, although they usually had some cocoa trading experience as sub-agents of other Licensed Agents. To facilitate the enrollment of farmers with the Company, local Seven-man Committees were formed, which were also to supervise the Farmer-Receivers. Apart from their commercial functions, the Seven-man Committees served as the grass-roots of the representative structure of the Farmers' Council, with District Committees and District Chief Farmers as the next step of the ladder. It is probable that some local groups had already been established during the days of the Farmers' Congress or earlier. Individual farmers and farmer-traders in particular who had shown themselves sympathetic to the CPP in the 1951 election campaign provided other points of departure for the new organisation. Party functionaries in the small rural towns, even when not farmers themselves, normally had close family ties with the farming community and could help in the recruitment of committees and Chief Farmers.

No commitment to the political aspirations of the CPP was required in order to realise the advantages of selling cocoa to the Cocoa Purchasing Company or to join as a member of the Seven-man Committees of the Farmers' Council. Several direct and material benefits were involved. Firstly and most immediately, the Company offered generous advances against future deliveries of cocoa. These off-season advances were an established part of the competitive weaponry of the cocoa trade. They were meant to tide farmers over the lean period preceding the main crop. The Cocoa Purchasing Company was well equipped to make a good show in this particular field of competition. Unlike for the foreign firms, the problem of profits was not pressing. The Cocoa Marketing Board provided operating capital free of charge. It was only necessary to ensure that excessive amounts of the funds required for actual payments were not tied up in outstanding advances. On the other hand, the need for buying cash was limited as a result of the 90 per cent advance paid by the Board on the declaration of purchases by the buying agents. Only the remaining ten per cent had to wait until the cocoa was delivered at the ports. Although this applied to all the licensed agents, it meant in the case of the Cocoa Purchasing Company that the interest-free capital initially supplied by the Board could be used with greater freedom. The subsequent inability to recover outstanding advances and the use of the Loans Agency to convert such short-term, off-season advances into credits for the "relief of indebtedness", were some of the major points raised against the Company.

The Purchasing Company as well as the cooperatives had another competitive advantage over the foreign firms in the payment of commissions to the farmers on the basis of tonnage sold. A commission was included in the cost schedule of the buying allowance. In the case of the firms, it had been used largely to remunerate brokers who in turn had passed on a part to their own sub-agents. By also drawing on the element of profit in the allowance, first the cooperatives and now the Purchasing Company were able to pay substantial commissions to ordinary farmers. The total commission offered by the Company was roughly three per cent of the producer price. Less than half, however, was received by the farmer. In 1955, the Working Committee of the Farmers' Council proposed that the £3.17/8 commission per ton offered by the Company should be distributed with £1.10/—to the receivers, the same amount to individual farmers, and the balance to the Farmers' Council.[39] In spite of the deductions, it is probable that the figures were quite competitive, at least as far as the foreign firms were concerned. But the latter, as well, were forced to increase or pass on commissions at the expense of profit margins in the new market situation.

The third and most formidable resource of the Cocoa Purchasing Company was the Loans Scheme. In this case, neither firms nor cooperatives were able to compete. It was a powerful instrument for expanding the market share of the Company by reserving the loans for its customers. It also served as a vehicle of political patronage. It could be used to win important farmers for the CPP and to reward those who were prepared to come out in open support of the party in the current political struggle. The ceiling for individual loans was £1,500, a very substantial sum by local standards.[40] The primary official purpose was to relieve indebtedness but the scheme lent itself to a much wider range of purposes. The formal requirements laid down were largely ineffective. Written documents on the original debt to be redeemed were requested. But village credit transactions were rarely undertaken in written form. The documents processed by the Loans Agency were necessarily largely fictitious. The scheme was wide open to manipulation by prospective beneficiaries, as well as by administrative officers. An elaborate and time consuming process would have been required, if the genuine nature of the debts were to have been checked. Much time would also have been needed to check the securities offered. The distribution of the funds does not seem to have been inhibited by such considerations. As already noted, almost £2 million were released in the first year of operations. By the end of the 1954/55 season some 15,000 farmers had received loans.[41]

In the administration of the Loans Scheme, the Seven-man Committees of the Farmers' Council received their most important assignment. According to an instruction from early 1954, the approval of the particulars concerning debts to be redeemed was to be sought from them.[42] They should inspect the securities offered by the farmers. The detailed directives issued by the Company's Loans Manager emphasised further the role of the committees in the scheme. It was specifically stated that the loans were to be given out through the Farmers' Council. The committees were to recommend genuine farmers in their group for loans, and serve as guarantors during the repayment period. Applications for loans should be delivered to the committees which "select the genuine cases of hardships for approval". They were jointly responsible with the debtor-farmer for the repayment of the loan. They were also to take charge of the farms pledged as security and make sure that the cocoa was used to liquidate the debt.[43] It is not clear to what extent these instructions were actually implemented. It does not seem as if the joint responsibility of the committee members was taken seriously. It is also clear that many transactions by-passed the Seven-man Committees altogether. Repeated appeals were made by Farmers' Council leaders requesting the Company not to grant loans to non-Council members. The Company was asked to inform their Supervisors to direct farmers who came to them with their securities first to register their names with the Farmers' Council.[44] The request seems to imply a much lesser involvement by the Council than that suggested by the official instructions.

In an attempt to refute the allegations raised by the Jibowu Commission, i.e. that loans were given to Council-members only, the General-Secretary maintained that it was only in April 1956, that is, after the granting of loans had ceased, that the Directors of the Company had accepted a Council proposal that loans should be given to its members only. It was felt "that those who were not members ... would not pay heed to the advice of the Council to make good the refund of their loans".[45] Loans had in fact been given to "any farmer irrespective of his membership of the Farmers' Council". In view of other evidence quoted, this claim seems somewhat exaggerated. It is quite likely, however, that the Council never managed to establish the degree of control over the loans scheme which it desired.

The involvement of the Farmers' Council in the Loans Scheme was nevertheless sufficiently intimate to have a major impact on the organisation. The scheme was launched almost at the same time as the Council. The administration of the loans funds coincided with the rapid expansion of the market share of the Cocoa Purchasing Company and with the multiplication of the Seven-man Committees. Loans, actual or prospective, were a major incentive for farmers to sell cocoa to the Purchasing Company and to register as members of the Farmers' Council. The provision of discriminatory services, in exchange for formal or real part loyalties, was an essential aspect of the social origin of the Council and the recruitment of its cadres. The real resources available to the joint Company-Council organisation greatly facilitated enrollment. But the manipulation of the genuine social grievances associated with indebtedness was also counter-productive. The discriminatory administration of the Loans Scheme was one of the major points raised when the Ashanti farmers rallied against the CPP after the 1954 elections.[46]

According to the General-Secretary, the existence of the Farmers' Council "would have been impossible but for the love and cooperation of the management of the Cocoa Purchasing Company".[47] The five functions of the Seven-man Committees listed by him were all directly linked to the activities of the Company: (1) to recommend farmers in their group for loans, (2) to supervise repayment, (3) to collect cocoa for the Company, (4) to distribute seasonal advances and (5) to collect seasonal advances. The Company was the senior partner. Areas covered by a Company Supervisor were declared to be the Districts of the Council.[48] The management of the Company was involved directly in organising the Council. Constitutional changes were undertaken on the recommendations of the Company.[49] The Company and the Council were two instruments by which the CPP sought to extend its political base in the cocoa-farming communities. The Council's claim at the Jibowu Commission, namely that the "CPP had never had any hands in the Farmers' organisation", is rather curious, especially against the background of the many statements by the General-Secretary and other senior Council officers about the role of the party and Nkrumah in creating the Council.[50] It reflects an ambivalence in the role assigned to the Council as an organisation established to ensure support for the CPP, but simultaneously aiming to become an all-national farmers' organisation.

In using cocoa trading as a vehicle of political mobilisation, the CPP built on the organisational experience of the cocoa farmers. For several decades farmers, in most cases led by cocoa traders, had sought to establish organisations capable of cutting in on an export market entirely controlled by foreign firms. These organisations had also been effective in mobilising a wider support in opposition to foreign control. The aspirations of some of these farmers and farmer-traders were successfully channeled into the officially-supported cooperatives. Others were drawn into the orbit of the nationalist movement. The attempt to tie these organisations more closely to the CPP met major difficulties, which had their main origin in the contradiction between the limited, commercial aspirations of the farmer-traders and the wider political aims of the party. Continued and expanded state control over cocoa income was an essential element in this conflict. The CPP created its own organisation with the help of CMB funds, which after 1951 had been brought within the reach of the party. The commercial side was entrusted to a separate, non-farmer organisation. This was very different from the earlier pattern where farmers and farmer-traders had organised themselves. They were now largely organised by non-farmer elements. Farmers' committees were established mainly as adjuncts to a trading organisation controlled by administrators who received authority from above.

The party could in this way draw on its predominantly non-farmer cadres in its attempt to establish political control in the cocoa-farming areas. The choice of solution, however, reflects not only a contradiction between the national aspirations of the party and the sectional interests of existing farmer-trader groups. It was also an expression of a contradiction between two different types of sectional interest within the cocoa trade itself, represented by the farmer-traders on one hand, and by the trader-clerks of the Purchasing Company of the other. The latter were primarily recruited from the core groups of the nationalist movement, the aspiring administrators and the small businessmen, whose opportunities for social advance were restrained by continued foreign domination. These groups had little interest in a farmer-based and decentralised type of market. They looked to a nationally controlled public sector for employment or for support against foreign capital.

The desire for a solution which facilitated the transfer of resources from the cocoa farmers to the State for the purpose of national development was reinforced by the sectional preferences of party cadres for a trading system in which their own aspirations could be most easily satisfied. The contradicting and reinforcing factors which shaped the joint Company-Council system of the 1950's were essentially the same ones that later manifested themselves in the monopoly organisation of 1961.

The political struggle 1954–1956

The successful expansion of the joint operations of the Cocoa Purchasing Company and the Farmers' Council during the 1953/54 cocoa season were brought to a sudden halt in the second half of 1954. The young Council organisation as well as the party itself were thrown into a major crisis. In the general elections of June 1954, the CPP government had been confirmed in power. Constitutional changes had increased its control over the colonial bureaucracy, in line with the policy of gradual devolution of power, now accepted by the colonial government. The CPP had obtained 55 per cent of the vote. But in an highly unsettled political situation, this figure did not mean that the party was safely in control. The elections had demonstrated some major weaknesses. While rival national parties had been shown to be largely ineffective, a substantial non-CPP vote had gone to independent candidates, in many cases persons who had first sought but failed to win CPP nomination. In Ashanti almost one-third of the votes were cast for independents. A party with an exclusively regional appeal had been highly successful in the North.[51] Only a few months after the elections a new powerful party was formed in Ashanti, the National Liberation Movement (NLM). It constituted a major challenge to the CPP and the intensive, frequently violent struggle which followed threatened to disrupt the smooth process of decolonisation. The NLM failed to develop its basically regional and separatist appeal into a national one. It also failed to unite the Ashantis. The CPP was again confirmed in power in the 1956 elections, but the party had been seriously shaken. The experience of the 1954–56 struggle had important consequences for the development of politics in independent Ghana.

The political crisis was precipitated by government cocoa policy. The first platform of the new anti-CPP movement was opposition to the suppression of local cocoa prices and the appropriation by the government of an increasing share of total cocoa income. In August 1954, two months after the elections, the government introduced the Cocoa Duty and Development Funds (Amendment) Bill. The decision was taken to freeze the cocoa price for a period of four years and to raise the rate of progression of the cocoa duty sharply. World market prices had been rising strongly and a substantial increase in the local producer price had been widely expected.[52] In chapter 7 I shall return to the important role of the 1954 Cocoa Ordinance in the context of the tax policies of the government, as well as to a discussion of the position of the cocoa price issue in the pre-independence political struggle. In the present context I shall confine myself to examining the impact of the political crisis on the development of the Farmers' Council.

In fact, the crisis began within the ranks of the Farmers' Council itself. The cocoa price revolt was led by Kofi Buor, the Regional Chief Farmer of the Council in Ashanti. A "Council for Higher Cocoa Prices" was formed in Kumasi, the Ashanti regional capital, with Buor as its chairman.[53] Buor appeared on a joint platform with leading members of the established opposition parties and was expelled from the Farmers' Council.[54] The cocoa-price issue put maximum strain on the unity of the organisation and on its ability to "represent" the farmers and the government simultaneously. The Council had been established only one year before the crisis and had hardly been able to consolidate the gains achieved during the rapid expansion of the 1953/54 season. Its resources and staff were very limited. Emphasis had been placed on its commercial side and little had been done to create a political commitment among its members, capable of resisting the claim for higher cocoa prices.

In June 1954, two months before the Cocoa Ordinance, the Farmers' Council had itself petitioned the Cocoa Marketing Board and the government for a higher price.[55] In August, the Supreme Council, the governing board of the organisation, was invited to a meeting with the government, where the decision to peg the price was explained. The Council continued to maintain, according to the General-Secretary's account, that the current price was too low.[56] It petitioned again for an increase. "While awaiting the Governor's reply, some of the farmers employed the help of Opposition Members to combat the Government for an increase in the price". In this situation, "the Council found it expedient to discontinue its legitimate demand for an increase". The break-away leaders were accused of sabotaging the Council's efforts in this direction.[57] The Council canvassed the cocoa areas to explain and defend the Cocoa Ordinance. This was its first major assignment in political mobilisation on behalf of the CPP government. It claimed that it was its determination and steadfastness which killed the demand for higher cocoa prices.[58] More important, however, was probably the government's retreat on the price issue in combination with a sharp decline in world market prices. Other questions also soon overshadowed that of the cocoa price in the strategy of the anti-CPP groups.[59]

The Farmers' Council continued to be deeply affected by the struggle in Ashanti. The organisation was threatened with complete disruption. The Second National Conference held in Okadjakrom in November 1954 was exhorted by the National Chief Farmer to remain loyal to the CPP government to which the Council owed its very existance.[60] Most of the delegates from Ashanti were absent and the threat of further defections was real. In districts not fully controlled by the "Cocoa Opposition", attempts were made to replace rebel Chief Farmers with loyal ones. Difficulties grew as the political struggle became more violent. According to the Council's chronicle, its officers "were attacked on and off duty, booed, stoned and jeered at by the political fanatics of the National Liberation Movement".[61] The files of the Council contain numerous references to violence of a more serious kind: people killed or wounded, farms, houses and other property burnt or damaged on both sides in the struggle. One Chief Farmer stated that "my Compound House here was dynamited and my car burnt to ashes as I took refuge at Kumawu".[62] Another local Chief Farmer claimed that the NLM "took possession of my farms and wilfully and maliciously peeled the barks of my cocoa trees which damage has rendered most of the trees dead".[63] The headquarters of the Council had to be moved from Kumasi, the Ashanti capital, to Accra.[64] By early 1955 the loss of control in Ashanti seemed virtually complete. It was reported at one meeting of the Working Committee that no districts in Ashanti had sent in their estimates and the regional executive had refused to form new district councils as requested.[65] The fortunes of the Council in Ashanti reached their lowest point when the National Chief Farmer himself was convicted and imprisoned on charges of arson.[66]

Efforts were made to prevent the further disintegration of the Council by those who administered the public funds, which were channelled into the joint operations of the Purchasing Company and the Council. Over one million pounds (more than one-third of the total amount released by the Cocoa Marketing Board for the Loans Scheme) were received by the Company during September 1954, immediately after the outbreak of the cocoa crisis. The delegates of the southern regions of the Council, meeting in Agona Swedru that same month, coupled their support for government cocoa policy quite bluntly to an increased release of such funds: "For reason that Government would issue more loans for farmers, we hereby assembled do vehemently resolve and be it resolved that the Cocoa Duty Ordinance and Price are reasonable and accepted".[67] In a memorandum on the "Revitalization" of the Council by the General-Secretary, the Loans Scheme was seen as a key element in the strategy to defeat the cocoa opposition: the granting of loans should be liberal, and they should be given on favourable conditions. "As it is our desire to win all the country's farmers, it would mean some sacrifice of funds on the side of the Government".[68] Trading advances were also considered important in the attempt to retain the loyalties of the cocoa farmers. As the Farmers' Council was entirely dependent on the Cocoa Purchasing Company, a minimum objective was to prevent farmers from selling their cocoa to other buying agents. The Farmers' Council pressed the Company for increased advances to the farmers, although the collection of outstanding advances was already seriously in arrears.[69] The Council also asked for increases in the commissions payable to the farmers at the end of the season.[70] An increase was granted in 1956, in spite of the impending bankruptcy of the Company.

Many southern cocoa farmers may have voted against the CPP in the 1956 elections. But an organised cocoa opposition failed to establish itself on the same scale as in Ashanti. Regional cleavages and the way in which these were manipulated by the CPP contributed to this result. Tensions between Ashanti and the South had deep roots both in historical conflicts and in present differences in social and economic structure. The way in which the Farmers' Council split in 1954 reflected these cleavages. The tensions were clearly felt as early as at the inaugural conference in 1953.[71] They were later to play an important role in the break-up of the cooperative movement, thus preparing the way for the Farmers' Council monopoly in 1961. Once it was evident that the cocoa opposition was closely allied with Ashanti traditional leadership and Ashanti resentment of southern political domination, its effectiveness in mobilising the southern cocoa farmers was greatly reduced.

The political struggle released by the Cocoa Ordinance of 1954 brought the Farmers' Council into open conflict with the aspirations of the farmers for higher cocoa prices. Such aspirations had been the major basis of previous attempts by the farmer-trader groups to mobilise the farmers. The link with this tradition, still represented in important respects by the Farmers' Congress, was now broken. The political function of the Council as an instrument of government policy was at this point clearly established. The ability of the Council to survive as a farmers' organisation depended in this situation less upon its backing among the farmers than upon the support provided by the CPP government. This also meant that the survival of the Council depended on the ability of the CPP to remain in control of the government. Unlike the party itself, the Council had no significant social base of its own to fall back upon if defeated in the parliamentary struggle. The direct stake of the Council in a CPP victory increased.

In the 1954 elections, that is, before the outbreak of the Cocoa Ordinance crisis, no particular campaign was undertaken by the Council on behalf of the party. Involvement in the 1956 elections was more direct. The Council ordered its members to vote CPP. Council officials at all levels were engaged in campaign work.[72] Earlier that same year, the Council participated in the party campaign for the Togoland plebiscite.[73] It later claimed that it was the stand taken by the Council which had "swung the issue in favour of the Party".[74] Officially, the Farmers' Council sought to preserve at least a semblance of an independent posture in relation to the CPP. The Minute Book reports that the Working Committee in 1956 "discussed all the major parties with special reference to their aims and ideologies" before it recommended to the National Conference to pass a resolution in support of the CPP.[75] On the same occasion, the Committee rejected a proposal made by the CPP headquarters for an exchange of representatives at the executive level between the two organisations. The reason given was that "it would be out of order to affiliate with a political party", as the Council was a non-political organisation. These recorded demonstrations of independence must be seen against the background of the recent public hearings by the Jibowu Commission, enquiring into the affairs of the Cocoa Purchasing Company, where allegations were made that public funds had been used for partisan political purposes. In this situation, the Council sought to show that there was "no truth whatsoever" in the claims that it was a CPP organisation. It maintained that it only—for the time being, as it were—supported the government "because of its policy favourable to the farmers".[76]

The Commission threatened the financial and organisational base of the Council, as founded on the operations of the Cocoa Purchasing Company. The real threat, however, lay less in the revelations of the Commission, which primarily repeated points already widely known, than in the actual financial crisis of the Company. It had greatly exceeded the financial frame laid down by the government and the Marketing Board. The Jibowu Commission added to the forces within the government, still ultimately under colonial control, which pressed for action against the Company. The apparent political hegemony of the CPP and the ineffectiveness of the anti-CPP parties had permitted the colonial government to adopt a more cautious attitude. The growing strength of rival political forces simultaneously made the CPP more susceptible to colonial pressures.

Collapse of the Cocoa Purchasing Company

The government statement accompanying the publication of the Jibowu Report in 1956 noted that it had become evident by the end of 1954 that "the various warnings and suggestions made to the Cocoa Purchasing Company through the Cocoa Marketing Board were insufficient to effect a change of policy".[77] Changes in the membership of the board of directors were made in August 1955 and the Managing Director (Djin) was removed ("though his services were still retained by the CMB"). The board chairman, who was also chairman of the CMB itself, was also removed. A police investigation in the affairs of the Company was ordered and the Committee (later Commission) of Enquiry was appointed early in 1956. Auditors reported that repayment of off-season advances and loans under the Loans Scheme was heavily in arrears and that large amounts of commissions had been paid out despite the losses incurred by the Company. Outstanding off-season advances had been transferred to the Loans Scheme, and the operating capital, intended for cocoa purchases, had been used for loans. These were improperly granted and no adequate arrangements were made for their recovery.[78] The ministerial directives to liquidate the Company were issued in May 1957.[79] The financial crisis was obvious as early as the end of the 1955–56 season: operations were grinding to a halt; no fresh addition to the operating capital had been issued since October 1954 and the last funds for the Loans Scheme were received in September 1955.[80] The Company made losses on current trading and part of the operating capital was locked up in outstanding advances. Repayments under the Loans Scheme were insignificant and provided no basis for continued lending. The CMB demanded that the Company should "correct the impression gained by some farmers that loans were not to be repaid".[81] The chairman of the Company explained to the Working Committee of the Farmers' Council in August 1956 that the Company could not pay any off-season advances to the farmers. They needed "the little money they had left in hand" as buying cash.[82] The Council in its turn had to cut down expenditures. It was a crisis for the Council as much as for the Company itself.

The Farmers' Council had little direct influence on the trading operations of the Company, although the Seven-man Committees were supposed to supervise and assist the local receivers. The Council now joined the critics and expressed concern about the "huge outstanding balances due from the Company's receivers". As a solution, it proposed that the latter should be made directly responsible to the committees.[83] The proposal was part of an attempt by the Council to establish control over the Purchasing Company, which so far had been the senior partner. This new ambition was most clearly revealed in a memorandum where the General-Secretary suggested that the Company should be transformed into a buying agent of the Council.[84] The latter should in this manner be given an opportunity to save the Company from any future failure "as its failure would affect the Council considerably". A sharp attack was made on the "unscrupulous actions" of the staff of the Company, whose numbers were also said to be vastly excessive. The Council would at the local level form "societies", which were to employ Secretary-Receivers for the local trading operations. It suggested that the buying allowance granted by the Marketing Board should be shared fifty-fifty between the Company and the Council. The official chronicle of the Council claims that the Council itself demanded the abolition of the Company.[85]

The declining fortunes of the Cocoa Purchasing Company led the Farmers' Council to look for new sources of revenue. Contacts were made with the Agricultural Development Corporation, another state organisation. It was arranged for the Council to establish a trading agency to buy palmnuts.[86] The Farmers' Council Agency subsequently undertook to buy other commodities as well, for the Corporation. Unlike the Cocoa Marketing Board, the latter had limited resources. No advances were given. The Council had little capital of its own and this new partnership turned out to be unrewarding. The Agency is of interest in the present context as a step in the transformation of the Farmers' Council into a business organisation, a process which already commenced before the final collapse of the Cocoa Purchasing Company. When the Company was brought into liquidation in May 1957, the Council was also virtually forced to close down. It was unable to pay either its officers or its rent.[87] But these were passing difficulties. That same month, at a Special Conference in the Accra Community Centre the President of the Council announced that the government had "thrown the guantlet to them": the Council was to take over the functions and duties of the Cocoa Purchasing Company. This was done, according to the General-Secretary, in return for the assistance given to the party in the struggle for independence.[88]

The Cocoa Purchasing Company was liquidated, but it reappeared as the Ghana Farmers' Marketing Association, a subsidiary of the Farmers' Council. Direct control by the Cocoa Marketing Board was eliminated. The Purchasing Company had been a subsidiary of the Board. The position of the new organisation was in this respect not basically different from any other private Licensed Buying Agent. The direct involvement of the Board in the affairs of the Company had caused much friction. The Board was held responsible for irregularities committed by the Company but complained that its directives were persistently ignored.[89] A close union was at the same time apparent, at least in terms of overlapping directorships.[90] The new organisation, on the other hand, was controlled exclusively by the Farmers' Council.

The competitive strength of the Purchasing Company lay in its access to capital supplied by the Marketing Board. How then was the new organisation to be financed? The starting point was humble. The minutes of the first board meeting recorded that the chairman was entrusted to raise a private loan in order to be able to pay the £30 registration fee for the new company.[91] But help was forthcoming. The government gave it a £250,000 loan and additional credit was provided for the purchase of the assets of the Purchasing Company from the Marketing Board. Ministerial directives fixed the price to be paid.[92] The total loan of £437,000 was later transferred from the government to the Marketing Board when it was realised that the government could not lend money without the prior approval of the National Assembly.[93] The Council was also able to share in credits channelled through the Cooperative Bank. The government guaranteed a bank overdraft of £1 million in 1958.[94] The Loans Scheme, which was no longer a matter of lending money but of trying to recover what was outstanding, was taken over by the Marketing Board where it continued to run heavy deficits for many years. Receipts were not sufficient to cover the cost of collection.[95]

Much of the marketing staff of the Company was taken over by the Council. The former Farmer-Receivers were renamed Secretary-Receivers to fit the new marketing societies established by the Council, partly on the model of the cooperatives. Continued central control over the receivers and the absence of financial "autonomy" at the society level made the system rather different from that of the cooperatives. Its central features were largely the same as those which characterised the monopoly system of 1961. The discussion of the new structure may therefore be postponed until the monopoly itself is examined in subsequent chapters. Instead, we shall turn at this point to the consolidation of the Farmers' Council as a national farmers' organisation in the pre-1961 period.

Official recognition

The period from independence in March 1957 to the introduction of the new, republican constitution in 1960 saw the consolidation of the state powers of the CPP.[96] The bitter struggle with various opposition groups was increasingly held outside the parliamentary arena. Acts of repression and subversion mutually provoked and reinforced one another, preparing the way for a one-party state. State powers could now be more directly applied to the largely unsolved problem of establishing an organised political base among the agricultural producers. Opposition to the cocoa policies of the government had disrupted the attempt to bring the country's farmers together in one, CPP-affiliated organisation. The Cocoa Purchasing Company had exhausted itself as an effective source of revenue for rural party work and patronage. But the party was now less dependent on such indirect methods of making funds available for organising the farmers. The liquidation of the Purchasing Company, although primarily a consequence of financial mismanagement, also reflected changes in the structure of state power. Apart from providing a convenient way of disposing of the undesirable financial legacies of the Company, the new arrangement expressed more adequately the political monopoly aspirations of the party.

The Farmers' Council was declared in 1957 the only officially recognised farmers' organisation in the country.[97] No other organisation was accepted as legitimately representing the farmers. The origin of this particular form of monopoly representation is not clear. A similar phenomenon existed in the colonial political system itself, where the authority of a chief to speak on behalf of "his people", although based in theory on a mandate from below, depended on official recognition by the colonial government. The official recognition of the Council was followed by instructions to the Marketing Board to pay it an annual subsidy of £100,000.[98] The General-Secretary also announced that the government had authorised the construction of a £100,000 headquarters for the Council.[99] The Council was thus able to expand its organisation considerably beyond the levels achieved in the pre-independence period, despite the poor trading results of its first two years.[100] Districts which had been dissolved were reestablished and the allowances and salaries of Chief Farmers and other staff were raised.[101] Gone were the days when the officers of the Council "had to work long hours without pay".[102]

Reviewing the past year, the General-Secretary reported to the National Delegates Conference in Nsawam, January 1959, that it had been a year of important achievements: The Council had used the official recognition effectively. The foundation of a permanent farmers' organisation had been firmly laid.[103] He emphasised that the future progress of the Council continued to depend on CPP control of the government. To mark this unity, it was proposed that the Council should openly align itself with the party and that the constitution of the Council should be amended accordingly. A resolution was passed by the Conference to that effect.[104] The new constitution stated among other things that it was the duty of the Council to keep the farmers informed about the government's good intentions, to support the government and to accept the leadership of the CPP.[105] The political situation after independence made it no longer necessary to keep up earlier pretensions of non-partisanship. The Conference was assured by

Gbedemah, the Minister of Finance and a leading member of the party, that "as long as they supported the party in power all their needs constructively submitted" would be considered.[106] The importance attributed to the Council by the government was emphasised at the following National Delegates Conference at Kpandu, January 1960. The Conference was opened by Nkrumah in the presence of a number of important government and party leaders, including Botsio, Gbedemah, Asare, A. E. Ofori-Atta, and Edusei.[107] On this occasion, there was a special reason for the government to show its appreciation and support of the Council. The latter had agreed on behalf of all the cocoa farmers of the country that 12 shillings per load, 17 per cent on the current producer price, were to be deducted as a "voluntary contribution" to the Second Development Plan.[108] The direct association of the Council with the fiscal policies of the government demonstrated the new confidence of the party as well as the Council in their ability to maintain political control in the cocoa-farming regions. The situation was very different from the days of the Cocoa Ordinance and cocoa revolt of 1954. The importance of the Council was also acknowledged by making the General-Secretary, Martin Appiah Danquah, a minister with "Cabinet rank".

Relations with the party were not without frictions. The party was anxious to extend its direct control over the organisation but met with opposition from Council cadres who wished to protect their own independent authority against incursions by party officials. The Bibiani Conference in June 1960 declared itself unable to accept a merger of Council and party district organisations, as proposed by the General-Secretary in the name of the Supreme Council.[109] The desire of Council officers to protect their independence had also been demonstrated at the 1959 Conference, which revolted against a decision by the CPP Central Committee, regulating the procedure for the election of Chief Farmers. The Ashanti Regional Chief Farmer declared on this occasion that he had the mandate of delegates present to challenge the party order. The Council, he claimed, was an autonomous organisation which could not be dictated to in this manner. After a heated debate the party line was accepted by the Conference in all important respects.[110] A similar incident was recorded in 1957, when Nkrumah temporarily suspended the election of Chief Farmers. The Working Committee of the Council noted that the compliance with these directives was a "special favour", but that they were not prepared to accept similar interventions in future.[111] The General-Secretary was not present at this meeting, and at the following one the minutes were amended, removing the unduly offensive section.[112]

The United Ghana Farmers' Council emerged in the post-independence years as a semi-official national farmers' organisation with the financial and political backing of the CPP government. Its ability to function in this capacity was no longer restrained by the funds and administrative support provided by associated commercial activities. But its orientation continued to be commercial as well as political. A wide range of new economic and administrative objectives were included in its programme. These fast-growing aspirations were revealed in the General-Secretary's Annual Report to the 1959 National Delegates Conference. The Farmers' Council planned to establish a multipurpose cooperative organisation covering marketing of export crops and crops for local consumption, agricultural credit banks, insurance, processing of agricultural produce, supply of farm implements, and creation of consumer cooperative stores.[113] The Report spoke of the part which the Council was now to play in the agricultural extension services of the government. It had been invited by the Ministry of Agriculture to participate in a country-wide scheme for spraying against capsids, a new pest which threatened the cocoa industry.[114] The government had asked the Council to take over the Young Farmers' Clubs, a farm-settlement scheme, which was now renamed Future Farmers' Organisation.[115] The Council established separate fishermen, poultry-, pig-, and cattle-farmers' associations and a variety of projects were planned in these fields. Particularly high hopes were attached to plans for mechanised agricultural production. A first step in this direction was taken with the establishment of the Vakpo Cooperative Farm in 1960.[116] The chequered career of the Farmers' Council as a multi-purpose entrepreneur in the non-cocoa sector must unfortunately be omitted from the present study. Cocoa marketing, however, remained by far the most important enterprise in which the Council was involved. The cocoa economy continued to be the principal arena for its political performance. By 1959/60, the Council had regained the share of the market once held by the Cocoa Purchasing Company. Its ambitions nevertheless went further. A campaign was launched against foreign buyers and local middlemen. The final struggle for control over the local cocoa trade had begun. Before drawing further conclusions from this account of the origins of the Farmers' Council and its claims for political and commercial monopoly, I shall examine this struggle in the next chapter. At this point I shall merely summarise the main stages in the early development of the Council and related organisations.

Origins of the Farmers' Council: summary

The early post-war years in Ghana witnessed an upsurge of nationalist politics. The United Gold Coast Convention was formed in 1947 with a leadership drawn from the professional and commercial elite. Kwame Nkrumah, called in to act as the General Secretary of the Convention, broke with its conservative leadership in 1949 and formed the Convention People's Party, the CPP, which developed rapidly into a mass movement. The colonial government was obliged to revise drastically its time schedule for the constitutional devolution of power. Elections were held in 1951, resulting in a sweeping CPP victory. The CPP leaders were given limited ministerial responsibilities, extended to "full internal self-government" in 1954, and finally, independence in 1957, after some years of strong party rivalries and two more elections won by the CPP.

In its pre-1951 work, the CPP sought to make maximum use of existing organisations to reach out as broadly as possible. Especially the local and regional "youth" associations were tapped for their mobilising potential. These groups drew their members largely from the urban or small-town petty-bourgeoisie and wage earners, including many aspiring for wage employment. Many had some formal education which set them off from the non-literate masses. Many may have had a family background in the cocoa economy. Some were traders and farm-owners. On the whole, however, with their educated and urban bias, these groups did not reach far into the farming communities. In an attempt to enroll the latter, the CPP sought to ally itself with existing farmers' organisations. Many of these had a pre-war record of activities, especially in connection with the successful hold-up of cocoa sales and the boycott of European goods organised in 1937/38 to fight the monopolistic practices of the foreign trading companies. The effort to form farmers' organisations on a regional or national level was intensified during the latter half of the 1940's. The farmers' movement was particularly concerned with cocoa marketing; it was spearheaded by a stratum of big farmers and farmer-traders. The Ghana Farmers' Congress, established in 1949/50 was such an attempt to bring existing farmers' organisations together. The CPP was closely involved but it did not succeed in turning the Congress into a functioning party organisation. Three issues formed the basis of farmers' agitation: their aspirations to increase their share of the commercial profits of the cocoa trade; their opposition to the price policies of the public export monopoly; and their opposition to the methods used by the government to fight the swollen shoot disease through the compulsory cutting out of cocoa trees. The CPP used these issues in its own agitation and supported the farmers' case against the colonial government. But when the CPP leaders were given partial governmental responsibilities in 1951, the divisions between them and the farmers soon became evident.

The Cocoa Marketing Board had been established in 1947 with the main official purpose to stabilise prices paid to the producers. These prices were fixed annually by the Board, which held exclusive authority over the marketing of cocoa. The difference between the producer price, and the world market price was partly accumulated by the Board for stabilising purposes and partly paid to the government in export duties. A small portion was used by the Board for its own development programme. The surplus of the Board rose sharply during the first years' operations and it came under strong attacks from farmers' organisations. The CPP ministers supported a temporary increase in the producer price but were more anxious to protect public cocoa income for the purpose of the party's ambitious public-development programme. Confrontation with the farmers was inevitable and contributed to the break-up of the Farmers' Congress. But disagreement was also strong over the organisation of the local cocoa trade. Foreign firms, farmers' cooperatives, and a small number of local private traders had been appointed Licensed Buying Agents of the Cocoa Marketing Board to buy cocoa from the farmers at the stipulated producer price. They were remunerated by an official buying allowance. The Board also advanced part of the current capital of the trade. The cooperatives received official colonial backing. The farmer-trader groups which dominated the Farmers' Congress, in their turn sought support from the new CPP ministers, who were now in a position to influence the Cocoa Board. Realising the strategic importance of the cocoa trade for the purpose of political organisation, the CPP leaders made such support conditional on the acceptance of CPP leadership. Unable to reach agreement with the Congress leaders who feared party domination, the CPP government launched its own buying organisation, the Cocoa Purchasing Company, which was established as a subsidiary of the CMB. With financial support from the Board and with a loans scheme (to relieve farmers' indebtedness) as a special attraction, the Company managed in a few seasons to carve out for itself almost one-fifth of the market. It is not clear whether it was the creation of the Cocoa Purchasing Company that exasperated the relations between the CPP and the established farmer groups or if it was the exasperations of these relations which motivated the launching of the Company. It may have worked both ways. Anyway, the Company became part and parcel of the party's efforts to establish a new national farmers' organisation under its own control.

The United Ghana Farmers' Council was founded in 1953, partly based on the pro-CPP sections of the defunct Ghana Farmers' Congress. Until the liquidation of the Cocoa Purchasing Company in 1957, the Council depended almost entirely on that organisation for its revenues. The Council was organised at the local level around the trading and credit activities of the Company. Farmers' committees were formed, with limited advisory and supervisory functions. The new organisation had hardly been established before it was broken up by party struggles. A second election had been held in 1954 which confirmed the CPP in power, but opposition to the CPP was growing. The professional and commercial elite groups which had been robbed of their expected constitutional inheritance by the CPP upstarts were joined by regional and traditional interests which for varying reasons feared the centralising and modernising ambitions of the CPP. The party itself, despite its spontaneous mass following, was based on a loose network of personal alliances and contacts. It was not sufficiently consolidated to prevent numerous internal divisions over personal and local issues. The decision by the CPP government after its 1954 election victory to freeze the producer price of cocoa at a time of booming world market prices and to raise steeply the level of cocoa taxation triggered off an opposition movement in Ashanti, the most important cocoa region. Most of the Ashanti establishment of the Farmers' Council defected. The new party, the National Liberation Movement, developed into an essentially regional party which failed to establish itself on a major scale in the other cocoa regions. The cocoa issue itself subsided into the background as market prices fell and producer prices were raised. Regional contradictions of a more general kind came to the forefront. Although defeated in Ashanti, the CPP won the third round of elections held in 1956, which prepared the way for the final British withdrawal in the following year. It was possible for the Farmers' Council to reconstitute its organisation in Ashanti, basing itself on the strong CPP minorities in many districts.

The growing financial difficulties of the Cocoa Purchasing Company and the partisan use of its funds had stirred public controversy and prompted an official inquiry, the Jibowu Commission. The Company was liquidated in 1957. Its trading activities were transferred to the Farmers' Council, which established its own marketing organisation much on the same line as that of the Company, but with increased formal responsibilities for the farmers' committees. Unlike the Company (which had been a subsidiary of the Marketing Board), the new organisation was not subjected to government control, although it received government loans and subsidies. After three seasons the Farmers' Council had recovered the market share once held by the Purchasing Company. The Council demanded that it should be made the sole licensed buying agent of the CMB, that the foreign buyers should be expelled, and that the cooperatives should merge with it. The commercial and political monopoly claims of the Farmers' Council went hand in hand. In 1957, it was recognised by the CPP government as the only organisation entitled to represent the country's farmers. The CMB was charged to pay an annual subsidy to cover the non-commercial activities. For all practical purposes, the Council had been a CPP organisation from the start. It had canvassed in support of government price and tax policies and participated in political campaign work at election times. After independence, its new role as the only officially recognised farmers' organisation was combined with a more formal attachment to the party. In 1959, the Council exercised its representative functions by granting a major deduction from the official producer price as the farmers' "voluntary contribution" to the Second Development Plan. The campaign to expel the foreign firms had begun the same year. A closer look at this campaign and the subsequent struggle between various national trading interests yields considerable information about the class character of the CPP regime and its internal contradictions.

IV. The Struggle for Monopoly

At the time of independence, Ghana's local cocoa trade was still dominated by foreign firms within the framework of the export monopoly of the Cocoa Marketing Board. But public financial support to cooperatives and the Cocoa Purchasing Company had reduced their market share from virtually 100 per cent before the war to some 60–70 per cent in the mid-1950's. The cooperatives expanded rapidly immediately after the war, but stagnated at the time when the Cocoa Purchasing Company entered the market. They continued to buy roughly one-fifth of the country's cocoa during the rest of the 1950's. The distribution of market shares can be studied in the following table:

Cocoa Purchases by Licensed Buying Agents, 1951–1961: Percentages

Year	Foreign Firms	Co-ops	Purchasing Company	Farmers' Council	Other Local Agents
1951–52	84	14	—	—	2
1952–53	74	19	6	—	2
1953–54	61	20	18	—	2
1954–55	61	19	19	—	1
1955–56	62	18	18	—	2
1956–57	68	18	13	—	1
1957–58	73	19	—	7	2
1958–59	66	21	—	11	3
1959–60	57	24	—	17	3
1960–61	37	30	—	28	5
1961–62	—	—	—	100	—

Source: deGraft-Johnson Report 1967, p. 59.

The foreign firms built their control of the market on a complex hierarchy of local agents, of which only a few were employees. Most were tied to the firms directly or indirectly through advances and commissions. (See Chapter 2.) These local cocoa buyers, frequently cocoa farmers as well, constituted an important section of a rural petty-bourgeoisie, operating under the umbrella of foreign capital. The independent local Licensed Buying Agents, that is, private Ghanaian traders buying directly for the Cocoa Marketing Board, controlled only a few per cent of the market.

The rise of the Farmers' Council as a monopoly buyer of Ghana's cocoa is discussed in this chapter in relation to the three major groups in the market: the foreign firms, the local buyers (in most cases buying for the firms), and the cooperatives. A first phase was characterised by the Council's campaign, partly in alliance with other groups, for the expulsion of the foreign firms. The campaign met little resistance and the United Africa Company, one of the largest buyers, withdrew from the market at an early point. Here I shall discuss why the Council was successful and how economic nationalism in this case was compatible with continued foreign penetration. The second phase, the struggle for the inheritance, overlapped with the first, especially in the Council's attacks on brokers and middlemen, that is, local traders buying on behalf of the firms.

Organised opposition to a party monopoly in the cocoa trade came in the first place from the "old" cooperatives. The commercial aspirations of farmer-traders had partly been channelled into the cooperative movement, a result of government support as well as restrictions. The "merger" of all cooperatives under the Farmers' Council meant for all practical purposes the liquidation of the old cooperative movement. The defence of the cooperatives was closely linked to the defence of free enterprise against the expanding role of the state and the party in the economy. Party and government were divided over the monopoly claims of the Council. Repeated reversals of decisions preceded the ultimate confirmation of the new scheme. The chapter concludes with a review of the social forces which cooperated and opposed each other in making the monopoly.

Expulsion of the foreign firms

Early in 1959, Ghanaian papers published reports that the Supreme Council of the Farmers' Council demanded that the cocoa buying licences of all foreign firms should be withdrawn by the Cocoa Marketing Board. In case the Board failed to comply the Council threatened that 10,000 farmers would march to the Board's offices.[1] The Board was taken by surprise. No serious discussions either between the Council and the Board or the Council and the relevant ministry seem to have been undertaken before this radical initiative was launched. The Borad decided to issue a press release making it clear that it had no intention of withdrawing any licences or of succumbing to pressures from "unauthorised outside bodies". A petition submitted by the Council asserted that the continued presence of foreign firms in the marketing of agricultural commodities had "a Colonial and Capitalist background". Farmers in Independent and Socialist Ghana could no longer respect such a system.[2] Much of the detailed argument was directed less against the foreign firms as such, than against their local agents, the brokers, who were said to exploit the farmers. Emphasis was also laid on the deteriorating effects of brokerage on the quality of cocoa.

A long tradition of agitation against the foreign cocoa-buying firms existed in Ghana. Before the Second World War it was primarily connected with the question of prices paid to the producers. The firms were accused of manipulating prices to the disadvantage of farmers as well as local traders. Agitation over prices helped aspiring farmer-trader groups to establish a wider following. The introduction of statutory price control and a public export monopoly cut the link between the price issue and continued foreign domination of the local cocoa trade. The focus of agitation was directed at the distribution of commercial profits within the framework established by the Marketing Board. Although occasional attacks were made on foreign participation in the local trade in general, no concerted campaign was levelled against the foreign cocoa buyers before independence. The cooperatives and the Cocoa Purchasing Company provided outlets for local commercial aspirations and acquired substantial shares of the market.

In the 1954–56 period, commercial competition between local and foreign buying establishments was drawn into the general political struggle. Foreign firms, and Cadbury in particular, were accused of assisting the political opponents of the CPP. It was claimed that they "agreed to deal with farmers who were dismissed from the Farmers' Council for rebelling against the Government" and that they provided them with loans.[3] One local group reported that farmers who had been disappointed with the Cocoa Purchasing Company had been approached "by the underground workers of the National Liberation Movement in the name of Cadbury, offering strong promises to pay their loans".[4] The oppositional Ashanti Farmers' Union was granted a Cadbury loan of £70,000 according to another source.[5] The company has denied allegations of political involvment which it claims were based on misconstructions of normal commercial operations.[6] Whatever their factual basis, the allegations added to the animosity in CPP quarters against the firms. Their commercial practices were also criticised. On one occasion, the Council threatened that no farmer was going to sell any cocoa to the firms unless they started to pay commissions directly to the farmers, as the cooperatives and the Purchasing Company were doing.[7] The criticism successively increased in pitch. In 1956, the General Secretary asserted that the Ghanaian farmer had now found out "how economically murderous the foreign firms who buy his cocoa have been to him".[8] Concrete proposals, however, were as yet not concerned with the elimination of the firms but with measures making their performance more acceptable. The main attack was directed against the middlemen, who bought cocoa on behalf of the firms. Only six months before the actual campaign to expel the firms started, the General Secretary of the Council proposed that they should be asked to extend their network of employees down to the village level, in order to eliminate the brokers.[9] Such a reform would have strengthened rather than weakened the control exercised by the firms.

The distinction between firms and middlemen was no longer maintained in the policy statement made to the National Delegates Conference at Nsawam in January 1959.[10] The existence of middlemen in the marketing of export crops was seen as the primary cause of the economic distress of the rural population. The middlemen were firms like United Africa Company and Cadbury which gave advances to local brokers who in their turn cheated the farmers. The money provided by the firms was the instrument with which the farmers were exploited. The statement claimed that the government had accepted the principle that the farmers eventually should market their cocoa through their own organisations. But there was "a school of thought" which assumed that this should be achieved through competition. In this way were the firms to be forced out of the market. The Farmers' Council did not accept the idea that the firms should be allowed to continue to operate at the expense of the farmers. The firms provided no services which local organisations could not give. The farmers should rather be protected against wasteful competition. The Council demanded legislation to prevent "any Company which does not belong to farmers" from participating in the local marketing of export crops. A resolution in support of this stand was passed by the Conference.[11]

The campaign did not have official government backing. The Cocoa Marketing Board expressed fears that the withdrawal of the firms would lead to a breakdown of the local trade.[12] In July 1959, however, the Board was officially informed that the United Africa Company had decided to withdraw from the cocoa-buying field with immediate effect, that is, it did not intend to participate in the 1959/60 main crop purchases.[13] The situation changed radically. UAC was one of the largest foreign buyers. It handled about one-fifth of the crop according to one source.[14] The basis for continued foreign presence in the local cocoa market was greatly weakened. Both the Farmers' Council and cooperatives expanded their share of the market in the 1959/60 season, but the main immediate beneficiaries from the withdrawal of the UAC were other established foreign firms. Staff and equipment were largely transferred to them.[15] Foreign capital still controlled over half the market. But the Farmers' Council was encouraged by the spectacular initial success of its campaign. Agitation increased. The General Secretary's Report to the National Delegates Conference in January 1960 was confident. The United Africa Company had left the field honourably. Other firms like Cadbury & Fry and Lyons were advised to follow the example "for as the United Ghana Farmers' Council has decided to get them out, they will by all means go". If they did not leave, the Council was going to use "any means at our disposal" to expel them during the 1962/63 season.[16] There were also calls for more immediate action. At the local level, a branch of the Council in Sefwi-Wiawso passed a resolution declaring that "we all farmers have also sworn the Sefwi Oath Tikwa and Asantewa that anybody found sending his or her cocoa to any of the foreign firms, has then gone against the Oath and this also means . . . (that) he or she should be sent to the Ahenfie (the Chief's palace) or the nearest Police station for prosecution".[17] The role of oath-taking in this context was of course primarily symbolic. The insistent stand taken by the Council was noted with concern by the Cocoa Marketing Board.[18] The Board was assured by a representative of the Ministry of Trade that the attitude of the government had not changed. A statement was drafted for the approval of the Minister, warning the Farmers' Council against seeking the premature removal of the firms. But it was evident that the CMB now saw the withdrawal of the firms as a matter of timing and administrative convenience. Local arrangements must first be satisfactory.

The official position was clarified in the Budget Statement of July 1960.[19] The government intended to "take the necessary steps to ensure that by the end of this development plan period, i.e. in 1964, all the purchasing of the Ghana cocoa crop shall be done for the Marketing Board only by Ghanaian farmers' co-operatives and unions recognised by the Government". The announcement was part of a general declaration of a moderately nationalist economic policy. It was one of nine measures for "winning the economic battle", including action to ensure that an increasing proportion of retail trade passed into Ghanaian hands and new laws and regulations restricting the operations of foreign banks and insurance companies. The package also included government participation in the marketing of mineral exports, increased protection for the local manufacturing industry, and incentives for import substitution and production for the local market in agriculture.

The subsequent debate in the National Assembly contained few comments on the government's plans for the local cocoa trade. There seems to have been wide agreement, including the parliamentary opposition, that the foreign firms should be expelled. The only substantial comment concerned the timing. One speaker maintained that there was no need to wait until 1964, as there already existed competent and experienced local organisations which could take over immediately.[20] This speaker was the owner of the largest private Ghanaian cocoa buying company. The reference in the Budget Statement to "co-operatives and unions" was obviously considered wide enough also to include his—entirely private—company. The question of who was to succeed the foreign firms was still largely unsettled. But there was no doubt that the days of the latter were numbered. More firms decided to leave without waiting for their licences to be withdrawn. Two general trading companies, the CFAO and Paterson Zochonis, announced their withdrawal in August 1960.[21]

At this point, the official timing was drastically revised. The Budget Statement had mentioned 1964 as the year when the Africanisation of the local cocoa trade was to be completed. Only a month after this statement, the Minister of Trade proposed in a draft Cabinet Memorandum that the transfer should already be effected from the 1961 mid-crop.[22] The memorandum was endorsed by the CMB. The fears so recently entertained seem by this point to have been abandoned. Timing was only slightly revised in an official announcement by the Minister of Agriculture at the National Delegates Conference in Bibiani, September 1960. The new marketing scheme, yet to be worked out, was to commence from the main crop 1961/62.[23]

The campaign waged by the Farmers' Council against the foreign cocoa-buying firms was swift and successful. It had at first no immediate official support, and the authorities most directly concerned, the Cocoa Marketing

Board and the Ministry of Trade, expressed disapproval. The change in the official position was facilitated by the decision by the United Africa Company to leave the market. The cautious time-schedule announced in the 1960 Budget Statement was soon abandoned in favour of more immediate solutions. The pace by which the positions were advanced reflected also a growing radicalisation of government economic policies. General declarations in favour of economic independence and decolonisation, as manifested in the Budget Statement, were increasingly followed by direct public intervention in the economy, restricting the field of foreign capital and creating new national economic institutions. A plan submitted by the Ghana Chamber of Commerce to the Government envisaged the complete nationalisation of trade in Ghana within five to seven years, according to press reports.[24]

Why did the United Africa Company withdraw? The question touches on important problems of power in the post-colonial context. What was the scope of national political power within a framework dominated by foreign capital? Were nationalist movements able to bring about fundamental changes in the pattern of domination? The Africanisation of Ghana's local cocoa trade is part of the historical experience within which these problems may be examined. Why did foreign capital put up so little resistance to nationalist advances? Was it due to the strength and determination of the nationalist movement or was foreign capital after all not such a formidable obstacle to national development as often envisaged? The United Africa Company, perhaps more than any other organisation, incarnated the might of European capital in West Africa.[25] Was it forced to abandon the local cocoa trade because of nationalist agitation? It is suggested here that the withdrawal must be seen as part of a wider process in which foreign capital was redeployed, leading to increased rather than reduced commitment.

As discussed in Chapter 2, the acceptance by foreign capital of the export monopoly was facilitated by changes in foreign participation from retail to wholesale trade, and from general merchandise to more specialised lines of trading. These changes were primarily induced by the growing effectivness of local enterprise in expanding the market for European merchandise at the retail level. On the produce-buying side, local traders were also increasingly effective in catering to the raw-material requirements of European and North American industry. Growing competition and declining profits could temporarily be countered by restrictive practices, as was done in the 1930's. The Cocoa Marketing Board temporarily ensured a resonable return on capital already committed. It reduced current capital requirements through its advances to the Licensed Buying Agents. A spokesman of one of the firms has suggested that little fresh capital was invested in the local produce trade in the post-war period.[26] But post-war developments in Africa brought other and more promising opportunities for foreign capital. In Ghana, the United Africa Company embarked on a major programme of investment in the late 1950's, including a large timber processing mill, a brewery, a soap and detergent factory, a textile printing mill, a car-assembly plant, and modern department stores.[27] Nationalism certainly contributed to this redeployment of resources but less by blocking established lines than by providing infrastructure, tariff protection, tax concessions and expanded markets for more advanced forms of penetration. In this context, the up-country purchasing of cocoa was a minor commitment of declining relevance to the overall strategy of foreign capital.

The structural changes were reinforced by the way in which the Marketing Board was increasingly brought under nationalist control in the post-1951 period. Profits depended on the Buying Allowance fixed by the Board. The allowance specified what were considered reasonable costs for each transaction. It is difficult to say how the official schedule reflected actual costs.[28] Public financial support for cooperatives as well as the Cocoa Purchasing Company increased competition and reduced the ability of the firms to claim the full profit to which they were entitled according to the Buying Allowance. A growing portion of profits had to be passed on to sub-agents and to some extent, to the farmers in the form of increased advances and commissions, in order to protect market shares. With national control over the CMB, firms met difficulties when claiming compensation for increased costs. In 1957, repeated petitions by the Licensed Buying Agents resulted only in minor adjustments.[29]

A distinction must be made, as earlier, between foreign cocoa buyers representing general merchant interests as the United Africa Company, and those, in the first place Cadbury & Fry, which were agents of cocoa manufacturers. We have seen that the former were the first to withdraw. Unlike the United Africa Company, Cadbury was not involved in a process of redeploying local resources. Its West African interests continued to be primarily directed to the cocoa trade and its first commitment was to ensure a satisfactory supply of cocoa both through direct participation in marketing and through contacts with local public agencies. Profits improved substantially after the withdrawal of the UAC but, on the whole, Cadbury's post-war performance was affected by the same constraints which led the UAC to consider the trade to be no longer profitable.[30] In Cadbury's case, however, there was no question of merely abandoning the market, as did the UAC. As a major customer of Ghana's cocoa, Cadbury was anxious to foster market arrangements capable of expanding supplies, reducing costs and improving quality. The company was at first strongly sceptical of the Farmers' Council and sought to postpone the transfer. Ultimately it decided to cooperate.[31] It was closely involved in designing the new marketing scheme. Managerial staff was provided to serve in advisory capacity during the first two years of monopoly.[32] When the organisation was dissolved after the 1966 change of government, the company which continued to follow the local cocoa market closely, expressed appreciation of the Council's performance, although this was hardly an opportune view to hold at the time.[33] We shall return to the role of manufacturing interests when evaluating the monopoly in the next chapter.

From the point of view of foreign capital—and here no distinction need be made between the various types of firms—Africanisation of the local cocoa trade seems not to have been a cause for great controversy or conflicting interests. The transfer could be effectuated in considerable harmony, in spite of the occasionally strong language of the nationalists. When the last firms left in 1961, it was possible to recall their "merciless exploitation", while on the other hand, they were officially thanked for their "devotion and efficiency".[34] The European-controlled local trading system, which once had performed an essential role in expanding the market for manufactured goods and securing the supply of raw materials, had largely exhausted itself as a necessary or profitable outlet for European financial and managerial resources. Investment opportunities lay in other directions; local enterprise, private or public, was increasingly capable of handling the elementary marketing functions on behalf of European industry. The withdrawal of the firms was part of a process of readjustment within the colonial or post-colonial economies, leading to new and more advanced forms of engagements by foreign capital. At the level of the individual firm such readjustments could of course be painful. A gradual transition may have been preferred, but the Farmers' Council campaign probably accelerated the process. Attempts to withstand such pressures in order to delay the process, although perhaps desirable in a short-term perspective, were unwise as they may have provoked nationalist forces in other fields, fields far more dangerous for longer-term interests. Local entrepreneurship had outgrown the bounds set by foreign management. The prevailing system of foreign financial and managerial control in the colonial produce trade was obsolete also from the point of view of basic metropolitan interests, concerned with the supply of raw materials. It was therefore possible to challenge this particular aspect of foreign presence without threatening the fundamental relationship between European industry and African primary production.

The campaign against the middlemen

The campaign against the foreign firms was at the same time directed against the traders who bought cocoa on their behalf. In fact, arguments advanced in favour of the elimination of the firms largely concerned the undesirable trading practices of the traders. Why were the anti-foreign and anti-broker arguments so closely linked in the strategy of the Farmers' Council? Was it simply a natural consequence of the actual integration of the operations of foreign capital and local traders? It may be argued that the latter merely constituted the weakest link in the system of foreign domination and were therefore singled out for the brunt of the attack, although their foreign principals were the real target. An effective political campaign against the firms, on the other hand, could have been based on the nationalist argument alone, that is, on the claim that foreign capital no longer was performing any legitimate function which local enterprise could not undertake. For this purpose there was no need to attack the local traders. There was no inevitable connection between the anti-foreign and the anti-broker argument.

The traders constituted a broad stratum of experienced private entrepreneurs. They were natural candidates for the succession, once the firms were to withdraw. The strength of the trading system controlled by the firms lay primarily in the commercial network established and maintained by the brokers. The United Africa Company could withdraw without causing general disruption of the trade, despite its large share of the market, because the local traders were still at their posts. The brokers were therefore also the natural rivals of the Farmers' Council in its effort to establish a political and commercial monopoly in the cocoa areas. In attacking the brokers the Council appealed to resentments against trade malpractices harboured by farmers and by those concerned with the development of production. Most of the points raised were part of the established criticism.[35] Manipulation of scales and weighing above authorised weight were the most common allegations.[36] The question of prices as a source of conflict between farmers and traders had largely been removed as a result of statutory price control, but even the official prices left a margin for manipulation and bargaining. Deductions from the price were in many cases made to cover local transport and storage costs before the produce reached the official buying points of the Cocoa Marketing Board. In a petition to the Board, the Farmers' Council maintained that brokers bought low-quality cocoa at reduced prices and mixed it with good cocoa to make it pass the quality mark prescribed by the Board.[37] This, too, was a well-established criticism from the pre-war trade, but the uniform pricing system of the Board, with no price differentials for quality, was of course a special invitation to such practices. The Council's argument was concerned with the undesirable effects of brokerage on the quality of cocoa. To protect quality and to free the farmers from exploitation was the declared purpose of the Farmers' Council's first petitition for the removal of the firms and brokers.[38] Competition at the village level was seen as a cause of the deterioration of quality, as well as a means of exploitation. The mixing of bad and good cocoa kept down the over-all quality of Ghanaian cocoa. Competition made it possible for farmers to receive full price even for inadequately fermented and dried cocoa.[39] The immediate beneficiary of competition was the farmer, who was able to pass on preparation, storage and transport costs to the buyer. If competition were so intense as the argument seems to presuppose it should also have provided the farmer with some protection from exploitation through the manipulation of prices and weights. The logic of the quality argument, on the other hand, was that higher standards could only be enforced if the farmers were prevented from taking advantage of competition. The emphasis on quality served to justify the reorganisation of the local market in terms acceptable to the manufacturers and the Marketing Board, which represented the Ghanaian cocoa industry vis-à-vis the former. Without relevant price differentials either in the international or local market, the enforcement of higher quality was in the interest of the manufacturers, not the primary producers.[40]

In attacking the brokers, the Farmers' Council paid particular attention to "Lagosians" and other marginal, primarily non-farmer, non-Ghanian groups whose position in the cocoa-farming communities was particularly vulnerable.[41] These "aliens" had less opportunity and were less motivated to find outlets for surplus funds in land and other fixed property in the areas where they were operating. They were therefore more dependent on trading and money-lending. They were also less protected by kinship against local resentment. In this sense, they were a natural target. By focusing on this "pure" type of middlemen instead of the more numerous and more typical farmer-traders it was possible to steer attention away from the internal contradictions among the cocoa farmers themselves. But attempts were also made to force farmer-traders to take sides in the conflict. The National Delegates Conference at Bibiani "strongly urged that arrangements be made for Farmer Brokers who posed as middlemen to be punished by law, the penalty of which should be £100 fine plus a year's imprisonment".[42]

The connection between brokerage and exploitative money-lending was emphasised. Brokers were alleged to charge excessive interest rates on off-season advances.[43] This was contrasted with the liberal terms offered by the cooperatives and the Farmers' Council. Other evidence suggests that competition had in fact forced firms and brokers to grant advances on similar conditions.[44] The line between trading advances and general money-lending was fluid, and higher rates of interest were probably charged for larger and more long-term advances, corresponding more to terms prevailing in the local money-market. The Council also claimed the superiority of its own system of buying when it came to the question of paying commissions to the farmers. The official buying allowance granted by the Cocoa Marketing Board did not provide for such payments. But it was possible to use the profit element of the allowance for this purpose, including whatever savings which could be achieved within the cost schedule. In theory, at least, there was room for such savings as the schedule was adapted to meet the high overhead costs, including foreign management, of the firms. Cooperatives as well as the Farmers' Council societies clearly had an advantage over the brokers in this respect. To remain competitive, many brokers were obliged to give commissions, at least to their more important customers, at the expense of their own profit margins. But the non-payment of commissions or their inadequate size was used as a major argument against the brokers. Farmers were prevented from reaping the full profits from marketing their own cocoa.[45] Profits should go, according to the Council, to the farmers themselves through their own organisations and not to middlemen, local or foreign. At this level, the case against middlemen was not concerned with specific malpracties. The participation of private commercial capital as such was rejected.

The elimination of the brokers did not follow from the desire to expel the foreign firms. Private Ghanaian traders could continue to operate after the firms had withdrawn, although the system of finance would have had to be reorganised, providing direct access to banks and to the advances issued by the Board. By demanding the elimination also of private Ghanaian traders, the Farmers' Council pressed for a "non-capitalist" solution. This was facilitated by the actual integration of local and foreign capital. Local traders seemed to have no separate identity. The nationalist argument gave support to a "non-capitalist" one. In this way, the monopoly aspirations of the Farmers' Council obstructed the aspirations of an important section of the national trading class which had successfully established itself within the colonial economy. The show-down was achieved with little effective opposition at the national political level. In the National Assembly, spokesmen of entrepreneurial interests attacked the tendencies of government and party organisations to monopolise business. The main concern was with the fate of the old cooperatives. I will argue below that the defense of the latter was in fact closely related to the defense of "free" enterprise in general. But the little direct support given to the brokers may be contrasted with the concern over the future of the much smaller number who were directly employed by the firms. They were to be guaranteed continued employment in the new organisation. The brokers, on the other hand, were "mostly farmers, Lagosians and Goa men and they could take care of themselves", according the Farmers' Council.[46] The brokers were many and handled a large business but as a social and political force they were weak. This weakness was no doubt related to their financial dependence on the firms. It reflected the declining interest of foreign capital in the local cocoa trade. But that decline was in itself partly caused by the growing effectiveness of local trading organisations. Why did the brokers not succeed in freeing themselves from financial dependence on the firms? Why did they not become independent Licensed Buying Agents? To understand this failure it is necessary to examine the way in which participation in the trade was regulated.

Before the war the supremacy of the firms depended on a combination of privileged access to trading capital and export channels. The Marketing Board removed much of the basis for this supremacy. It took over a major share of the trade financing, using the system of 90 per cent payments on the declaration of purchases. The amount of private capital required was greatly reduced. Independent access to the world market was blocked altogether. Much of the effective local market knowledge was in the hands of the brokers. Favourable conditions should have existed for their transformation into independent Licensed Buying Agents. This did not happen. The average share in the market held during the 1950's by Ghanaian private buyers, not including the cooperatives, was only two per cent. (See table at the beginning of this chapter) Two factors contributed in particular to prevent the brokers from establishing themselves on their own: firstly the restrictions enforced by the licensing system itself; secondly, discrimination in favour of the cooperatives and other state-supported buying organisations in the allocation of trading capital. A Licensed Buying Agent had to meet certain qualifications. The candidate should have a guarantee from a commercial bank (foreign-owned) showing financial capacity to handle a minimum amount of cocoa. This was first set at 50 tons but was later raised to 100.[47] A letter was required from a shipping firm (foreign-owned), allocating storage space at port. Finally, a certificate had to be issued by the Government's Produce Inspection, showing that up-country storage facilities owned by the candidate were suitable.[48] The bank guarantee, in turn, presupposed access to securities in the form of fixed property, primarily urban housing. Banks did not usually accept cocoa farms as security. As the majority of the brokers were farmer-traders based in the village economy, it is easy to see that the bank guarantee alone could be a major obstacle.

The number of licences issued was restricted. The 1952 Annual Report of the CMB gave the total number of Agents as 28, of which 14 were African.[49] A restrictive practice was in line with the circumstances which led to the establishment of the Board. The purpose of the pre-war market-sharing agreements was to restrict competition and to establish closer control over the brokers. As the firms offered their cooperation with the post-war scheme, it was reasonable to expect that their sub-agents were not unduly encouraged to set themselves up on their own. The special access of the firms to banks as well as shipping firms facilitated a restrictive interpretation of the formal conditions regulating fresh entries to the market. But other forces also worked in favour of restrictions. In 1957, the desirability of limiting the number of Licensed Agents was discussed both by the Board and between the Board and the government. A decision to restrict the number to 32 had been taken earlier. It was noted that there were no formal grounds for the decision.[50] The justification put forward was concerned with the need to "eliminate the small buyers who usually resorted to malpractices". It was also considered desirable to minimise the administrative work of the Board.[51] The Board voted in favour of a slight increase in the number but the Minister gave directions that the existing limit should be kept.[52] It was later announced that the minister's approval was required before any new licence was issued by the Board.[53] The records show examples of candidates whose formal qualifications were accepted by the Board but where the application was turned down by the Ministry.[54] Some of the controversy over individual licences clearly concerned the political rather than the commercial qualifications of the candidates. We noted in Chapter 3 how a group of farmer-traders associated with the Ghana Farmers' Congress failed to obtain a licence because, according to its leaders, they refused to accept CPP control over their organisation. In 1957, this company, the Ghana Producers Association, was actually accepted as a Licensed Agent at a meeting of the Board. The decision was reversed when it was later realised that the company was an agency of the National Farmers' Union, an organisation engaged on the side of the political opposition.[55] The failure of the company to buy the stipulated minimum amount of cocoa was stated as the formal ground for the reversal, although in fact only the small mid-crop had been available since the licence was granted. Restrictive licensing was reinforced by the desire of the CPP government to use the licences as an instrument of political control.

More important in obstructing the growth of independent Ghanaian cocoa buyers was the way in which commercial opportunities were heavily weighted in favour of publicly supported organisations, first cooperatives, then the Cocoa Purchasing Company and finally the Farmers' Council. This was done primarily through the supply of trading capital from public sources but also through other forms of administrative support. Farmer-traders were encouraged or compelled to pursue their commercial aspirations within these organisations in order to be able to benefit, for example, from funds supplied by the Cooperative Bank. Traders and their clientele could reconstitute themselves into cooperative societies and unions. Less reorganisation was needed to get access to the trading capital of the Cocoa Purchasing Company. It is not known how profitable these forms of business were, as compared with pre-war brokerage but the scope for individual entrepreneurship was considerable within the new organisations, as well.

Without the financial and administrative backing of either the government or foreign capital, the private Ghanaian Licensed Buying Agents were not able to establish themselves as a major force in the local market. It is possible that inefficiency, mismanagement, and other factors contributed to this failure. The obstacles created by public policy, however, were major. This must be recognised if the strength of local commercial forces in the cocoa economy is not to be underestimated. A number of factors cooperated in blocking their growth in this sector: foreign firms, anxious to maintain control over their sub-agents; the colonial government concerned with achieving a "constitutional" transfer of power in the cocoa trade by encouraging cooperatives; the nationalist leaders wishing to use the cocoa trade as a platform for their political aspirations; and the marketing board system, supported by colonial as well as nationalist interests, which restricted the scope for competitive enterprise in general through its regulations and control of prices and profits.

The CPP position on the role of Ghanaian private enterprise in the cocoa trade contained certain contradictions. In spite of official support for the Farmers' Council in the post-independence period, the strategy was by no means clear. In 1957, when the Cocoa Purchasing Company was liquidated, its former chairman, a Kumasi businessman and CPP parliamentarian, established a cocoa trading company of his own, strongly backed by the government. At first, a buying licence was refused by the Marketing Board because of his indebtedness to the Cocoa Purchasing Company, and thus, indirectly to the Board itself. But pressure was applied by the Ministry and a licence was issued.[56] The following year, the new company (Agricultural Producers Ltd.), applied for a £200,000 loan from the Board. This was a substantial sum and can be compared with the £400,000 lent to the Farmers' Council for the same purpose. The Minister pushed the issue but the Board took a cautious attitude. It decided first to examine the securities offered. They were found to be largely fictitious. This, however, did not deter the Minister, who directed that the loan should be granted, against the strong reluctance of the Board. Securities were to be provided afterwards. The minutes noted seven months later that no securities had yet been submitted.[57] The new company soon bought more cocoa than all the other Ghanaian private Licensed Agents together. In 1960/61, it bought 11,000 tons, or 3 per cent of the total crop.[58] Encouraged by the loan to the Kumasi company, some of the Ghanaian Licensed Agents set up a joint company composed of individual Agents "who for financial and administrative reasons wished to operate as a team" and applied for a £500,000 loan. The Marketing Board, however, declared that it did not accept the principle of granting loans to its Buying Agents.[59] No ministrial support for the applicants was forthcoming this time.

Government backing of the Agricultural Producers Company was political patronage primarily on an individual, rather than a class or group basis. As such, it did not necessarily indicate that the government was prepared, as a principle, to push private Ghanaian enterprise if controlled by persons loyal to the party. The support given this particular company may merely be a case of personal exercise of ministerial powers. If that was the case, it still demonstrated the lack of official commitment to a "non-capitalist" solution at the time when the government was forced to wind up the affairs of the Cocoa Purchasing Company. It was possible for government agencies simultaneously to pursue different and partly contradictory solutions. It reflected the ambivalent class character of an anti-colonial movement with petty-bourgeois leadership, vacillating between "state capitalist" and private capitalist means of emancipating itself in a situation of colonial economic dependence.

Cooperatives and free enterprise

Colonial institutions, private and public, obstructed the growth of independent Ghanaian cocoa traders. Cooperatives, on the other hand, received official support. They were first introduced by the government in the late 1920's, mainly as an instrument for improving the quality of cocoa.[60] They were subjected to close official control. Individual cooperative societies and unions were not free either to by-pass the foreign firms or to bargain with them on their own. Their cocoa was sold to the firms by colonial officials through a system of tenders. After the war, the official approach to the cocoa cooperatives changed. They were now to provide an outlet for the farmers' commercial aspirations, which had threatened to disrupt the trade, as had been dramatically demonstrated in the 1937/38 cocoa hold-up. Cooperatives, properly supervised, could facilitate a peaceful and gradual transfer of power in the local market. A Department of Cooperatives was established in 1944 to provide administrative assistance and financial control. A Cooperative Bank handled the growing capital resources of the "movement". A government guarantee made it possible to borrow current trading capital from the commercial banks up to a maximum amount of about £2 million.[61] Under these favourable conditions, the cooperative share of the market grew from a few per cent in the late 1930's to almost 20 per cent in the 1952/53 season.[62] Public financial support was vital. But the cooperatives were also able to draw heavily on the commercial experience and resources of the established farmer-traders. Accumulated farmer resentment against the foreign firms added to the rapid success of cooperative marketing.

Expansion was halted when the Cocoa Purchasing Company entered the market. Why did the CPP create a new cocoa-buying organisation instead of building on the cooperatives? One answer was suggested in the previous chapter: cooperatives, as well as other established farmer-trader organisations, were less adaptable to the wider political aims of the party, including increased state control over the distribution of cocoa income. I do not know whether an attempt to enroll the cooperatives was actually made at this stage. We have seen how sections of the Ghana Farmers' Congress group of farmer-trader organisations resisted CPP and government control. But the cooperatives were already in important respects dependent on the government. Would they not for this reason have been more susceptible to party control? The close participation of colonial institutions in the administration of the cooperative sector may have made it less attractive as a springboard for the CPP in the cocoa economy. The leadership was able to use its special relations with the colonial government to protect the cooperatives against attempts of CPP control.

Whatever the earlier relations between the cooperatives and the CPP, once the Purchasing Company had been established a conflict was inevitable. The Cooperative Marketing Association, the national organisation, petitioned the government against the Company scheme. DeGraft-Johnson mentions that plans were developed by the government to protect the interests of the Association and to prevent it from being destroyed through the activities of the Company.[63] It is not known what these measures implied. The cooperatives succeeded in maintaining their share of the market but further growth was inhibited. With the collapse of the Cocoa Purchasing Company, the cooperatives resumed their role as the leading Ghanaian buyers. Their prestige was enhanced by the failure of their rival. The Farmers' Council, as well, advocated at that point a cooperative type of solution to the difficulties of the Purchasing Company. Greater powers were to be given to the Seven-man Committees in supervising and controlling the Farmer-Receivers, or the Secretary-Receivers, as they were now to be called. The General Secretary of the Council told a conference of District Chief Farmers in Kumasi to refrain from propaganda against the cooperatives, because "their methods of operation are appreciated by the Farmers' Council".[64]

In 1957, when faced with the problem of reconstructing the badly damaged organisation inherited from the Purchasing Company, the Farmers' Council approached the Cooperative Marketing Association with the purpose of establishing closer links between the two organisations. The plans were abandoned after a most "unfortunate and unfavourable" meeting with the leaders of the Association.[65] The invitation was rejected. Nor was the Department of Cooperatives willing to accept the local units of the Farmers' Council as cooperative societies in the sense prescribed by the Cooperative Ordinance. After pressure from the Minister of Labour and Cooperatives, the Council succeeded in having its buying agency as a whole registered as one "society".[66] A Council memorandum claims that the old cooperatives and their supporters sought to prevent the registration and that "the battle was tense and difficult" before the Council could make its "triumphant entry into the country's Cooperative Movement".[67] The Department of Cooperatives began surveys of the Council's branches to find out which of them could be established as "autonomous" cooperatives.[68] This exercise was discontinued as the government proceeded to dismantle the old cooperative structure.

The first move was taken at the 1959 Congress of the Cooperative Movement, when the Minister responsible for the cooperatives outlined a proposal for a unified structure, including the new Farmers' Council "cooperatives" and those set up by the Trade Union Congress. A new central organisation was to be established and the existing apex body, the Alliance of Ghana Cooperatives, was asked to reconstitute itself for that purpose. It refused. In this situation, the government launched a National Cooperative Council, and withdrew official recognition and grants from the Alliance. Atta Mensah, a man with a strong party record, was put in charge of the new organisation.[69] The following year, the government announced its intentions to reserve the local cocoa trade for cooperatives and officially recognised unions. The old cooperatives sought to safeguard their position in the new order. The Ashanti, Brong Ahafo and Sefwi unions of the Cooperative Marketing Association broke away and made a separate deal with the Farmers' Council. According to an account given by the Registrar of Cooperatives in 1966, the leadership of the Farmers' Council and the newly-established National Cooperative Council instigated the split, "using every trick and artifice" in order to convince the break-away unions that the southerners were cheating them.[70] It is probable that southern dominance of Association leadership contributed to the split. Some of the southern leaders were active members of the political opposition against the CPP. ABASCO, the organisation established by the seceding unions, voted for a merger with the Farmers' Council. Its leaders were invited to the Bibiani Conference in 1960.[71] The President of ABASCO "thanked the Council for accepting his cooperative organisation in principle".[72] The terms of the merger, however, were far from settled. ABASCO expected to retain its own organisation, while accepting the political leadership of the Farmers' Council at the national level. The Council, on the other hand, aimed at control at the local level, as well. Facing the prospect of being manoeuvered out of the market altogether, the rump Cooperative Marketing Association dismissed those leaders whose presence was most offensive in the eyes of the CPP, namely the chairman Mark Addo and the secretary George Oteng.[73] A third cooperative-based opposition politician, K. B. Ntim, was removed with the abolition of the Cooperative Bank, another step in the drive to reorganise the cooperatives under party control. The business of the cooperatives was transferred to the newly established state-owned Ghana Commercial Bank.[74]

A committee under the chairmanship of the Minister of Labour and Cooperatives met in November 1960 to advise on the new cocoa marketing scheme. High officials of relevant government and public bodies and the contending cooperative organisations participated. At the first meeting, a proposal was outlined by the General Secretary of the Farmers' Council.[75] The Council was to be the only Buying Agent of the Board, but during the first two years the old cooperatives were to continue as buyers at the local level under the auspices of the Council. "If after two years of trial there are two or more such operating cooperative units in one locality arrangements would be made to bring them under the ambit of one executive body." In addition, the scale-buyers, that is, the local buying staff of the firms, would be given an option to remain at their posts, at least during the first year of operations. The autonomy of each cooperative society, as laid down in the Cooperative Ordinance, was emphasised. "Assets of any cooperative society must be collectively owned by all members" and the decisions of these autonomous societies should be "final and conclusive". On the other hand, there would be only one national organisation and its decisions would also be "binding as well as final" at all levels. The formula presented by the Council sought to combine a degree of temporary independence for the old cooperatives at the village level with the effective supremacy of the Council in the overall leadership of the combined operations of the three major groups of buyers. The formula contained important contradictions. How, for example, was local autonomy and Farmers' Council supremacy to be combined? The committee adopted the outline as the basis for its future work. A sub-committee was entrusted with the task of working out the details.[76] But the sub-committee failed to report. Instead, the Principal Secretary of the Ministry of Labour and Cooperatives and the Acting Registrar of Cooperative Societies worked out a scheme of their own, which they presented to the committee.[77] The minutes do not indicate the contents of their memorandum but it probably was largely along the lines of one prepared by the Registrar and separately documented.[78] It supported the approach taken by the old cooperatives. The Farmers' Council was only to provide a "summit organisation", with each subagent operating its business and accounts separately, dealing directly with the Marketing Board.

The unity maintained so far by vague formulas was ended. The representatives of the Council did not attend the following meeting of the committee. Still chaired by the CPP Minister, R. O. Amoako Atta, it developed in fact into a general attack on the way in which the Farmers' Council was pursuing the merger.[79] The Ministerial Secretary for Labour and Cooperatives, a parliamentarian, reported that he had recently returned from a tour of the cocoa areas. A serious situation had arisen as a result of statements allegedly made by the General Secretary of the Council. According to these, the old cooperatives were to be abolished altogether and their

Secretary-Receivers would have to reapply for employment with the Council. All assets were to be taken over by the Council and no member of the old cooperatives would be eligible for any post in the new organisation. Judging from the internal records of the Council, the reported statements may well have given an accurate picture of the Council's stand at this particular time. The minutes of the General Purpose Committee noted, for example, that the cooperatives had obstructed the struggle for independence and that "it would not be safe to allow them to retain their respective identities". Leaders of the cooperatives should not be considered for executive posts in the Council, "thereby having the opportunity to sabotage the cause of the Council which they have for a long time sought to wreck".[80] The reported statements deviated strongly from the picture given by the Council's spokesmen to the committee. Its members, including the Minister, "showed great surprise at these startling statements". Spokesmen of the Department of Cooperatives considered an early merger "highly impossible and very dangerous". Each society should be allowed to sell its cocoa through its own apex body. The trial period had to be at least two to three years. The Farmers' Council, they claimed, had no proper cooperative organisation. Only 19 of the 350 Farmers' Council societies had as yet qualified to be registered as cooperatives. A motion was unanimously adopted by the committee confirming its earlier decision to recommend separate buying units at the local level but now adding that separate apex bodies should also be allowed. Any society participating in the new scheme should be registered under the Cooperative Ordinance. The implication was that the bulk of the Farmers' Council societies had to be reorganised in order to qualify. The supervisory role of the National Cooperative Council, the new party-controlled body, should not cover the old cooperatives. A "reasonable grace of time" would have to be allowed before any merger was considered.

The Ministry of Labour and Cooperatives and the Department of Cooperatives had come out strongly in support of the line advocated by the old cooperatives. Particularly noteworthy was a comment made by the Minister himself that the Farmers Council was "appropriately responsible for the organisation and welfare of all farmers" and should therefore not "conduct specialised business operations as the sale of cocoa". This attack on the way in which commercial and political functions were integrated went beyond the specific critique of the Council's trading scheme raised at the meeting. It was agreed that the Minister should take action on the reported statements, which had "resulted in unrest in Ashanti and Brong Ahafo". It was suggested that the Cabinet should be approached "before the present alarming position gets out of hand". It was also suggested that the Minister should make an early pronouncement over the Radio "correcting the erroneous impression created" by the Farmers' Council.[81]

This seems to have been the last meeting of the Amoako-Atta committee. Four days earlier, a rival meeting was held under the chairmanship of the Minister of Agriculture, Kojo Botsio, also Minister of State for Parliamentary affairs. The Ministry of Agriculture, the Cocoa Marketing Board, and the

Farmers' Council were represented by leading functionaries. The local Manager of Cadbury & Fry was in attendance. No representatives of the Ministry of Labour and Cooperatives, the Department of Cooperative, or the cooperatives themselves were present. The General Secretary of the Farmers' Council reported on the new organisation. Nothing was mentioned, as far as may be seen from the minutes, about separate buying organisations at either the local or any other level. The impression given was of one integrated organisation. No opposition to the scheme was recorded.[82]

The government was obviously strongly divided on the appropriate strategy to choose. Opposition to the planned reorganisation of the local cocoa market grew stronger as more details about the new scheme became available. Questions were asked in the National Assembly about the statements allegedly made by the General Secretary of the Council.[83] The government spokesman was unable to confirm or deny the reports. A major occasion for voicing misgivings about the pending monopoly of the Farmers' Council was provided by a debate on the progress report of the Ministry of Labour and Cooperatives in February 1961.[84] The debate concerned not only the Farmers' Council but the new cooperative policy in general. This was seen by speakers as a move towards the monopolisation of economic opportunities by party organisations. Far-reaching plans had been announced by the Trade Unions Congress for union-owned cooperative enterprises in a number of sectors of the economy. Resentment was expressed against the growing power of the TUC and the Farmers' Council. Their leaders had recently been awarded the rank of Cabinet ministers. How could monopoly powers be given to a "voluntary organisation" like the Farmers' Council which was not part of the state and not controlled by the National Assembly?[85] All the critics except one were CPP members. "Official" opposition members in the Assembly were at this time less than a dozen. The efficiency and integrity of the Farmers' Council was questioned. The Council was not thought capable of handling the entire Ghanaian cocoa crop. Delays and inefficiencies would occur. The quality of Ghanaian cocoa would be affected. Other speakers claimed that farmers would have to wait for payments and the distribution of equipment such as cocoa bags would not function satisfactorily.[86] It was maintained that many of the Council employees could not be trusted. They had cheated the farmers in the days of the Cocoa Purchasing Company. According to one speaker, farmers were grumbling and crying: "We have sent you at long last to Parliament, we have voted for the CPP but we are still being exploited by people who call themselves supporters of the CPP". Those who had cheated the farmers in the past should not be allowed to do it again. "Nobody should grow fat at the expense of the farmers who toil to produce the cocoa."[87] Reference was made to a theft case involving a large sum and a senior Council official and to losses already said to be incurred by the Council during its first years of operation.[88]

The performance of the Council was contrasted with that of other buyers, especially the cooperatives. As these had been reasonably efficient they should be allowed to continue. One speaker suggested that there ought to be at least six or seven buying establishments to cater to the interests of the farmers. It was proposed that those who had been buying cocoa for Cadbury should be allowed to turn themselves into cooperative societies. A special plea was made for the Ghanaian owned Agricultural Producers' Company. The principle owner was himself a CPP member of the Assembly. Worries were expressed about what would happen to all the individuals presently engaged in the cocoa trade. The monopoly would cause unemployment. Not less than 10,000 would lose their jobs, according to one speaker. The various cooperatives and the Agricultural Producers' Company should therefore maintain separate establishments "so that each organisation will be able to employ labour".[89] While several speakers expressed concern for the employees of the cooperatives as well as the firms only one referred specifically to the consequences of monopoly for the wider group of farmer-trader.[90] Much of the argument, however, concerned the evils of monopoly in general. The benefits of competition and free enterprise were elaborated. Attempts were made to use a language in consonance with the prevailing, socialist-oriented ideology. Farmers should be free to sell to anyone they liked. Competition would ensure that the farmers' interests would be adequately protected. The right to form cooperatives should not be reserved for some people only. Without competition it would be difficult to "foster the growth of socialism and the country's economic emancipation". Monopoly, it was claimed, would lead to misuse of power and irresponsible business practices: "if I am the only man on the road I can freely swing to either side . . ." "It is always the inefficient man who does not not want competition." Not only cooperatives but also private Ghanaian business should be encouraged: "When the wolves—that is the imperialists—were here, we were asked to close our ranks, but now that they have gone our Government should encourage Ghanaians to come up". Business opportunities should not be monopolised: "We are in a Socialist state and everybody must contribute equally to the economic emancipation of the state".[91]

The debate in the National Assembly on the progress report of the Ministry of Labour and Cooperatives was a manifestation of opposition within the CPP against a strategy which increasingly restricted the scope for private Ghanaian capital. In the context of the cocoa economy, the defence of competition and free enterprises was at this stage primarily a matter of supporting the independent cooperatives against the Farmers' Council. Private entrepreneurship had already to a considerable extent been channelled into cooperative forms.

The attacks on the TUC and the Farmers' Council met little opposition in the National Assembly itself. The state-owned press, however, reacted sharply. The *Ghanaian Times* maintained that the attacks on the party wings were attacks on the party itself and on its socialist strategy. The principle speaker of the Assembly debate, a CPP businessman, was accused of acting on selfish grounds because he feared competition from a TUC-owned enterprise. The critics, according to the paper, were ideologically bankrupt and failed to realise that the party had chosen a socialist line.[92] The Secretary General of the TUC was reported in the press to have told a large public meeting that the attacks in the Assembly were directed against the "socialist bastions" of the country, but that the President and the party had the appropriate machinery to deal with the critics. Particularly the threat implied in the last formulation caused much excitement and concern in the Assembly, where a counter-attack on the press was supported by a ruling by the Speaker that "contempt of the House" had been committed. Several prominent party members, who had not participated in the original debate, joined on this occasion in the condemnation of the press. One minister asked another minister, who was officially responsible for the state press, to resign. The session seems to have ended in tumult.[93]

The debate involved wider problems of strategy than those raised by the monopoly plans of the Farmers' Council alone. These plans, however, contributed to bringing the latent conflict between liberal and statist or corporative aspects of the CPP strategy into the open. The divisions demonstrated by the committees set up to reorganise the local cocoa trade are placed in their political and ideological context by the Assembly and press debates.

The State Agency scheme and the triumph of monopoly

An official statement on the cocoa purchasing scheme was read to the Assembly in March 1961 by the Minister of Labour and Cooperatives.[94] A major change of government policy was announced. The planned Farmers' Council monopoly was abandoned. The Council was assigned a clearly subordinate position in the cocoa trade. The dominant role was instead to be played by a State Buying Agency "in order that the State should retain a share of the profit accruing from the purchase of cocoa". The new organisation was to be a division of the Cocoa Marketing Board itself. The Agency, not the Council, was to take over the staff and equipment of the departing foreign firms. The Farmers' Council, the Cooperative Marketing Association and other Ghanaian bodies which "in the past purchased cocoa successfully" would be permitted to participate in the cocoa trade but only on a limited scale and until further notice. A special concession was made for the Agricultural Producers Company, the company owned by a CPP parliamentarian. Other small private Ghanaian buyers were not to be allowed. All cooperatives were to be supervised by the National Cooperative Council, the party organisation.

Why did this sudden change of policy occur and what did it signify? First, it should be remembered that the foreign buyers had already been asked to leave by the end of the 1960/61 season. The decision had been preceded by a nationalist campaign which probably had wide support in the party. It was difficult from a political as well as a practical point of view to ask the firms to stay on at this late point. The most important ones had made new arrangements and would not be particularly anxious to remain in the prevailing atmosphere of uncertainty and controversy. It was therefore not possible to further postpone a solution to the problem of how the dominant part of the crop, still handled by the firms, was to be bought in the 1961/62 season. Serious doubts had developed about the ability of the Farmers' Council to replace the firms or, for that matter, to take on the entire crop. Various formulas were discussed whereby the old cooperatives and the scale-buyers of the firms were to continue to operate more or less independently during a transitional period. A solution on these lines was complicated by the conflict over the amount of overall control to be exercised by the Council. But the conflict involved not only different views about the Council's ability. More importantly, it concerned government strategy and the future of independent enterprise, private or cooperative, in the Ghanaian economy. The party was divided and the political risks were considerable in case the Council failed after its control had been enforced against strong opposition. If, on the other hand, the old cooperatives were permitted to compete with the Council for the market vacated by the firms, their position would probably be further strengthened. The future consolidation of the market under government or party control would be made even more difficult.

A State Agency, under the direct control of the Marketing Board, seemed to provide a way to increase control, while at the same time sidestepping the conflict. The contending organisations would be left largely in the position where they were before, while the vacated market was directly reserved for the state. The government and the party could in this way jointly achieve control of over two-thirds of the market without forcing a show-down with the old cooperatives and the wider commercial interests which had been engaged on their side. But the State Agency scheme was also a partial victory for the parliamentary critics who had attacked the growing political and economic powers of the wings of the party. The ambitions of the Farmers' Council were rebuked. By preserving competition and an independent cooperative sector, the demands of the private entrepreneurial interests within the party had at least been partly met. The State Agency, in taking over the foreign-controlled sector, could continue to provide an umbrella for the cocoa brokers. The decision on what place this group should hold in the future market was postponed.

It has been suggested that the State Agency scheme was actively pushed by some of the departing foreign firms.[95] The foreign share of the market was now dominated by firms directly controlled by the cocoa manufacturers. It is understandable if these showed a keen interest in future market arrangements. As buyers in the international market they were in continuous contact with the Marketing Board. Direct participation by the Board in the local trade would provide better conditions for influence and communication. From the point of view of the firms, the Board was more likely to ensure "responsible" trading practices than the Farmers' Council. The political and non-commercial functions of the latter could be expected to interfere with the commercial ones. But the Marketing Board itself may also have supported the State Agency scheme. It was less concerned with the political organisation of the farmers than with marketing cocoa. Together, the firms and the Marketing Board were in a position to exercise "expert" influence on the government, particularly when evaluating the capacity of the marketing organisation of the Farmers' Council.

The State Agency scheme was short-lived. A month after the official statement of policy, the government reversed its position. The President, in his *Dawn Broadcast* (April 8, 1961), announced that the new scheme had not been favourably received by the farmers. He had therefore finally instructed that the Farmers' Council should be given the sole responsibility for organising the purchase of cocoa.[96] The Ministry of Labour and Cooperatives, including the Department of Cooperatives, was abolished. The staff of the Department was transferred to the National Cooperative Council. By this step, the main institutional source of resistance against the monopoly within the government itself was eliminated. Public control over the Council was to be exercised by the Auditor-General. Accounts would be examined in the same way as those of the government itself. This proposal sought to meet the criticism of those who had claimed that a monopoly of this kind was much too important to be managed by an organisation which stood outside the normal channels of public control. The emphasis on the effective integration of the Council and the other party "wings" with the party itself was another way to meet the demand for public control.[97]

The declared purpose of the *Dawn Broadcast* was to seek to re-establish the unity of the party, which had been threatened by the conflict over the new roles which the party organisations demanded for themselves in the economy. This involved the question of the place private enterprise was to take in public strategy. The speech sided primarily with the press and union critics of the parliamentarians. Paradoxically, the State Agency scheme had in important respects signified a victory for a more "liberal" approach. The *Dawn Broadcast,* on the other hand, marked a definite anti-capitalist posture. An important aspect was the ban on the private business activities of CPP members of the National Assembly and the government. Private business commitments had been attacked by the press as a main cause of the "anti-socialist" position of certain parliamentarians. The wider ideological context of the President's speech underlined the "anti-capitalist" aspect of the monopoly scheme, now confirmed in unequivocal terms. The rehabilitation of the Council, the image of which had been damaged by the State Agency proposals, was also part of an attempt to enforce the supremacy of the party and the unity of leadership. When appealing against the Agency scheme, the Farmers' Council had stressed its serious political consequences. A petition submitted contained nothing on the commercial advantages or feasibility of its planned monopoly. Monopoly was necessary, it claimed, if the Council was to be able "to win all farmers for the CPP".[98] The State Agency scheme would leave the farmers in other organisations outside the control of the party. The prestige not only of the Council but of the party itself was at stake. The unity of the party and the government would also be seriously damaged if the Council, as a CPP organisation, were drawn into open competition with an agency established by the government. The petition stressed the vital function performed by the Council when organising the farmers for the party. It was the aim of the Council to get farmers to support government economic policies and to help the government develop the rural areas by making the farmers contribute "towards the development of their own villages and towns". Monopoly trading profits would be used for this end instead of being shared among individuals.

The petition demonstrated the primacy given by the Council to the wider political functions of the trading monopoly in its attempt to pursuade the government to reverse its position. Strong doubts had been raised about the commercial feasibility of the monopoly. Little could have happened in the month between the statement issued by the Minister of Labour and Cooperatives and the *Dawn Broadcast* which could have removed the basis of such doubts. It is therefore probable that decisive weight was given to political considerations, of the type raised by the Council, when the monopoly was finally confirmed. The importance of the political argument was emphasised by the old cooperatives when, following the *Dawn Broadcast,* they in their turn petitioned the government against the threat which they thought had been averted.[99] ABASCO strongly denied allegations that they were opposed to the CPP government. They claimed that their members had voted solidly for CPP in the presidential elections of 1960 and that they had donated £2,000 to the construction of the regional headquarters of the party. A CPP member of the National Assembly maintained in a letter to the President that the top managers of the old cooperatives had been bad politically but that the organisation was sound and should be allowed to continue. Public opinion, he claimed, was opposed to the government's new cooperative policy. He quoted an informant as saying that there was a general feeling among the public that personal properties would one day be confiscated by the government because of the way certain people talked about the President's "Socialist Policy" in conjunction with cooperative farms and cooperative everything.[100] The General Secretary of the Council had in fact been obliged to deny "wicked rumours" that the government intended to confiscate cocoa farms.[101] One petition in favour of ABASCO was submitted, surprisingly, by the Commissioner of Police (Special Branch) who certified the political loyalty of the present leadership.[102]

The decision of the government was this time final, although both opposition and CPP politicians continued to demand that it should be reconsidered.[103] On May Day 1961, the Farmers' Council organised an Emergency National Delegates Conference and a farmers' rally at the Accra

Young Pioneer Centre to listen to the official announcement by the Minister of Agriculture.[104] The Council was now the sole buyer of Ghana's cocoa. The detailed plans for the new scheme were presented. After the Conference farmers dressed in white calico, a sign of victory, paraded through the streets of Accra. The offical report recalled the "fierce but silent battle" which had been fought. "Every day brought along with it wild and unfounded rumours that the Government intended to change the decision." The Council had achieved a "victory over their political opponents who were mainly in the fold of the ABASCO". The elimination of the middlemen from the cocoa trade was also celebrated.

The old cooperatives were absorbed by the Farmers' Council. The General Secretary announced the new arrangements to which the ABASCO and the Cooperative Marketing Association were said to have agreed.[105] There was to be only one society and one buying centre in a given area. All farmers in the area would belong to that society. Existing societies would absorb farmers who had sold cocoa to foreign or Ghanaian private buying agents. If in one area there was only a society belonging to the old cooperatives, this would be made the sole buyer but a new executive would be elected under Farmers' Council supervision. If there were both a Farmers' Council society and another cooperative in the centre the two executives would merge. There would be a majority of Farmers' Council members, including the Chief Farmer, the Vice Chief Farmer and the Treasurer. A similar arrangement was made at the district level. As the district organisation of the Council covered the entire cocoa area, the formula ensured that all posts as Chief Farmers and Treasurers at that level were automatically retained by existing Council functionaries. This was also the level where the Council recruited representatives to the regional and national bodies. The Council undertook to absorb the employees of the old cooperatives as well as the foreign firms. Significantly, only the supervisory and clerical staff were covered by this measure. The bulk of the employees—the labourers, drivers, and watchmen—were not.[106] The relevant assets of the firms were bought by the Cocoa Marketing Board on behalf of the Council at their full value. In the case of the cooperatives, no payments were made for assets taken over. This was in accordance with the logic of the merger; assets and liabilities were merely pooled. The savings and investments of the old cooperatives were in this way absorbed by an organisation over which they exercised no control.[107]

The roots of monopoly: summary

In this and the preceding chapter the historical roots of the Farmers' Council and its commercial and political monopoly have been examined. In the following chapters the monopoly will be discussed in terms of how it functioned. At this point, the process leading to its establishment can be summarised. At one level, the monopoly was the outcome of competition between rival groups for economic opportunities in the cocoa trade. In this respect, the monopoly was in the first instance a national solution whereby European economic control was eliminated, to the benefit of local groups. But it was only one among several possible national solutions. Alternatives with political backing in the Ghanaian situation were mainly of three types; one was a competitive market where any Ghanaian organisation or company would be free to participate. It implied in the first place the transformation of the brokers who bought cocoa for the foreign firms into independent traders, selling directly to the Marketing Board. A second type was a farmer-based solution, building on the traditions of public support for cooperative marketing. A third type, finally, was based on direct state participation; an extension of the export monopoly of the Marketing Board to the local cocoa trade. Various combinations of these models were possible. The Farmers' Council solution contained elements from the cooperative model, although mainly at the ideological level. On the whole, it came closer to the state monopoly type, but in this case, the administration of the monopoly was entrusted to a party organisation.

Why was this model chosen? It was the outcome of a process where the political aspirations of the CPP and the commercial aspirations of a group of traders connected with the party mutually reinforced each other. A number of factors contributed to shape this process, including the historical link between cocoa trading and political organisation, the limited access of the CPP to the farmers, the conflict over the distribution of cocoa income between the farmers and the state, the weak position of the independent capitalist class, and the colonial tradition of state intervention in the cocoa trade.

The CPP, in its early attempts to organise the farmers, relied heavily on the tradition of farmer-trader organisations. Farmers and local traders, in spite of internal antagonism, had been united in opposition to the foreign firms. In the pre-war period, the latter were natural targets for farmers' dissatisfaction over poor prices. The commercial aspirations of farmers and traders alike were frustrated by foreign domination of trading capital and export outlets. Many farmers were traders and most traders were farm-owners. They were united in a belief that if they could by-pass the foreign firms it would be possible, not only to secure trading profits but also to strike better bargains in the world market. It was widely assumed that the firms combined to hold down producer prices as well as the profits of the local traders. The basis of this unity between farmers and traders was weakened by the introduction of the state export monopoly and statutory pricing during the second world war, consolidated after the war in the Cocoa Marketing Board. The question of the producer price was separated from that of foreign control over the trade. Independent access to the world market was blocked for firms and farmers alike. Official colonial support for cooperative marketing offered at the same time a new avenue for the gradual increase of the Ghanaian share in local trading profits.

Other grievances, especially opposition to the swollen-shoot control scheme and the price policies of the Marketing Board, activated the farmer-trader groups and brought them in contact with the anti-colonial movement of the late 1940's. The Ghana Farmers' Congress was a loose federation of such local and regional groups. The alliance between farmer-traders and the nationalist leaders was never strong. The Congress was divided on the issue of its political affiliation. The pro-CPP leaders formed the Farmers' Council. But this organisation also split, as the CPP showed no tendency to discontinue the price and tax policies of the colonial government. Most of the Ashanti cadres of the Council defected in opposition to the 1954 Cocoa Ordinance.

The Congress leaders, true to their background, showed a primary interest in cocoa trading. The Farmers' Council in its turn was closely associated with the Cocoa Purchasing Company, the CPP-controlled subsidiary of the Marketing Board. The Council depended financially and administratively on the Purchasing Company, its senior partner. Political organisation was an extension of cocoa trading. The merger of the political and commercial functions was completed as the Council took over the buying activities of the Company when the latter was liquidated in 1957, the year of independence.

The political monopoly of the Farmers' Council was consolidated after independence. The Council was awarded official recognition as the only organisation entitled to represent Ghana's farmers. The Cocoa Marketing Board was authorised to pay an annual subsidy. From this authoritative position the Council sought to establish its control over the cocoa trade. It called for the expulsion of the foreign firms. Foreign capital, represented above all by the giant United Africa Company, made little resistance to the nationalist moves. The UAC withdrew soon after the Council's campaign was launched. The withdrawal, however, was less a result of nationalist threats than of the fact that more profitable opportunities were available in new directions. Far from indicating disengagement, the step was linked to increased participation by the company in the Ghanaian economy. Large investments were made in manufacturing. Cadbury, the other main foreign buyer represented the European processing industry. As such it was particularly concerned with regular supply and quality and less with commercial profits at the African end of the business. It cooperated—after some hesitation—with the monopoly scheme and provided advisors to the Council.

Foreign control in the local cocoa trade was based on privileged access to capital. Most of the actual buying was done by agents and sub-agents, operating largely as entrepreneurs in their own right, although financially dependent on the firms. The experience and market knowledge of these agents, the brokers, should have made them natural candidates for the succession, once the firms withdrew. Their growth into independent traders, selling directly to the Marketing Board, had been obstructed by the licencing system and, above all, by public support for marketing cooperatives, the Cocoa Purchasing Company and the Farmers' Council. Brokers as well as firms were hard pressed by competition from these new organisations. But the rapid expansion of the latter was also a result of their ability to enlist traders who had previously sold cocoa to the firms.

The Farmers' Council made the local brokers a main target in the campaign against the foreign firms. In this way, the largest group of potential rivals in the market could be identified with foreign and colonial domination. In its attacks on the "middlemen", the Farmers' Council appealed to the antagonism between farmers and traders. It also invoked the more recent cooperative tradition among the farmers. But a farmer-based cooperative solution fitted poorly the type of organisation which had been created by the Purchasing Company and the Council. This organisation was characterised by central control over capital and staffing, with no autonomy at the local level. Changes in a more "cooperative" direction undertaken by the Council did not alter the general pattern. The organisation was created not by farmers but by clerks, traders, and store-keepers, belonging—unlike the farmers—to sections of the society from which the CPP cadres were mainly recruited. From the point of view of these groups, the Company-Council type of organisation provided better avenues for advancement and entrepreneurship than a farmer-based cooperative movement.

CPP preferences for a "bureaucratic", rather than a cooperative solution were linked to its choice of strategy for organising the farmers politically behind the party. The existing cooperatives were unreliable in this respect. Their leadership had to a certain extent joined the parliamentary opposition to the CPP government. Attempts to establish CPP control over the cooperatives met strong resistance. But more importantly, the old cooperatives like other farmer-trader groups represented the sectional interests of the farming and trading communities and were therefore naturally opposed to CPP aspirations to control the surplus of the cocoa economy for government and national development. The Farmers' Council offered a solution where centralised commercial and political control were closely integrated. The organisational experience of the cocoa farmers was primarily linked to trading. The combination of trade and politics was therefore a natural starting point for efforts to organise farmers for the nationalist movement. The close identification of political mobilisation and cocoa trading meant by implication that the amount of cocoa bought also became a measure of political success. A commercial monopoly did in this context appear as a logical consequence of the desire for political monopoly.

The choice of trade as a vehicle of political mobilisation and control, although historically rooted in the tradition of farmer-trader organisations, emphasised nonetheless the limited direct access of the party to the farming community. Initially, this may have been a natural consequence of the politically more advanced position of the non-farming groups in the growth of the nationalist movement. But access did not improve with the consolidation of CPP power. Instead, it deteriorated in the face of the growing conflict over the distribution of cocoa income. The repressive aspects of the party's attempt to organise the farmers were brought to the fore. The commercial monopoly served in this respect to eliminate cocoa trading as a platform of organised opposition. The importance of the repressive aspects was enhanced as world market prices fell at the same time as the ambitions of the CPP government greatly increased. (For the role of the Farmers' Council in relation to the price and tax policies of the government, see Chapter 7.)

Forces backing the monopoly found support at the ideological level in the socialist aspirations increasingly espoused by sections of the party leadership and the press. Plans for far-reaching nationalisation of trade were discussed. Cooperatives, controlled by the party through the newly-established National Cooperative Council, or by the party wings, were assigned a key role in the new strategy. The Trade Union Congress had advanced plans of its own for union-owned enterprises of a wide variety. The cocoa-trading monopoly of the Farmers' Council was, however, the only substantial result of this new cooperative deal. In other directions, the state itself, not party organisations, dominated the new strategy. A common denominator, however, was the rejection of private capitalist solutions in a growing sector of the Ghanaian economy.

These developments were by no means unopposed. Although anticapitalist elements were included in the ideology and policies of the CPP even in the early 1950's, private enterprise had, on the whole, been encouraged, clearly in accordance with the aspirations of large sections of the party cadres. A number of leading party men were important private entrepreneurs. In the National Assembly, they were able to speak for the growing class of businessmen, within or outside the party. Private Ghanaian enterprise in the cocoa trade was weak. Only one independent company had succeeded in establishing itself on a substantial scale, and even then only with the financial support of the government. On the whole, the major "private" alternative to the monopoly claims of the party and the government in this sector was represented by the independent cooperatives. These also had much support within the government itself. The cocoa monopoly of the Farmers' Council was opposed as part of a shift in official strategy, reducing the scope for independent enterprise in general.

The social cleavages underlying the ideological conflict within the CPP were probably small. Private, state and party enterprises may have provided largely substitutable or complementary avenues of economic advance for the social groups from which the party recruited its cadres. These groups represented no doubt a variety of experiences in the field of private enterprise, but access to office and wage employment was for most of equal importance. It was by no means certain that income from capital was a superior source of wealth when compared with income from public office in an underdeveloped capitalist economy of the Ghanaian type. The conflict over public versus private enterprise was therefore partly an expression of rivalry among members of the same class. Confronted with the superior resources of foreign capital it was natural for aspirant groups to seek financial support and legal protection from the government. The commercial monopoly of the Farmers' Council was in this perspective a case where one section of this class succeeded in realising their monopoly aspirations by offering at the same time to perform certain political functions on behalf of the government.

But the conflict over the cocoa trade involved at the same time very real social cleavages between those social forces which dominated the CPP and those which so far had played a leading role in cocoa trading. It was only for the members of the urban based petty-bourgeoisie with its degree of education that a "non-capitalist" solution could hold up a credible alternative as a means of private advance. Only they could compete for the offices of the new party and state bureaucracies. The real losers were not the party businessmen who had fought in the National Assembly for free enterprise and free competition. Few of these had any private interests in the cocoa trade. The losers were the farmer-traders who had handled the bulk of the trade either as brokers or as cooperators. Recruited mostly from the upper stratum of the farmer community, these were the people with the most experience in organising the farmers and those most likely to turn farmer organisations against the central government and its heavy appropriation of cocoa income.

The political functions were fundamental in tilting the balance of forces in favour of a "bureaucratic" monopoly solution. At a general level this involved the political monopoly claims of the CPP and its strategy for organising the farmers behind the one-party state. But more specifically and acutely, it concerned the financial basis of this state and its programme for national development as built on the surplus of the cocoa economy. The commercial monopoly aspirations of the Farmers' Council cadres and the political aspirations of the CPP leaders mutually reinforced each other.

V. The New Middlemen

This chapter examines the performance of the Farmers' Council as the buyer of Ghana's cocoa from the establishment of the monopoly in 1961 until its dissolution in 1966. Nevertheless, this is not an attempt to evaluate the overall success or failure of the Council as a commercial enterprise. It did well in some respects and rather poorly in others. How the overall balance is struck depends on what aspects different observers consider to be most important. Nor is this chapter concerned with the general argument for or against monopoly marketing, which clearly is an important issue, and one which continues to preoccupy those concerned with the cocoa trade.[1] I am aware that much of the evidence presented in this chapter could be used to underpin a case against monopoly. Nevertheless, a balanced judgement in that respect also requires a proper enquiry into the evidence against "competitive" buying, both as it existed in pre-monopoly days and as it was when partly reintroduced in 1966. Such an examination is not attempted here. But the case in favour of a non-competitive, farmer-based marketing organisation does not stand or fall with the outcome of a comparison between the Farmers' Council system and what came before or followed it. A system controlled by farmers can take many different forms, as can one based on private traders and a competitive market. It may be noted, for example, that the Ghana Cocoa Marketing Association, the "old" cooperatives and the strongest opponents to the Farmers' Council, demanded in 1966 that they in their turn be made the sole buyer of Ghana's cocoa.[2] It is probable that if this had been granted their monopoly would have taken on a distinct character of its own. In more recent years, the dominance of the Produce Buying Agency, a subsidiary of the Cocoa Marketing Board, represents another type of monopoly situation with its own advantages and difficulties.

I believe that there exists a strong case for a system of marketing based on the cooperative enterprise of the producers. But the problem of producers' control cannot be solved without considering at the same time the stake which Ghanaian society as a whole has in cocoa production. This is a question of finding a form of organisation which best fulfills the necessary balance between the interests of the producers and the interests of national development. This was the type of problem which faced the leaders of the Farmers' Council and it provides the context within which the present examination of the Council's commercial performance should be seen. The aspects of this performance reported here are those which I believe are particularly relevant in judging the manner in which the Council served as such a crucial bridge between sectional and national interests.I am concerned in particular with the growth and consolidation of a new class of bureaucratic middlemen, in pursuit of their own class interests at the expense of the farmers as well as the wider national interests defined by the CPP leadership. I shall return to these wider interests in subsequent chapters. At present, however, our focus is on the dealings of these new middlemen with the farmers.

The chapter opens with a brief account of the substantial achievements claimed for the Council, particularly in raising the quality of the cocoa at the same time as it successfully moved a greatly increased crop from village to port, to the full satisfaction of the Cocoa Marketing Board and overseas customers. From the farmers' perspective, however, these achievements were overshadowed by other obstacles to the smooth flow of trade. They complained about frequent delays and irregularities in the handling of the crop at the village buying centres. There was no compensation for the enforced higher quality standards. The difficulties and delays may by themselves have been minor, considering the overall size of the commercial operation involved. Their significance in the eyes of the farmers, however, was greatly magnified, as they were coupled with widespread malpractices, corruption and private profiteering by Council officers. The Secretary-Receivers, the clerks in charge of the local buying centres ("societies") were the chief targets of farmer resentment, as they were the officers with whom they were most directly in touch. To numerous farmers, these clerks personified the oppressive and exploitative nature of the new party state.

Their list of grievances was long, including such features as the manipulation of scales and weights, the imposition of unofficial levies; favouritism, extortion and corruption in the handling of crop and cash; misappropriation of local funds and profiteering in the distribution of farm implements. Collusion between senior and junior officers and between employees and the elected officers contributed to the farmers' distrust for the new system. Council officers were accused of accumulating private wealth at a rapid rate. I have looked in particular at the connection between their official and private business activities. There were several profitable sidelines of the cocoa trade, such as road transport and the construction of cocoa sheds, which offered opportunities for such "joint" enterprises. While some officials may have drawn much of their income from unofficial sidelines, salaries, wages and allowances were no doubt more important for most. The Farmers' Council was the largest commercial establishment in the country. It was able to offer a wide range of employment opportunities, with yearly increments and promotions.

The official argument for eliminating foreign firms and the old middlemen was that the profits of the trade should be reserved for the farmers themselves. But the expansion of the new bureaucracy left little room for farmers' profits. The part of the surplus specifically reserved for this purpose, the commissions, was absorbed and used to finance overall operations. Certain monopoly savings seem to have been achieved but the bureaucratisation of the trade added at the same time new types of claims on the surplus, including the costs of maintaining excessive staff and "uneconomic" buying centres. The concluding section of the chapter summarises the commercial performance of the Council and what it adds to the overall picture of the class character of the organisation.

The achievement

In 1966, the military government appointed a *Committee of Enquiry on the Local Purchasing of Cocoa.* The chairman was J. C. deGraft-Johnson, of the University of Ghana. The task of the Committee was to investigate the effects of the monopoly and to recommend a new system for the internal marketing of cocoa. The most immediate problem was to find an interim solution for the 1966/67 main crop. There was therefore little time either for investigating past performance or for long-term plans. The Committee reported in July 1966. It declared itself unable to evaluate the economic performance of the Farmers' Council and recommended that a full Commission of Enquiry should be appointed. This recommendation was accepted by the government but never implemented.[3]

Considering the strong anti-CPP mood of official public opinion at this time, the Report was remarkably favourable in its assessment of the commercial administration of the monopoly. It spoke of the firstclass machinery created for the collection of the cocoa crop. The organisation had worked: it had moved the country's record crop to the ports while maintaining very high quality standards. Overseas manufacturers had been satisfied and claims for rejected and disputed parcels had been virtually eliminated. The improvement in quality was seen as "an excellent performance by any standard".[4] The achievement of the Farmers' Council was praised in a statement submitted to the Committee by the Managing Director of the Cocoa Marketing Board.[5] Not only had the quality of cocoa improved, but farmers had also been saved from exploitation by middlemen. Monopoly had put to a halt the stealing of farmers' cocoa by labourers and others: a farmer could sell cocoa only in his own locality, where his production and performance were well known to everybody. He claimed that abuses resulting from direct and indirect competition had been eliminated. Waste and inefficiency had also been reduced and the farmers had obtained means of collective representation and negotiation. Commenting on allegations of various malpractices committed by Council officials, he maintained that such would always exist but that they were more easily investigated and punished under a monopoly.

Cadbury, a major customer and a former Licensed Buying Agent, was asked for its views. The company was represented on the Committee and appeared moderately satisfied, at least to the extent that it did not argue in favour of a change of the system. When a monopoly buying organisation is fully efficient, it claimed, there was no reason why it should not be "the correct medium for this particular purpose".[6] Although the company did not suggest that the Council had been fully efficient, neither did it support the reintroduction of competitive marketing. It emphasised the role played by the staff taken over by the Council from other buying organisations in ensuring the "early success" of the monopoly scheme. Senior Cadbury advisers had worked with the Council during the first two crop years. On completing his assignment in 1963, one of the advisers wrote a favourable review of the new organisation.[7] He spoke of the benefits from "the virtual elimination of the broker", of the very substantial improvement in quality, the high standards of shipping weight, and of the construction of more and better storage. The crop was graded and sealed faster than ever before. In 1966, this same Cadbury adviser submitted a memorandum where the Council's successful handling of the 1964/65 bumper crop was quoted as evidence that its system of marketing had worked to the benefit of the country.[8] The magnitude of the task, he claimed, was not always understood. Cadbury may have been biased in favour of an organisation which it had helped to build, but it had also been well placed to observe actual performance.

In 1961, misgivings about the Council's ability to handle the entire cocoa crop were widespread among government and Marketing Board officials, as well as in trade circles. Disorganisation in the cocoa trade could cause serious damage to the national economy. These misgivings were reflected in the last-minute attempt to launch the State Agency scheme. Oppositional farmers appealed to trade interests, warning against inevitable delays and a rapid deterioration of crop quality. The actual performance of the Council was therefore a positive surprise to officials and manufacturers alike. The flow of the trade was on the whole uninterrupted and quality was greatly improved, in spite of the fact that these far-reaching changes in the organisation of the trade coincided with heavy increases in production. Particular emphasis was given to the improvement in quality. Cocoa is graded according to the proportion of defective beans per consignment. In 1960–61, Ghana's main crop was classified as 73 per cent first grade. In 1964–65, the official figure was 99.8 per cent. Although these figures were doubted by the Committee, it was generally accepted that a very substantial improvement had been achieved.[9] The method used by the Farmers' Council was quite simple: it accepted only first-grade cocoa.[10] An upward revision of the quality standard for grade one was also introduced.[11]

The Farmers' Council leaders could claim with some justification that they had succeeded in realising the basic objectives of the organisation. National control had been established over the local cocoa trade; the "middlemen", local and foreign, had been successfully replaced.[12] But the degree of satisfaction expressed by the spokesmen of the CMB and the manufacturers was not shared by the large number of farmers who gave evidence to the deGraft-Johnson Committee. The Committee toured the cocoa regions and received oral and written evidence. The general tendency was highly critical of the way in which the monopoly was run. The report itself gives a brief summary of the critique and a few examples of concrete allegations. The committee files contain a wealth of additional documentation. It is obvious, given the political situation of the immediate post-coup months, that the dissatisfied and oppositional farmers were more likely to come forward and give their views on an organisation which had been overthrown and which was condemned by the new government. It is also probable that farmers who had been cocoa traders or active cooperators in the pre-monopoly days were overrepresented among those who gave evidence.[13] The allegations of malpractice, theft, misappropriation, corruption, and extortion were none the less numerous and specific and lend support to the Committee's assertion about widespread hostility to the monopoly among the cocoa farmers. The Committee noted that even if few of these allegations could be substantiated, the observable effects of the monopoly were such as to explain the hostility.[14] The internal records of the Farmers' Council at the same time abundantly demonstrate that the allegations were by no means post-coup concoctions invented to blacken the reputation of the organisation by those anxious to take over the trade. Its own petition files contain a mass of evidence of similar kind. National Assembly debates, press reports and harsh criticism also from government ministers show similarly that the public reputation of the Council in these respects was seriously tarnished already before the military coup.[15]

The Council's success in maintaining an uninterrupted flow of trade was an achievement which the farmers could not be expected to appreciate for its own sake. The enforcement of higher quality, that is, of an increased productive effort on the part of the producers, may possibly have resulted in some competitive advantage in the international market, although even this could be disputed. It has been suggested that grade differentials find "little or no reflection in the world markets" and that improvement in quality therefore represents a "substantial misdirection of effort".[16] Be this as it may, it is clear that farmers gained no visible advantages on that account. They received no compensation for the increased labour required to bring the quality of their produce in line with the new higher standard.

With the question of prices removed from the immediate domain of the trade, the farmers' concern when judging the monopoly was primarily with the quality of the marketing services provided. It involved, first of all, questions about how and when cocoa was received, weighed, graded, and paid for. It also involved off-season advances and such other services as the buyers might provide. Farmers were concerned about the regularity and reliability of these transactions, and the way in which they were treated by the buyers. They were of course particularly concerned with the problem of cheating.

The general framework within which these transactions took place was drastically altered by the monopoly. The multitude of intermediaries between individual farmers and the Cocoa Marketing Board was replaced by a uniform structure. Previously, a farmer might have sold his cocoa to a neighbour, a village farmer-trader, a Nigerian shopkeeper, a petty-trader approaching him in the field, or to a by-passing lorry. He might have taken it to the nearest market town or to a more distant buying centre where he could expect to receive the full control price. He could have sold his cocoa to independent traders, company agents, to a cooperative or a Farmers' Council society. Under the monopoly all farmers in a particular area were members of one organisation. They were to sell their cocoa in one place only, the buying centre of the society, and to one person, the Secretary-Receiver appointed by the Farmers' Council headquarters in Accra. Let us look at some of the farmers' grievances with the new system.

Obstacles to the flow of trade: the farmers' perspective

From the farmers' point of view, the flow of trade was by no means as smooth as total crop movements suggest. Obstructions occurred in different forms. At an elementary level, there was evidence of delays at the Council's buying centres. In 1963, the CPP executive in Chiraa, Brong Ahafo, met with other local authorities to pass a resolution of non-confidence in the local Farmers' Council staff. They spoke of week- and sometimes month-long delays in weighing and payments.[17] A memorandum submitted to the deGraft-Johnson Committee in 1966, also from Brong Ahafo, claimed that the handling of cocoa was held up for over a month as a result of inadequate staff and storage facilities.[18] The exceptionally large 1964/65 crop in combination with the hold-up of exports, a result of an international boycott organised by the Cocoa Producers' Alliance, caused major congestion and delays in many areas. In other seasons, complaints about delays in weighing exeeding a few days were rare. But even short delays caused problems: the distance to the buying centre from outlying villages and farms could be considerable. If the cocoa was not taken care of by the receiver the same day it might have to be brought back to the village. Alternatively, a private arrangement had to be found for storing and guarding the cocoa at the buying centre. Both alternatives involved inconvenience, extra costs and certain risks. The deGraft-Johnson Report referred to a case where a farmer lost a large consignment of cocoa which he had left over night with a private watchman.[19] A chief in the Bekwai area claimed in 1966 that farmers in the villages under his jurisdiction had frequently been forced to bring their cocoa back because the Secretary-Receiver was not at his post.[20] But delays were more frequently caused by congestion at the buying centre. The need to queue up for long hours to have the cocoa weighed, graded, and paid for was particularly resented. It was found degrading and insulting. An Ahafo farmer called it a "disgraceful affair".[21]

Payments were occasionally made by promissary notes while the farmers waited for actual cash (the "chit-system"). The extent of the delays involved is not clear. The deGraft-Johnson Report claimed that the evidence at their disposal indicated that they "normally ranged from two to six weeks".[22] This referred probably only to delays actually reported and not to the Council's normal buying routine. The Report mentioned extreme cases where final payments were delayed for four months and more. Memoranda in the Committee files give evidence of such long delays only from the 1965–66 season.[23] At that time, however, trade was disrupted by the military coup and the dissolution of the Farmers' Council. In the Council's own records there were occasional references to month-long delays in earlier seasons, but on the whole, these may have been exceptional.[24] During the mid-crop season of 1961, the Marketing Board discussed rumours that the Farmers' Council was using chits instead of cash payments. It was alleged that private traders were taking advantage of this situation by buying cocoa from the farmers at a price below the official one. The rumours were confirmed by Board members who had been asked to investigate the matter.[25] They spoke of delays of one week or more. The Council denied the allegations, which, in its opinion, were based on the "unfounded fear that the UGFC wanted to take the farmers' cocoa for nothing".[26] These may have been initial delays, not necessarily typical of following seasons. Repeated instructions from the Council headquarters to the buying staff that "farmers must not be kept waiting for payment" nevertheless indicate a continuing problem.[27]

Negligence or absenteeism on the part of the Secretary-Receivers was sometimes mentioned as the cause of irregular payments: one receiver was accused of ignoring the farmers waiting for cash. He would keep the money in his safe and leave for his farm "until such time that would suit him".[28] But the smooth flow of cash to the buying centres involved genuine difficulties of planning and timing. In order to hold down financial costs it was desirable to limit the amount of "idle" cash at the society and district centres to a minimum. As far as possible proceeds from early purchases in the season should supply the cash for the later ones. It was difficult to tell how large the crop would be in a particular area and when it would be delivered by the farmers. The Ghana Commercial Bank, which handled the distribution of cash on the Council's instructions, mentioned the disruptive effects of unexpectedly large or early crops.[29] Delays were not caused by Secretary-Receivers, others claimed, but by those who controlled the central allocation of cash at the Council's headquarters.[30]

The use of chits and irregular cash payments caused uncertainty and anxiety among the farmers. Usually non-literate, they were suspicious of paper transactions. Even quite minor delays could cause serious difficulties, beside the nuisance and expense of repeated visits to the buying centre before finally being paid. It was alleged that payments were often made by instalments, "thus reducing a farmer's entitlement to such bits and pieces that budgeting was impossible".[31] The deGraft-Johnson Report referred to an extreme case where a farmer was said to have been burnt to death by labourers who were enraged by his failure to pay them.[32] One farmer claimed that as a result of the delays the labourers had abandoned the cocoa farms.[33] There were other types of commitments which could not be easily put off. Some farmers spoke of indebtedness caused by the delays. One, for example, claimed that he was forced to take a £20 loan with £10 interest from a moneylender to be able to settle a doctor's bill before he was actually paid for his cocoa.[34] The insecurity caused by delays at the weighing scales and cash counters of the village buying centres was of a new type with no direct counterpart in the pre-monopoly system. The multitude of buyers in the past had in most cases allowed the farmers to sell their cocoa whenever it had been prepared for the market, or even before. Quick service may have been linked to deductions from the official price but competition provided at least a certain guarantee that such charges conformed roughly to "normal" trade standards.

The reorganisation of the market also affected the way in which short-term advances were granted. These had been an important feature of the competitive system. The Council had claimed in 1961 that the monopoly would make it easier to recover outstanding advances with the farmer and thus increase the total resources available for such short-term credits.[35] Farmers complained, however, that the advances granted by the Council had been less than those of the foreign firms.[36] In June–July 1962, a £3 advance was authorised per ton sold in the previous season. It was considered insufficient by the National Delegates Conference and it was doubled the following season, according to the General Secretary's Report.[37] The most numerous complaints in connection with the short-term loans, however, concerned the rate of interest charged by the Council. The Eastern Region Steering Committee of the CPP appealed in 1963 against the one shilling per pound charged.[38] This rate on half-yearly loans corresponds to the official 10 per cent rate announced by the Council in 1962. The Council itself had made arrangments to borrow the money at 6$^1/_2$ per cent from the Ghana Commercial Bank, which allowed for a nice profit.[39] Many petitioners claimed that no interest had been charged in the pre-monopoly days.[40]

The enforcement of higher quality standards was another obstacle to the smooth flow of trade, as seen from the farmers' perspective. They were concerned about what to do with the grade II cocoa which was no longer accepted.[41] Numerous complaints were submitted about the refusal of Secretary-Receivers to accept what farmers themselves considered to be perfectly good cocoa.[42] Council officials had nothing to lose from being insistent. The pre-monopoly traders could not afford to be as strict. They were often pressed by competition to accept cocoa of doubtful quality. If nothing else, they stood the risk of losing outstanding advances.[43] But they could also compensate themselves or even add to profits by mixing cocoa of varying quality to make it pass the required minimum standards.

The zeal of the Council in its campaign for higher quality was great. Field officers were threatened that "drastic measures" would be meted out to them if low-grade cocoa was bought.[44] At the buying centre, the zeal could certainly take drastic forms. According to a letter to the *Daily Graphic,* forwarded to the Council—with a note that it would not be published—a Secretary-Receiver might shout at the farmer: "Hey, old man! How dare you bring such inferior cocoa beans for sale! You are going to answer for it in the police station".[45] According to the deGraft-Johnson Report, it was arranged for subgrade cocoa to be confiscated to prevent it from being used for mixing.[46] It was to be paid for only when and if it was sold in the world market. The Committee found no records of such payments either to societies or to individual farmers. The actual extent of such confiscations is not known and may not have been important. The threat of confiscation, however, added to the pressures behind the quality drive.

The trade in cutlasses (matchets), the principal tool of cocoa farming, and gammalin, the insecticide used for spraying against capsids, was not a part of the marketing of cocoa as such. In the monopoly period this trade was partly taken over by the Council. Its widely dispersed, permanently staffed buying centres seemed to provide a suitable network for the distribution of these implements essential to the cocoa economy. Shortages in combination with serious irregularities in the system of distribution soon turned the Council's participation in this field into a major liability. Few issues were so frequently raised in the petition files of the Council and in the memoranda submitted to the deGraft-Johnson Committee. The complaints were given much publicity, even before the military coup. The Council was obliged to defend itself against "fantastic and unfounded allegations by some people, including sections of the press".[47] A public enquiry was demanded in the National Assembly.[48] At one point, a special government committee was set up to ensure fair and equitable distribution.[49]

A Secretary-Receiver complained in 1966 about the difficulties facing the local field staff in this respect: "The then General-Secretary always assured the farmers that he is sending them plenty of Gammalin and cutlasses but after the speeches they do not get them".[50] How, he asked, could a Secretary-Receiver possibly distribute four cases of gammalin and 100 cutlasses fairly in a society of 400 farmers? The General-Secretary, on the other hand, pointed at the insufficient supplies available to the Council. He reported in 1965 that only 86,000 cartons of gammalin had been received during the past 17 months, while the requirement at the recommended spraying rate was 800,000 cartons a year.[51] Effective demand, even at subsidised prices, was certainly not that high, but it was obviously far above the amount actually supplied. Reports by field staff of the Ministry of Agriculture in the 1964/65 season spoke of serious shortages in all regions: "In fact farmers everywhere still complain very bitterly about the critical scarcity of the gammalin".[52] The local manufacturers of gammalin, the Imperial Chemical Industries at Tema, in their turn referred to irregular orders and inadequate import licenses for raw materials.[53]

Large scale spraying had been initiated by the Ministry of Agriculture in the late 1950's, first in the form of a free spraying scheme undertaken by the Ministry itself. Trial results had been very impressive and large increases in production were expected. Official estimates indicated that production might rise by over 50 per cent within the current five year plan period.[54] A campaign was launched in 1959 to encourage the farmers to spray their own trees. Gammalin and spraying machines were heavily subsidised.[55] The farmers were quick to respond to the new opportunities which, at least in the short run, increased earnings.

The Farmers' Council was associated at an early point with the spraying scheme. It provided supervisors and instructors and participated in the distribution of the spraying machines.[56] The distribution of gammalin was taken over from the Ministry of Agriculture in 1963.[57] According to the General Secretary's Report for 1962/63, a dramatic shortage developed soon after the Council's take-over.[58] He said that he did not want to discuss the alleged causes but maintained that they were now remedied. But a really serious shortage does not seem to have developed until the 1964/65 season. After much public pressure, the Farmers' Council agreed to hand back responsibility to the Ministry of Agriculture.[59] Statistics submitted on actual distribution to the districts reveal a very irregular pattern.[60] The allocations cannot be explained by local differences in demand. Uncertainties and miscalculations of the supplies at the centre probably contributed to the irregular pattern.

Participation by the Farmers' Council in the distribution of cutlasses begun in 1962. An agreement was made with the state-owned Ghana National Trading Corporation that cutlasses were only to be sold in the GNTC stores in Farmers' Council district centres. They were also to be sold only to "legitimate farmers" introduced by Council officers. A uniform price was to be charged.[61] At this time shortage had already developed and the Council was to make sure that the implements reached those in need of them, without prices and supplies being manipulated by unscrupulous middlemen. The Minister of Agriculture announced that three million cutlasses had been ordered "to get rid of the greatest obstacle facing farmers at the moment".[62] The shortage was one of many which arose in the wake of the government's 1961 decision to introduce stricter import control. As in the case of gammalin, a black market rapidly developed, with prices far above the official one.[63] The cutlass was a principal tool not only in cocoa production but in the entire forest zone. The failure of a system of distribution based on the cocoa trade alone should not have been surprising. At a meeting of the Cabinet in 1965, the Minister of Agriculture raised the question of the shortage and of reports of cutlasses selling at over six times the official price.[64]

The bulk of the farmers' complaints and allegations concerned irregularities by individual Secretary-Receivers at the society level.[65] As in the case of delays in weighing and payments, much of the difficulties originated in inadequate resources and genuine problems of allocation and administration. Much may be credited to inefficiency and inexperience. Farmers, however, stressed other causes of the irregularities, and above all, the criminal and corrupt activities of the Receivers, alone or in collusion with senior officers and members of society committees. Such allegations were raised in relation to every step in the marketing process from weighing and grading to payments and the distribution of advances, cutlasses and gammalin. General difficulties in the administration of the monopoly, shortages and poor planning, added to the opportunities for private profiteering by individual officials, opportunities which were inherent in the new market organisation. Let us turn to this important aspect of the Council's performance.

Public monopoly and private profiteering

The manipulation of scales and weights by traders has been a problem in many different types of societies. It is a long-established grievance of the Ghanaian farmers. The 1938 commission on the marketing of West African cocoa claimed that much evidence had been submitted to show that farmers were defrauded by false weighing.[66] Spring-balances used in "bush-sales" were easily tampered with but the platform scales of the large brokers could also be manipulated. The need to prevent such malpractices was put forward by the Farmers' Council as a major reason why private cocoa buyers should not be permitted. The monopoly would improve security in the market place. The spring-balances were to be eliminated. The platform scales at the buying centres were to be regularly checked by senior officers. All transactions would be supervised by members of the executive committee of the society.[67]

Reports about widespread rigging of scales and manipulation of weights by Council officials were received by the deGraft-Johnson Committee.[68] The 1966 Report gave details of one case in the Nsuakwate society where farmers had discovered that a piece of lead had been attached to the scale. Inspection by a District Officer revealed a large surplus in the society's stores.[69] One report in the files of the committee was submitted by an Ashanti farmer who himself owned a scale and was therefore able to challenge the false weighing.[70] In this case, a piece of lead attached with the help of soap was discovered. In another instance, a piece of wood was found inside the scale.[71] An internal auditor of the Council reported in 1964 from Sunyani that he had discovered scales manipulated by metal applications.[72] The 1966 report also mentioned allegations that scales were sometimes placed in a way which "enabled the labourers to play tricks with their feet and thus interfere with the correct weighing".[73]

At a large farmers' rally in Kumasi in 1966 it was claimed that Council officers used to prevent farmers from checking the way in which the cocoa was weighed.[74] "Abuses were showered on you if you tried to watch the weight during the course of weighing". A petition from Akroso claimed in 1965 that the farmers were told by the committee "that the General Secretary had announced that no farmers should weigh his cocoa with an ordinary handing scale".[75] The farmers were unlikely to be deceived by such pretensions. But there were other methods. Farmers from Nyinahin reported in 1964 that the Secretary-Receiver and committee members did not allow them to approach the scale to verify weights.[76] Some farmers claimed that they were subjected to threats and actual reprisals if they dared to challenge the method of weighing. Even quite small adjustments and overweights could produce a surplus of several bags by the end of the season. To individual farmers such pilfering was less noticeable. They would hesitate to challenge the method of weighing as they depended on the cooperation and services of the Secretary-Receiver.

Overweight could also be arranged without recourse to the manipulation of scales. A bag of cocoa was supposed to have a certain standard weight. Numerous reports maintained that receivers demanded additional weight above the official figure.[77] It was reported in 1965 from Kukurantumi that for three years the Secretary-Receiver had weighed the cocoa at 148 pounds per bag although "it was well known to all farmers" that it should have been weighed at 144 only.[78] The excess, it was alleged, had been shared between the receiver and the committee members. The Akroso petition, quoted above, claimed that the farmers were told that a bag should be weighed at 146 pounds and not 144, "as an order from the General Secretary".[79] The farmers in this case had found out from the District Officer that there was no such order.

This particular method of profiteering was greatly facilitated by the confusion surrounding the use of surplus weight as an officially sanctioned method of financing certain expenses at the central as well as the local level.[80] Such deductions varied from season to season and it was not easy for the farmer to know which were authorised from above and which were not. Deductions decided by regional, district and local party and government authorities added further to the confusion and widened the scope for manipulations by marketing staff and committee members. In the 1962–1963 season, centrally proclaimed deductions amounted to $2^1/_2$ pounds per bag, including one pound to meet commercial risks for which there "was no insurance cover" and for the maintenance of landrovers.[81] One pound was allotted for allowances to local committee members and Secretary-Receivers and to cover "incidental society expenses". The $^1/_2$ pound was to cover shrinkages (reductions in weight) at society level. In the 1963–64 season, a further one pound deduction was authorised for the construction of feeder roads.[82] The landrover and feeder road-deductions were abolished in 1964, after considerable public criticism.[83] The Eastern Region branch of the CPP demanded the abolition of the extra weight because it encouraged the Secretary-Receivers "to dupe the illiterate farmers".[84] A much-elaborated petition to the President was submitted by J. G. Edusei of Sunyani.[85] He emphasised that transport costs (landrovers) and committee allowances and expenses should not be separately charged to the farmers, as they were already covered by the buying allowance granted by the Marketing Board. The construction of feeder roads, he claimed, should be financed from the general resources of the government, to which cocoa, of course, contributed the major part.

The deduction for society allowances and expenses was authorised by the National Delegates Conference in 1962. The General Secretary's Report noted that it sparked off "a great deal of trouble in the villages".[86] The deductions were strongly opposed. At an Emergency Meeting of the Central

Region division of the Council he assured committee members that "they would by all means be furnished with their share of the surpluses and sample residues as agreed upon".[87] Some farmers, he said, were merely envious of these benefits and were therefore trying hard to dispose of the present committees by using the influence of local politicians and District Commissioners. A letter from Dodi Papase in 1966, signed by a chief and 49 farmers, spoke contemptuously about "some handful of persons whom they themselves call Society Committee Members" who shared the farmers' money between themselves, using "a queer name as surplus weight for their selfish ends".[88] Receipts from this particular deduction were labeled the "Members Cocoa Fund", of which the society treasurers normally received one-third.[89] The rest was divided between the Secretary-Receiver, the Chief Farmer and other committee members. One part was kept for society expenses. In addition, the committees could use the special deduction for "shrinkages" and other expenses. A farmer claimed that this was also shared among the committee members after a part had been reserved for expenses in connection with visits by the General Secretary, the Regional Secretary and Chief Farmers.[90] Part of the Members' Cocoa Fund was claimed more directly by higher officers. The Regional Executive of the Council in Ashanti decided in 1963 to transfer one part to cover expenses at the district level.[91] In 1964, for example, the headquarters authorised a payment of £1,500 out of such funds to cover the cost of a van bought for the Party.[92]

Locally authorised deductions (surplus weight) covered a variety of purposes including development projects, party buildings and society expenses. A district meeting in Bogoso in 1962 decided to add one extra pound to the official weight, in order to finance the construction of district headquarters for the Council.[93] Elsewhere, farmers protested against surplus weight collected by the local chief.[94] In Kpedze, the District Executive Committee decided to use a 2 lbs deduction to finance an extra £6 monthly payment to the Secretary-Receivers.[95] The Auditor-General reported disapprovingly to the General Secretary in 1965 that a 2 lbs deduction was enforced in Central Region for the new regional headquarters being constructed by the CPP.[96]

The Cadbury adviser expressed concern in 1962 about the additional weight charged to cover society expenses. It did not appear in the books and was not accountable by the Farmers' Council.[97] The absence of accounts for the official deductions at the society level further increased the scope for private misappropriations. Farmers spoke of "unnecessary and compulsory deductions for developments which were never carried out".[98] In Kumawu in 1963, farmers demanded that any local surplus weight must be declared to them.[99] The local Chief Farmer had been approached in this matter but refused to give any information. He claimed, according to the petitioners, that he had been instructed by his superiors "to put everything into darkness". Similarly, a Peki farmer claimed that surplus weight had never been declared to the farmers and that the Secretary-Receiver, when questioned, had simply told him that the money had gone to the District

Office and that "it was not for us".[100] Another farmer from the Volta Region reported that when Secretary-Receivers were asked to account for overweight they refused, threatening the farmers with "CPP and Nsawam" if they insisted.[101] Nsawam, in this case, stood for a much feared prison.

Surplus weight was also extracted without reference to either officially or inofficially inflated "standard" weights. On a moderate scale, such extra weight could be expected to go unchallenged, as the farmers wanted to avoid further difficulties. They reported a variety of methods used in such open pilfering, such as the receivers' practice of refusing to take note of weights of less than 5 lbs.[102] The deGraft-Johnson Report noted allegations that labourers handled unsealed bags in such a way that cocoa was intentionally spilled and that farmers were prevented from collecting it again: "Did you not see cocoa on the ground when you came in? How can you distinguish yours from the lot on the ground?"[103] "Huge quantities" were dropped on the floor in this manner, according to a petition from Badu Society in 1963.[104] The methods could be even less subtle. According to a Mampong farmer "they take off some of the beans and use it as their own profits".[105] Petitioners from Nargode Society, Nyinahin, also claimed that cocoa was removed from the bags in this way.[106]

One problem was how to solve the bookkeeping aspects of these appropriations. The Ashanti Farmers' Union claimed that entries were made into the books of the society in the names of fictitious farmers.[107] The Regional Auditor of the Council in Volta Region complained in 1962 about this particular practice.[108] It has also been suggested that Secretary-Receivers handed over the surplus cocoa to confidants among the farmers, who in their turn resold it to the society.[109]

Congestion and delays at the buying centres provided a fertile ground for extortion and bribery. The conditions themselves may have been the result of factors largely beyond the control of the local marketing staff. Irregular timing of the farmers' demand for market services was bound to strain the capacity of the system, especially during the peak of the season. At such times, the prospects of long delays, day-long queuing and perhaps repeated visits to the buying centre created a potential demand for preferential treatment. Some Secretary-Receivers were reported to serve their relatives first. "He who spoke about this whimsical practice had his cocoa delayed intentionally or at times rejected and later burnt", according to one report from a group of Kumawu farmers.[110] But preferential treatment could also be bought and the step to discrimination and extortion was short. Farmers were refused or delayed unless something was offered in exchange, either a direct charge paid before the service was obtained or indirect ones, such as deductions made while the cocoa was handled.

The deGraft-Johnson Report suggested that delays were sometimes artificially created, in order to facilitate this type of profiteering: "Secretary-Receivers were known to watch a build-up of loads of cocoa at their sheds without making a move to attend to the cocoa."[111] This build-up, the Report claimed, often continued until the farmers were tired and hungry. "Just when the farmers were beginning to feel dejected", they were told that they would have to come the next day. In this situation, many preferred to offer inducements. As retold, the method may seem too crude and provocative to be a regular trick of the trade. There were no doubt other, more refined varieties. A farmer could count himself lucky, a report from Sunyani maintained, if he merely had to spend a few cedis to secure the attention of the receiver. The alternative was to spend days at the shed before it was his turn. Farmers had to "offer promises or pull wires to secure prompt services". As a result of these difficulties the farmer had to accept any weight declared by the receiver. "Any attempts to challenge weights meant further delays, more expenses and inconveniences, (and the) displeasure of the officials."[112]

Many farmers played according to these rules in order to get preferential treatment or to prevent discrimination. Others protested. A farmer of Akropong, Ashanti, appealed to the headquarters: what should he do? The Secretary-Receiver had refused to weigh his cocoa when it was his turn. Should he carry it back to his own village or leave it over night at the centre? "Many a time farmers have lost their cocoa at this same buying centre because it was not weighed and the farmers had to leave it".[113] A petitition to the Juaso District Office of the Council asked for the removal of the society treasurer because, among other things, he "usually collects money from the farmer before he instructs the labourers to weigh their cocoa".[114] Others objected to "the extra charges by the labourers, with the consent of the secretary, from farmers before their cocoa is weighed".[115] An ex-Chief Farmer from Nkawie district explained that he had sent five bags of cocoa to be sold but that the Secretary-Receiver demanded £5 before he would take them.[116] "As I refused to do so the clerk also refused to buy the said cocoa and it was put down until all became spoiled and I threw it away". A Kumasi farmer claimed that before the cocoa was weighed, "the farmer would be asked to stand in a corner with the person weighing the cocoa and farmer was told to give money as drink".[117]

The way in which cocoa was received and weighed was one primary cause of conflict between the farmers and the monopoly and a major source of private profiteering by Council officials. Grading was a second. We have seen how the enforcement of higher quality meant that more labour was extracted from the producers without compensation. Disputes over quality were heightened by bribery and extortion. Quality was measured by taking a sample of beans and counting the number of defective ones. These could be mouldy or damaged by insects. Grade I allowed for a maximum of five (later three) per cent defective beans.[118] Unlike weighing, grading depended on personal assessment. Measures were imprecise and the scope for arbitrary treatment considerable. The immediate task of grading rested with the receivers, but Produce-Examiners, employed by the Ministry of Agriculture, were to control the cocoa before the bags were sealed for transportation to the depots. They could also be called upon in the case of disputes between the farmers and the receivers. The deGraft-Johnson Report mentioned allegations about collusion between receivers and examiners.[119] First-grade cocoa could be classified as second or sub-grade. Farmers were forced to accept deductions for the purpose of reconditioning, that is, for adjustments of weights when the allegedly defective beans had been removed.

A letter to the *Daily Graphic* in 1961 spoke of "a certain game" played by the Council officials, refusing to accept good cocoa.[120] Another letter claimed that the farmers, "trembling with fear", bribed the grader after they had been threatened with various reprisals because of their "bad" cocoa.[121] A petition from Badu Society mentioned that cocoa which had been first rejected as bad was later accepted if the farmers agreed to share the proceeds with the clerk.[122] Two farmers from Akim Oda, described the situation in the following way:

> It may happen that an old man may bring his cocoa from his cottage to town to sell. The Receiving Clerk being anxious to have a share of it, quickly declares the cocoa to be slaty. This the old man has nothing to say as the clerk is the only buyer in town . . . This is a kind of stealing which is killing the farmers.[123]

They were afraid, the letter added, to report the matter to the executive committee of the Society, as this would place them at loggerheads with the clerk. In some cases, farmers may have accepted in good faith the claim that their cocoa was in need of reconditioning, only to find out later, as reported by one farmer, that "not a bean was thrown away as a bad one".[124] This farmer referred to one occasion when he had sent his wife to weigh 70 loads of cocoa at the Kenyase Ahafo centre. The receiver had demanded a deduction of six loads against "pickings", that is, for reconditioning. "If I did not agree he would refuse the cocoa". The example indicates the amounts which could be involved in this particular racket. The material does not allow any conclusions about "typical" fees charged. They may have been much lower than in this case. But there are also occasional reports about higher levels. One farmer maintained that his 15 loads of cocoa had been rejected because he had refused to offer the one-third share "always received" before the cocoa was accepted.[125] This was certainly an exceptional case, if at all true. It is difficult to imagine that margins of extortion could be maintained at such levels, considering the element of political control which still existed and which will be further discussed in the next chapter.

Apart from the charges of false grading, farmers claimed that cocoa taken for the sample checks was manipulated in a way to show a lower grade. It was alleged, for example, that labourers kept bad beans in their pockets which they used for the purpose of deflating the grade.[126] In the pre-monopoly period, brokers were accused of buying sub-grade cocoa at a reduced price and up-grading it to acceptable standards by mixing it with high-quality supplies. The same accusation was now made against the Farmers' Council officials. Especially in the first seasons of the monopoly, it was alleged that old brokers continued to buy sub-grade cocoa which they resold to the Farmers' Concil.[127] According to an internal audit report from the Central Region, senior officers were directly involved.[128] They bought low-grade cocoa and stored it privately. When first-class cocoa was received from the buying stations these private purchases were reconditioned to the required grade.

As in the case of weighing, delays in payments were sources of corruption. Shortages of cash at the buying centres meant that bribes were offered to secure preferential treatment and that fees were imposed on normal service, as well. A report from Sunyani mentioned rumours that District Officers received money from individual societies in order to secure priority in the allocation of cash.[129] Secretary-Receivers, in their turn, demanded "something" from the farmers for early attention when cash had been secured from the district headquarters. Delays were probably in the first place caused by inadequate resources and poor planning but it was also maintained that money actually available was misappropriated by Council officials. The postcoup enquiry reported allegations that trading cash was used for money-lending and for other private business. In some cases, large sums were said to have been used as down-payments for private trucks.[130]

Unlike manipulations directly involving the farmers, such allegations were more difficult to substantiate. Farmers reported what they believed had happened to money which did not reach them in time and they drew conclusions from the observed private business activities or spendings of the officials. Farmers could be specific about the delays but not about their causes. Similar claims were made about cash meant for short-term advances. Committee members had taken the greater part of the loans, according to a senior officer who investigated problems of how to recover outstanding advances in the Eastern Region.[131] Other reports maintained that the advances were given to the wives of officials for trading purposes.[132]

A common allegation was that money for short-term advances was used for private money-lending. A petition from Santrokofi Society in the Volta Region in 1964 reported that it was a general belief among the farmers that the committee and the Secretary-Receiver withheld the short-term advances for that purpose.[133] The issue was raised several times in the reports submitted to the 1966 committee of enquiry. "The mighty Lords of the Society Committees kept the money and lent it out at big interest for themselves", one Ahafo-Ano farmer claimed.[134] Loans, it was also alleged, were often given to nonfarmers.[135] The Chiraa petition already quoted asserted that loans were given out at 100 per cent interest which sounds high even by the standards of the local money market.[136] But effective rates of 50 per cent were by no means unusual in the latter case. The gap between these free market rates and the official rates prescribed by the Council was wide enough to allow a considerable scope for private profiteering by those in control of cash at the various levels of the organisation.

Shortages and irregularities in the distribution of insecticides and cutlasses were a major source of resentment. They were widely associated with private profiteering, nepotism and favouritism on the part of officials and committee members. The deGraft-Johnson Committee mentioned reports of District

Officers using local shopkeepers and sometimes their own wives to retail these goods at inflated prices. As in the case of delays in the allocation of cash, it was difficult for the farmers to know what was actually going on: were supplies being diverted, or were the shortages genuine? "I and the farmers here suspect some underhand dealings in the distribution of matchets", one farmer reported to the General Secretary in 1965.[137] Lack of accounts and information about supplies and the way they were distributed added to suspicions. It could even be a problem among the committee members themselves. In one society, the treasurer was accused of concealing supplies from the other members, and of selling privately at inflated prices.[138]

Sometimes the reports about how supplies were diverted could be quite detailed, as in the case of one petition from Kukurantumi: eight cases of gammalin had been supplied, one being given to the Chief Farmer, one to the Odikro (a local chief), and the remaining six cases to one committee member who allegedly sold them in another region at more than twice the official price.[139] On one occasion, the same committee and Secretary-Receiver were foolish enough to charge a greatly inflated price when selling to the Adontenhene (another local chief). They were made to apologise and refund the difference. In another society, it was alleged that "committee members take a case each while a farmer gets only one tin".[140] Field staff from the Ministry of Agriculture reported in 1965 that farmers complained that relatives of the Chief Farmers and other Council officials still had some gammalin but that they only sold it to close associates or at exorbitant prices.[141] Other farmers spoke of "some special people who get some always". They wanted a senior officer to investigate the activities of the receivers in this respect.[142] A chief of Manso-Mim, Bekwai, reported in 1966 that farmers in his area could get neither gammalin nor cutlasses, because "the Secretary Receiver in my town is having Farm and Families here, and he used to give the whole supplies of Materials to them".[143]

The farmers' views about the new marketing system were primarily influenced by what was observed or rumoured in connection with weighing and grading, and the distribution of payments, advances, and implements. These were all transactions in which they were immediately involved. But they were also influenced by reports and rumours of misappropriations and thefts of society funds and cocoa by marketing staff and committee members. In such cases, farmers may not have been affected as individuals, the losses normally being carried by the organisation as a whole, but their confidence in the system was further undermined.

A catalogue of various types of misappropriations is found in a report by the Council's security department in 1962.[144] It listed 20 cases which had been referred to the police, including cases of fictitious receipts, false entries in the accounts, cash shortages, theft of cocoa, theft of empty bags, misappropriation of surplus cocoa, and overdeclarations. The last item referred to cases where receivers had declared higher purchases than actually achieved, in order to get temporary access to cash for private use. A far greater number of cases were mentioned in the internal audit reports of the same period. It was the policy of the Council to solve these "disciplinary" problems as far as possible within the organisation itself, without involving the police and the courts, especially if the misappropriated funds could be recovered from the culprits.[145] An audit report showed cash shortages in 11 out of 22 societies in the Abuakwa District, ranging from £13 to £652 and totalling £3,345. In the Kufuor Camp Society, Bibiani, the Secretary-Receiver was reported to have failed to pay for £448 worth of cocoa. He was unable to account for £2,371 and had misappropriated £301 out of the Members' Cocoa Fund. In the Hwidiem District, overdeclarations ranging from 9 to 249 bags were reported in eight societies.[146] It is not possible to give a systematic picture of the disclosures of the internal audit reports as most of them had been removed from the Council files at the time of my investigation. Other evidence indicates that the number of reported cases may have been quite high. The standing orders issued by the Administrative Secretary of the Council in 1962 noted that numerous Secretary-Receivers had been found indulging in "fraudulent and unscrupulous means of living".[147] At a meeting in Koforidua of senior marketing and audit staff in 1964 better control over the receivers was urged:

(They) are the key officers constantly in touch with money and the farmers . . . their work is highly tempting and unless an officer is extremely socialist-minded anything might happen anytime.[148]

It is difficult to tell how much the auditors were actually able to detect. They complained about the poor quality of bookkeeping at the society level and the opportunities thus opened up for manipulations.[149] It was often hard to verify local accounts. The large number of transactions involved, the dispersion of the societies, and the inexperience of the auditors added to the difficulties. Local officers could also do much to sabotage effective inspection. A Volta Region auditor, suspecting fictitious entries into the accounts, reported that he was unable to ascertain the true facts, because the Secretary-Receiver responsible "played hide and seek".[150]

At a meeting in Sunyani the General Secretary warned District Officers against taking bribes from Secretary-Receivers.[151] It was alleged after the coup that the latter lavished gifts on senior officers in order to secure their cooperation.[152] Others maintained that the Secretary-Receivers were rather forced to compensate themselves by taking from the farmers what senior officers exacted from them.[153] The 1966 committee of enquiry recorded that it was the practice of senior officers to collect tributes during visits to the societies. In some cases, District Officers demanded a share in cash delivered to the societies.[154] Other officers, including the auditors themselves, similarly taxed the receivers. The latter claimed that in this situation they were left with no option but to recover the outlays from the farmers in order to be able to balance the books.[155]

Senior and junior officers were probably in many cases tied to each other by a complex relationship where protection was traded against a share in the locally extracted surplus. Cooperation and discretion was a condition for acquiring and preserving employment and certain levels of earnings at both ends of this relationship. A fascinating example of collusion between junior and senior officers was revealed in the so-called "Depot Keepers' Case". In 1964, 20 Depot Keepers in Ashanti petitioned the President against their recent dismissal from the Farmers' Council.[156] According to the petition, they had been made scapegoats for shortages which had in fact been caused by excessive "horning" by Produce Examiners on the instructions of the Regional Officer of the Council. The Produce Examiners were supposed to make sample checks by piercing the sealed bags with an implement called the "horn". Only a handful of beans were normally to be removed. The examiners however had been offered 30 shillings for each bagful of cocoa they were able to "horn out" for the benefit of their unofficial principal. The Depot Keepers reported that during the 1962–63 season 2,558 bags valued at £17,997 were extracted in this way at the depots under their control. The horning was done in the presence of weighing- and siding-supervisors and security officers. It is not known if officers other than the examiners were given a direct share in the surplus. The Depot Keepers themselves claimed that they had received nothing and that they did not know what happened to the cocoa after it had been removed. A certain reduction in weight was allowed for during transport from the depots to the port, but the "shrinkages" in this case were apparently excessive. The Cocoa Marketing Board sounded the alarm. The Depot Keepers—quite unfairly, according to them—were made to pay for the shortages and dismissed.

The Depot Keepers' story is not altogether clear, especially as far as their own involvement is concerned. Nor is it clear to what extent this was primarily a regional affair or whether the surplus was passed on further up. But the impression of corruption and collusion involving officers at even the highest level was enhanced by the official response to the petition. The Minutes of an emergency meeting of the Regional Working Committee of the CPP in Ashanti reveal that the Depot Keepers were in fact reemployed as Assistant District Officers. The General Secretary of the Council promised that their "scanty salaries" would be increased. They were also promised promotion to full District Officers as soon as there were vacancies.[157]

Illegal wealth and private entrepreneurship

The Secretary-Receivers were the principal target for the post-coup attacks on the monopoly. This does not mean that they necessarily were more exploitative than other categories of Council officials; they were simply more easily observed from the farmers' viewpoint. They came to incarnate the evils of monopoly, surrounded by a mythology where specific allegations and rumours were fused into a general picture of a new class of middlemen, replacing the pre-monopoly picture of the exploitative broker.

Such Secretary Receivers are very bad, cruel and notorious thieves ever since we have met (them) in our midst. Even they do not respect the farmers at all, and even when a very young boy is engaged as Secretary Receiver for about two years time, you see him riding a very promising Car and (have) mighty buildings with their scanty pay.[158]

Much of the post-coup material presenting this general picture, as opposed to more specific allegations, was clearly a result of an organised campaign. Many of the letters and resolutions submitted to the deGraft-Johnson Committee can be traced to the same areas, organisations and letter-writers. Similarly worded and serialised documents were produced in the Dormaa Ahenkro and Techiman districts of Brong Ahafo and in Ashanti-Akim.

Emphasis was laid on the wealth of the receivers, which was considered out of all proportion to their official earnings and consequently assumed to be ill-gotten. A meeting of farmers at Mangoase, Techiman, demanded an investigation into the assets of the Secretary-Receivers. The request was doubtless inspired by the enquiries which at this time were ordered by the military government into the private affairs of the ousted CPP politicians.[159] The receivers, they claimed, had become rich "by their misdeeds and malpractices", not from their monthly salary of £15. The Dormaa Cooperative Union maintained that receivers had put up "mansions" and owned "fleets of lorries" after only one or two years of service.[160] Large houses and private cars and lorries were the most commonly cited manifestations of the new wealth.[161] The speed with which wealth was acquired was particularly stressed: "Impecunious Secretary-Receivers became rich overnight".[162] A document signed by twelve Tepa farmers was very specific when describing this sudden transformation:

A Secretary-Receiver may arrive at a station with one six-yard cover cloth, one pair of khaki shorts, one pair of white shirts and one pair of shoes. He has no chair and no bed. In two years service as Secretary-Receiver you will meet him riding in a car and enjoying I.D.C. furniture in his room.[163]

I.D.C. (Industrial Development Corporation) symbolised in this case European-designed, factory-made furniture, in contrast to the simpler products of the local carpenters. The ethic of an agricultural community, where wealth is still associated with old age, was particularly offended by the young age of the prosperous clerks.

The prosperity of the receivers was contrasted with the poverty of the farmers. "These Secretary-Receivers are thiefs!!", one farmer exclaimed: "What they gain is more than the farmers own money".[164] The farmers had become "poor as church rats" while the clerks owned cars and buildings.[165] The anti-imperialist rhetoric of the CPP was applied to the relation between the farmers and the officials: "For farmers live in the midst of plenty yet we are the hewers of wood and drawers of water for the officials of the defunct UGFCC". The receivers rested comfortably in fine houses or rode about in luxurious cars, while the farmer who produced the cocoa "has no place to put his head and goes about with misery, hunger and tattered clothes".[166] The parasitical role of the receivers was emphasised. "Some of those clerks had never seen a cocoa pod before", but they enjoyed more income from cocoa than the farmers themselves.[167] The contrast was graphically illustrated by one Besease farmer: "The Secretary-Receivers have bought some lorries and the farmers who are plucking the cocoa to them could not buy a lorry's tyre".[168]

The ill-gotten wealth of the Secretary-Receivers was no doubt exaggerated. Much of the pilfering was probably on a scale hardly sufficient to finance either mansions or cars. But it is equally probable that *some* Secretary-Receivers actually bought private cars, built large houses and extorted or misappropriated money on a large scale. The comparisons between the prosperity of the receivers and the poverty of the farmers took no account of the large differences in wealth also among the farmers themselves. But the argument should be seen in relation to the real changes which took place in the relative fortunes of cocoa producers and wage-earners in the 1960's. The decline in producer prices stood out in contrast to the increments and promotions of the upper-level employees of an expanding public sector as will be futher discussed below.

A recurrent element in the allegations against Council officials was the emphasis on the *private business* aspects of how they handled their administrative duties. The impression is frequently given of private entrepreneurs, who not only put a price on the public service under their control, but also used public resources as a springboard for private investments in separate or allied lines of business. The relative prosperity of individual officers may have been the result of such enterprises, rather than of extortion and misappropriation alone. Frequent references were made to private trading and moneylending.

The transport business seems to have been particularly profitable. Transporting 400,000 tons of cocoa from all over the forest zone to the ports was a major operation. Only part of the total milage involved was handled by the railways. Much cocoa in the southern regions was carted by road all the way to the ports; cocoa from the interior was often transported long distances by road before it reached the railway sidings. The distances from outlying villages and farms to the buying centres could also be considerable. Private transport business had long been closely associated with export-crop production and was one of the most dynamic sectors of private entrepreneurship outside agriculture itself. It was an important field of investment for big farmers and cocoa traders alike, providing a bridge between emerging capitalism in rural and urban areas.

The cocoa-purchasing monopoly placed the Farmers' Council in a strategic position to exercise control over this well-established and prospering business. No attempt was made to take over the cocoa haulage directly. The Council continued to rely on private transport-owners carting cocoa on its behalf. This may partly have reflected the independent strength of the private transport industry in the Ghanaian economy. On the other hand, the Council's position as the sole Licensed Buying Agent of the country's cocoa provided ample opportunities for sharing, directly as well as indirectly, in the profits of the transport industry without formal ownership. It seems as if this share was primarily collected not by the organisation as such, but by individual officials, who were owners of lorries as well as those allocating contracts. Headquarters did make one attempt to centralise control. The Council files contain a proposed contract of August 1961, making the Amoakohene Company the exclusive agent of the Council in carting cocoa within Ashanti and Brong-Ahafo.[169] The scheme seems not to have materialised. In 1962, an agreement was signed with the National Cooperative Transport Society Ltd., which was to become the sole agent. Instructions were issued that "on no account shall officers load or permit to be loaded . . . lorries which are not registered as forming part of the fleet of the N.C.T.S. Ltd".[170] Special deductions were to be collected on the payments made to the lorry-owners: one shilling per ton for distances of 11 to 50 miles and then at an increasing rate. Objections to these arrangements were later raised by the Ministry of Labour, responsible for the cooperative sector. The Council replied that the agreement had never been ratified or fully implemented.[171] The intention was only that the Transport Cooperatives would make vehicles available "as and when it was necessary". The Ministerial report maintained that the N.C.T.S. was "everything but a cooperative". The lorries were owned individually and the basis of association lay in the right of members to participate in contracts secured by the leadership. As far as the Farmers' Council was concerned, this particular scheme seems essentially a method to secure a "tax" on private lorry-owners in exchange for supporting the monopoly aspirations of the Transport Cooperatives.

The field staff of the Council imposed their own fees in addition to whatever may have been prescribed by the headquarters. A petition submitted in 1962 complained that transport-owners were forced to pay a commission of six pence on every bag of cocoa to the Secretary-Receivers before they were allowed to cart cocoa, unless they had special relations to the receiver or the District Officer. The petition also claimed that Council officers monopolised the business by using their own lorries. "District Officers own fleets of vehicles . . . depriving Transport Owners of their rights". Reference was made to Nkrumah's *Dawn Broadcast,* which was interpreted to mean that those in the employment of the CPP government should not take part in private business. "Are the members of your Department practising Socialism?", the petitioners asked. They doubted whether it was compatible with socialism "to monopolise the work" in this way.[172] Similar allegations were raised after the military coup. It was maintained that District Officers and Secretary-Receivers had used their own trucks—bought with Council funds—and monopolised the carting of cocoa.[173] Farmers also complained that they were heavily charged by Council officials using private vehicles to convey the cocoa to the buying stations.[174]

The cocoa-carting business was the major example of how Farmers' Council officials successfully combined control over contracts with private entrepreneurship. Construction was another field in which this was practiced, although there the control of contracts seems to have been more important. The Council appeared as a builder primarily in its programme for new cocoa sheds. The programme was intended to cater for a much expanded crop and for the new network of stationery buying centres. By 1965, the General Secretary reported that a total of 616 sheds had been built or were under construction as part of this programme. The amount of money involved was considerable. The 91 sheds under construction were estimated at £213,068 and the 734 proposed sheds at over £3 million. A continued massive expansion was envisaged.[175] The Council's storage programme provided economic opportunities in the rural construction industry, which, along with transport, was a major sector of expansion for Ghana's local capitalist class. It was closely linked to the trade in building materials, where cement and corrugated iron sheets were the major main items. The control exercised by Council officers over the allocation of cocoa-shed contracts was an important source of private profit. Farmers mentioned cases of "partnerships" between officials and contractors.[176]

The complex relation which on occasion existed between those awarding and those receiving contracts may be illustrated by a petition from Praaso Society in 1963. It was claimed that a contract for a new cocoa shed had been given to an Accra contractor at £5,000. Individual members of the Society Committee, however, had taken the same work on contract from their own contractor at half the original sum. As members of the Praaso Town Development Committee, they had at the same time succeeded in securing communal labour, that is free labour, for this particular job![177]

There were frequent rumours and allegations about commissions being charged by Council officers on the contracts awarded for cocoa sheds. The CPP Propaganda Unit in Offinso reported in 1963 to the Council headquarters about such practices in their district:

> It is rumoured in the district that the District Officers and District Chief Farmers at Offinso, Nkenkasu and Techiman met on one Saturday at Offinso ... At the meeting they emphasized that Mr. Martin Appiah Danquah had instructed throughout the whole regions that they the officers should collect some money from the Contractors who may submit their names to them to be resubmitted to Accra for the award of the pending contracts. The District Officers, have therefore collected some amount ... with the view of maintaining the Party.[178]

The regional party authorities, in Kumasi in this case, denied knowledge of any authorised collections of this type. The report therefore concluded that they were probably undertaken "for the selfish ends" of the Council officers themselves. The directives from the headquarters may have been invented in order to dupe local contractors. But instructions of this type may well have existed.

Large lorries and trucks were the most conspicuous element in the picture of the private investments of the Farmers' Council officials. They demonstrated most visibly the strong entrepreneurial side of the new trader-clerks. The most important field of investment in Ghana's forest economy, however, remained cocoa itself. In the farmer-traders, cocoa trading had had a direct link to farming. Profits from cocoa farming were invested in trade and transport and the surplus of this wider enterprise was again partly reinvested in cocoa production. The Farmers' Council itself evolved from the tradition of farmer-trader organisations. This continued to be primarily reflected in the representative structure, with its farmers' committees and Chief Farmers, but the new trader-clerks were also largely recruited from the cocoa communities. Many officials of the Council invested in cocoa production, either under their own or family auspices. The impression given during my contacts with former employees of the Council was that this was a common element in their wider economic activities. They were often in a good position to obtain information about suitable land and rental conditions. Underemployed relatives could be engaged as caretakers or supervisors.

Such investments may be illustrated by the case of the General Secretary himself. The scale of operations in his case was probably exceptional but the pattern had wider relevance. His economic activities are documented by one of the post-coup commissions appointed to investigate the assets of leading CPP politicians.[179] After leaving school in 1945, Appiah Danquah was appointed part-time catechist of the Anglican Church at Nkawie. He simultaneously managed two cocoa farms given to him by his father. The gross farm income at the time was about £500 to £600 per annum, corresponding to a yield of five to six tons of cocoa. The net proceeds of these farms were said to have been largely reinvested in cocoa farming. From 1954 to 1966, the Commission estimated, partly with the help of the Council's own records, that Appiah Danquah's gross receipts from his farms were in the order of £20,000. Holdings in the Nyinahin Society were particularly large, yielding almost 20 tons in the 1965–66 season. The way output expanded indicates continued investments during the 1950's. For the 1957–62 period, the Commission also recorded investments of £4,000 in farms owned by Appiah Danquah's wife and in 1964–65, a further £2,000 in four farms owned jointly with a cousin. The picture is of a large and expanding enterprise, belonging perhaps to the top one or two per cent of the country's cocoa holdings, if judged by the 1963 purchasing records.[180] Investments on this scale were of course out of the reach of most senior (as well as junior) Council staff.

Large houses were singled out as a typical manifestation of the wealth of the Farmers' Council officials. The line between productive investment and conspicuous consumption was diffuse. The rise and decline of early cocoa wealth in Ghana is readily illustrated by the many magnificent, now greatly dilapidated buildings in the small towns of the Akwapim Hills, the home-area of the first major wave of migrant cocoa farmers.[181] These were "family houses" built in the home towns to which the migrants returned during off-season, for festivals and ultimately, for retirement. Although often highly elaborate, their commercial value was low because of the limited market for rental housing in these places and their character of personal manifestations associated with a particular family. But investment in housing could also be highly business-like, as the large properties acquired by Kwahu traders and farm-owners in the commercial quarters of Accra demonstrate. Many of the large houses of Kumasi, a commercial town more fully integrated in the cocoa economy, were built by cocoa farmers and cocoa traders. Even if not used for commercial purposes, the rental market in the cities gave the buildings an effective value which increased greatly as a result of rapid urbanisation. It was usually a safe way of accumulating wealth.

It is natural that the farmers paid particular attention to the activities of the Council officials in this field. We may again turn to the record of the General Secretary, as reported by the Jiagge Commission.[182] Between 1959 and 1965, Appiah Danquah spent a total of £36,000 on six different buildings, according to the Commission's estimates. Most expensive was a three-story house in Nkawie, his home village. Two houses were in Kumasi and three in Accra. Two of the six belonged to his wife. Two were under construction. The picture is one of buildings primarily acquired for private and family use. But they represented at the same time, with the possible exception of the Nkawie house, solid investment from which a high rate of return could be expected in a booming real-estate market.

It may be added that the Commission investigating Appiah Danquah's affairs estimated that his total expenditure during his period with the Council exeeded his "lawful" income, as defined by them, by 80 per cent.[183] The distinction between lawful and unlawful income was by no means clear. The close connection between the corrupt administration of official duties and private business activities makes such distinction particularly difficult. But the strong element of private entrepreneurship and unofficial profiteering should not be allowed to obscure the dominant, official way in which surplus was appropriated under the monopoly. The stake of the new middlemen was primarily in the wages and salaries which they were able to obtain for themselves. The monopoly provided a large number of people with an avenue for employment and careers. Let us examine the new administrative structure in terms of the opportunities created in this respect and the cost which it imposed on the cocoa economy.

Trade by bureaucracy: opportunities and costs

The Farmers' Council was by far the largest commercial establishment in the country and one of the single largest employers outside the central government itself. The recorded number of employees was at the time of the dissolution in 1966 over 13,000, excluding seasonal labour engaged during the peak months of the operations.[184] This figure may be compared with a total recorded wage employment in the mid-60's of about 400,000.[185] Some 95 per cent of the Council's employees were primarily engaged in the cocoa sector.[186] Labourers and watchmen comprised about half of the total. The Secretary-Receivers were the second largest group of employees: 2,268 according to the 1966 figures. Omitting the non-cocoa regions and other units not concerned with cocoa, the establishment list included some 230 Accountants and Book Keepers, 380 Auditors, 110 Clerical Assistants and Clerical Officers, 370 Depot Keepers and Store Keepers, 140 District Officers, 100 Messengers, 130 Tally Clerks, and 130 Typists, only to mention the largest groups of employees. The list contained a wide variety of functionaries, including 31 musicians and 21 actors employed by the Publicity Department.

In addition to the regular employees, there were those with fixed allowances or commissions on purchases. They primarily included Regional and District Chief Farmers, Vice Chief Farmers and the Treasurers at the district and society levels. These categories totaled about 2,000. Other committee members shared in the official or unofficial surplus. If this larger group is included, the total Council establishment in the cocoa sector involved about 30,000 people. It may be recalled that the total number engaged in cocoa production according to the 1960 census was about 520,000.[187]

The salary structure indicates the hierarchical pattern and the avenues of advance offered by the monopoly. The Eastern Region establishment at the time of the 1966 coup may illustrate the pattern at the regional level.[188] At the head was a Regional Officer at £1,850 per annum, followed by an Assistant Regional Officer and a Regional Internal Auditor at £1,255 and £1,160 respectively. District Officers ranged from £680 to £1,040, the typical pay being between £800 and £900. Next were the District Internal Auditors, most of whom were paid £680 per annum, followed by the Depot Keepers—also at district level—whose standard rate of remuneration was £530. Assistant District Internal Auditors and Assistant Depot Keepers received salaries ranging from £300 to £360 and from £240 to £270, respectively. Descending to the society level we find that the most common pay of the Secretary-Receivers was £200. Typists, drivers, and messengers at regional and district levels received amounts varying from £108 to £215. At the bottom of the establishment were the daily rated workers, the watchmen and the labourers, with wages corresponding to annual earnings of roughly £120 and £95, calculated on 300 work days.

The gap between the lowest—and the highest—paid in the region was considerable. The Regional Officer received an income twenty times that of the daily rated worker. A middle-school leaver, employed as a Messenger or an Assistant Secretary-Receiver at £108, could aspire to a career where, as a District Officer, he would have increased his initial earnings ten times. Such a career was by no means exceptional. At headquarters the distance between the top and the bottom of the hierarchy was of course even greater. The General-Secretary of the Council received a gross official income over forty times higher than that of the daily rated labourer. The gap remained almost as wide when net earnings are considered. Rates of taxation were very modest and large tax-free allowances were an important part of the earnings at the top. In 1956, the General-Secretary received a salary of £840 per annum plus an annual £120 housing allowance.[189] The salary was doubled at the time when the Farmers' Council was officially recognised as the only national farmers' organisation after independence. In 1960, when the government first announced the monopoly plans and the General-Secretary of the Council was awarded ministerial status, the salary was further increased to £2,500 plus a general allowance of £700 and a £600 housing allowance.[190] An additional allowance brought his total official income to £4,100 per annum after the monopoly had been established in 1961. Assistant and Deputy General-Secretaries, Financial and Legal Advisers and a small group of other top headquarters officers had similarly spectacular careers, although at a lower level.

Even quite ordinary employees experienced very substantial increases in their earnings. In 1963, for example, the staff of the Publicity Department had increases averaging about 50 per cent.[191] In 1964, the General-Secretary's Department offered increases varying between 15 and 55 per cent, or an average of about 30 per cent.[192] The new salaries can be compared with those received when these staff members were first employed in their respective capacities. In this case the average increase was about 75 per cent. Records are also available from the Shipping Department showing equally strong increases.[193] Comparing the proposed 1964 salaries with those received two years earlier we find that only one out of 34 employees received an increase of less than 20 per cent. More than half had increases of over 40 per cent and one-quarter had increases of 60 per cent and over. The 1962-63 increments are available for all headquarters departments.[194] On that occasion the average increase was a modest 8 per cent. Only a dozen people had increases over 20 per cent. It seems as if very large increases of the size recorded for certain headquarters departments were less common in the regions. I have examined the 1962-63 increments proposed for the Ashanti establishment.[195] The average increase was 11 per cent, but the bulk of the increase went to elected officers, the Chief Farmers and Treasurers. Few ordinary employees had increases of over 10 per cent, although the Regional Officer himself (the administrative head) had 38 per cent. The allowances of most of the Chief Farmers and Treasurers, on the other hand, were raised by at least 20 per cent. In many cases they were doubled and eight District Chief Farmers got a 150 per cent raise in pay.

The real value of these increases was of course reduced by inflation. The Monthly Index of Retail Prices for Accra, with June 1954 as 100, reached an annual average of 112 for 1960 and 136 for 1963. Towards the end of 1964, consumer prices were increasing more rapidly.[196] The general impression obtained from these scattered records is that at least up to this time, the salary and wage increases received by the Council's monthly paid staff were well ahead of price developments. The large group of daily-rated workers, on the other hand, had their real wages reduced already before the 1964-65 price squeeze.

The methods by which the monthly paid staff advanced were many. First of all, there were annual increments to which all such employees were entitled. According to a schedule from 1964 these ranged between four and six per cent per annum.[197] In addition, there were changes in the salary-scale attached to a particular post, up-gradings of posts from one scale to another, and of course promotions. Together, they offered a wide scope for individual advancement. Looking at the adjustment of salary scales alone, we find that District Officers, for example, were placed on three different scales ranging from £300 to £755 per annum in 1960, while in 1965 the same range was £480 to £1,020.[198] Many public corporations paid higher salaries than the central government. This was a major complaint within the civil service. The general salary structure of the Farmers' Council seems not to have deviated much from that of the Cocoa Marketing Board. The Board, on the other hand, was said to have among the most generous scales within the public sector.[199]

In a category of their own were the elected officers of the Council, paid by way of allowances or by shares in the proceeds of the trade. At the society level, the size of the official share in the surplus collected by local office-holders varied with the purchases of the societies. It may be noted that after the coup, society Treasurers were given a fixed allowance of £144 per annum, similar to the salary of an Assistant Secretary-Receiver.[200] This amount may have been considered roughly equivalent to their pre-coup receipts from the local surplus in typical societies. Society Chief Farmers were entitled to a share equal to half the Treasurers' part. The remaining committee members shared the same amount.[201]

District Chief Farmers and Treasurers were paid direct allowances by the Council. At one point they were graded according to the importance of their districts. In 1958, the official allowance of a District Chief Farmer varied from £36 to £120 per annum.[202] A year later the scale was raised to £180.[203] With the advent of the monopoly, it was increased by another £100.[204] It was now roughly equal to the salary of an Assistant Depot Keeper or Internal Auditor. Looking at the 1962–63 increments for Ashanti, we find that the allowances of some District Chief Farmers were increased by 150 per cent to £360, while those at the top of the existing scale had to be content with a 25 per cent increment.[205] On the same occasion, some District Treasurers had their allowances doubled to £348. The Regional Chief Farmer in Ashanti was paid £480 in 1959, £840 in 1962 and £1,200 in 1963. The National Chief Farmer experienced a similar rate of advance, receiving £2,500 in 1963.[206] I have no information on the increments received by the elected officers during the later years.

A number of "fringe benefits" were an important part of the official earnings of the senior employees of the Council. They included allowances for housing and transport, as well as the use of official houses and official cars. As we have seen in the case of the General Secretary, such allowances could be very substantial. The Auditor General complained in 1964 about the great number of official cars allocated to senior officers. He also noted that private cars owned by individual officers were bought by the Council and transformed into official ones. Both the General Secretary and his deputy had two official cars each (Mercedes 190 D and 220 S).[207] All officers above a certain income received interest-free long-term loans from the Council to buy their own cars, in conformity with the general practice in the public sector. Car maintenance allowances were paid to all those entitled to have a car. These ranged from £120 to £180 per annum in 1961, or roughly equal to the salary of a Secretary-Receiver at this point (up to twice the annual wage of a day labourer).[208]

Some of the "fringe benefits" for the senior officials of the Council, employees as well as elected ones, were less formalised. They included fees or gifts received from farmers and junior officers in connection with visits, inspections, and inaugurations.[209] When, for example, the General Secretary and the National Chief Farmer visited the Volta Region in 1964, the regional executive of the Council authorised contributions of £5 per society and £10 per district to the two dignitaries.[210] Special fees were charged when a new society or a new shed was opened. The 1966 Report mentioned levies of £40 to £100 on such occasions.[211] The practice of collecting "funeral donations" when senior officers or their close relatives died was another fringe benefit, which in the case of the most senior ones, could be quite substantial. It was rumoured after the coup that very large amounts had been collected for the funeral of the General Secretary's father.[212] The Council's own files record large donations by staff on the occasion of the death of the General Secretary's uncle.[213]

As a contrast to the opportunities created for the new middlemen it may be illuminating to look at the fate of the "profits" due to the farmers under the new marketing system. A major argument used by the Council when campaigning for the elimination of foreign firms and local middlemen was that all profits from the cocoa trade should go to the producers themselves. In the past, the question of farmers' share in profits had been associated with the payment of commissions and bonuses. The practice of paying commissions at the end of the season had spread, under the impact of the cooperative movement. The amount involved could be considerable. The Kumasi Cooperative Union claimed in 1966 that members of good cooperative societies used to get a commission of over £4 per ton.[214] Another group of Ashanti farmers maintained that commissions ranged from £3 and upwards.[215] Records show that in 1956 the Purchasing Company accepted a Farmers' Council request to raise its commission to £4.[216] The figures can be compared with the total Buying Allowance payed by the Marketing Board to cover costs as well as profits. In 1961/62 this was roughly £11 per ton. But most cooperatives and brokers were not able to pay commissions of this order. The portion of the commission which actually reached individual farmers varied. In the case of the Purchasing Company, for example, deductions were made for receivers, committee members and for the Farmers' Council. One witness in 1966 claimed that the committee members of the old cooperatives shared the commission between themselves and their Secretary-Receivers.[217] Others said that the cooperatives divided the commission in three parts, two-thirds being retained for "expenses" and only one-third paid to the farmers.[218] It is probable that practice varied greatly from place to place and from year to year.

The Farmers' Council promised that all farmers were going to receive the same commission irrespective of the profitability of a particular society. For the 1961-62 season, an amount of £3 per ton was set aside for this purpose. The amount was probably well above the average received in the past. The money, however, was not to be paid to individual farmers but to the societies, which were to use it for social amenties in their respective areas.

The idea that the profits of the monopoly were to be used collectively had a central place in the ideology of the Council. At the Bibiani Conference in 1960, the General Secretary spoke of "Social Development Committees", which would be formed in each society with the purpose of administering the common profits.[219] It was emphasised on other occasions that this was the essence of the socialist nature of the Farmers' Council cooperatives, as opposed to "Capitalist cooperatives in which dividends go only to the share-holders".[220] No details of how these funds were to be controlled and utilised were given when the monopoly scheme was presented in May 1961, but it was later announced that the money should be spent according to the recommendations of all the farmers in a society.[221] The instructions issued for the first monopoly season emphasised the role of the local executive committees in this respect.[222] It was envisaged that "by five years from now, the face of your villages might have changed completely", as a result of the collective use made of local profits.[223] The importance of these collective profits for rural development was also emphasised in the Council's petition against the State Agency scheme; If the monopoly plans were not implemented, "the development of the rural areas which is still at a snail's pace will remain unchanged".[224] The overall impression created by the various statements was that trading profits would be under local (society) control and be used for purposes normally identified as social amenities in rural Ghana, that is, for water, sanitation, roads, community centres and perhaps, electricity. Actual control and use of the funds turned out to be rather different. The Supreme Council decided that the farmers' commissions should be used exclusively for the construction of cocoa sheds. A circular announced that no other projects would be approved.[225] The need for investment in new storage capacity was no doubt great. The cocoa crop had increased dramatically and the new permanent buying centres required sheds. Much of the existing capacity was scattered throughout a number of small units. Many sheds were merely rooms in the houses of local traders. Large ones existed primarily in the principal centres. Although useful in itself, the shed-building programme of the Council was hardly a way of *distributing* profits, either for individual or collective purposes. This was rather a question of reinvesting part of the surplus in the trade itself.

Local control over the commission funds never materialised and no other local projects than sheds were authorised. Yet the amount allocated for that purpose was only a minor portion of what was officially reserved for farmers' commissions. Up to 1965, allocations for sheds were less than a quarter million pounds, while farmers' entitlements on the commission account reached one and a quarter million during the first monopoly season alone.[226] Had the rate of £3 per ton been maintained throughout the five years of the monopoly, a total of £6.7 million would have been available.[227] The Council, however, reduced the commission to £2 in 1962 and to £1.10 the following year.[228] Even so, a total of over £4 million could have been distributed. This was not done. Instead, the commissions, although specified as the "Farmers' Development Fund" for accounting purposes, were absorbed in financing the overall expenditure of the Council. The cut in the earnings of the farmers caused by the discontinuation of individual commissions was of course minor in comparison with the reductions in producer prices experienced during the first half of the 1960's. Nevertheless, it highlights the class character of the monopoly and the continued contradiction between producers and middlemen.

The Farmers' Council had maintained that the monopoly would not only place the profits of the cocoa trade in the hands of the farmers, but would also reduce the overall costs of marketing by pooling the resources of the trade.[229] There was much to say in favour of monopoly as an instrument to cut costs in the particular situation of the cocoa trade, as also noted by the cocoa manufacturers.[230] Competition between buyers was inefficient in keeping costs down, when prices and buying allowances were fixed by the government and the Marketing Board. The costs of overlapping and duplicating buying organisations were easy to see. Staff, equipment, sheds, transport and other resources were often employed ineffectively.

Did the Farmers' Council realise any monopoly savings? What costs, peculiar to the monopoly, were added? The financial records at my disposal do not permit firm conclusions about the way in which the total cost of cocoa marketing was affected by the monopoly.[231] It is possible, however, to identify certain general tendencies, related to the changes in the organisation of the trade.

Shortly before the 1966 coup, the Minister of Finance, Kwesi Amoako-Atta, sharply criticised the financial administration of the Farmers' Council in connection with a review of its 1964 accounts.[232] The organisation, he said, appeared to be top-heavy and saddled with heavy wage-bills and administrative expenses. The point of heavy over-staffing was also raised after the coup by the Managing Director of the Cocoa Marketing Board, who was otherwise highly appreciative of the Council's performance. He even suggested, according to the deGraft-Johnson Report, that it would be possible to reduce the staff by half.[233] The Council's Cadbury adviser reported as early as 1962 that the district staff was "in all cases excessive". He noted that during the four busy months of the cocoa season the staff did not appear hard worked and they could only be a financial liability during the remaining eight months.[234] Over-staffing may partly have been a result of weak administrative control. It was largely institutionalised, however, as a consequence of the introduction of a "bureaucratic"-type organisation into a seasonal trading economy. In the past, cocoa trading engaged for a few months a multitude of small traders who did much of the grass-roots buying on behalf of the major brokers and the firms. Cocoa trading was for most of them only complementary to other farming or non-farming work. The new buyers, although fewer in overall numbers, were mostly employed on an all-year basis. The waste of overlapping trading organisations was thus replaced by the waste of the seasonal underutilisation of the staff.

The formalisation of the trade introduced by the monopoly also brought other new costs. Much of the cocoa had been handled by traders with little expenditure on such things as supervision, accounting and auditing. Many were non-literate and expected only modest returns for their labour. The new organisation required complex administrative procedures and a variety of educated staff. It established a network of permanently staffed buying centres which were to be fully equipped with proper scales and storage facilities. Grass-roots buying in the past had been more mobile and involved less fixed capital. A spare room in an ordinary house could serve as storage. Only at a higher level had more advanced equipment been required.

The high costs of monopoly buying were largely caused by the proliferation of these permanently-staffed, fully-equipped buying centres. In 1963, a total of 1,457 centres (societies) were recorded by the Council.[235] Their size varied greatly in terms of the number of farmers enrolled, the area covered and the amount of cocoa handled. In dense cocoa areas, one village or town could have several centres while in other areas a single society might cater to a number of widely dispersed villages. The purchasing records indicate that the number of farmers in each society varied from as few as 30 to over 600. The typical number was in the 100 to 200 range.[236] In theory a society was expected to market a minimum of 200 tons of cocoa, but in practice many societies were below this mark.[237] The General-Secretary suggested in his annual report for 1964–65 that about two-thirds of all societies would have to be closed down if only those which were viable as separate units were to be retained. He said that he did not consider this to be in the interest of the farmers. He maintained that access to the market was a service which the Council was obliged to provide to all farmers "when and where it is needed".[238] Profits from areas of dense and high production were used to subsidise buying centres in less remunerative areas. The spread of "uneconomic" societies was criticised both within and outside the Council. The Chief Internal Auditor claimed in 1965 that "records at our disposal are unanimous that most of the societies have been showing big deficits in their operations year in and year out".[239]

But deficits at the local level did not exclude profits at higher levels. The overall financial records of the Council are not easy to interpret. The full accounts have been audited by the Auditor-General but not released. The deGraft-Johnson Committee maintained that the Council made large profits, or ought at least to have done so, considering the size of the Buying Allowance paid by the Cocoa Marketing Board, an annual subsidy from the Board, and other direct and indirect forms of government financial support.[240] The profit element in the official allowance was originally designed to meet the expectations of the foreign firms for a "normal" return on their capital. The very profitable business of Cadbury & Fry during that company's last years as Licensed Buying Agent, suggests that the allowance was not overly restrictive, especially not in a less competitive market.[241] In addition, there was the "commission" element which, as we have seen, was not passed on to the farmers by the Council. It also seems if the Council profited from a new transport item added to the allowance.[242] Moreover by lending money to the farmers it could add to its overall potential surplus.[243] That substantial profits were being made was clearly the view of the Minister of Finance who pointed in his 1965 *Budget Statement* at the "anachronism" that the Farmers' Council was not paying income tax.[244] He proposed a 30 per cent tax on profits and the Council was assessed to pay £300,000, corresponding to a £1 million profit. Judging from the unaudited accounts available to me, the Council may have made profits of this size in 1961/62 as well as in 1964/65 season.[245] In the latter season, however, these profits were clearly due to the "windfall" brought by the exceptionally large crop. On the whole we get a picture of rapidly growing overhead costs eating into generous official allowances.

The budgets of the Council allow us to compare the cost structure of the monopoly with the items of the buying allowance in a way which the accounts do not. I have looked at the 1964/65 budget, where total expenditure was estimated at £6.3 million with a net deficit of £400,000.[246] The budget was calculated on a large crop (490,000 tons). The deficit should be compared with the £1.1 million which was the official profit element included in the allowance. The potential profit, however, was consumed and turned into a planned deficit as a result of an expected £400,000 deficit on the regional account (including the society deficits) and another 400,000 allocated for head office salaries. The fact that the last item was charged against the "profit" element of the allowance is particularly significant. A smaller amount was the planned deficit of the Council's activities in the non-cocoa regions.

The deGraft-Johnson Report suggests that much of the cocoa profits were channelled into the non-cocoa activities.[247] The accounts do not give a clear picture of these transfers. The other operations seem to have been largely financed by bank loans. It is likely that the cocoa account was not only made to carry the current financial costs (bank charges) but also the growing deficits of this ambitious but unprofitable programme.

It was suggested by a CPP minister at one of the public enquiries held after the 1966 coup that the Farmers' Council made very large (£1 million) annual payments to the CPP. This report was not based on accurate information, according to the commissioners.[248] Payments to the party of a lesser order should not be excluded. The overall picture, however, is of an organisation where official profit margins as well as certain "monopoly savings" were increasingly absorbed by growing costs both at grass roots and headquarters level. The savings seem largely to have been a result of generous allowances and direct and indirect subsidies from the Cocoa Marketing Board rather than of a more rational organisation of the trade. The surplus available for transfer to the government or the party, into non-cocoa operations, or for that matter, back to the cocoa producers themselves, was rapidly diminishing.

The new middlemen: summary

The Farmers' Council monopoly meant that the profits of the cocoa trade were redistributed. The part which in the past had gone to foreign firms was now reserved for Ghanaian national enterprise. But only a small part of the earnings of the pre-monopoly trade was collected by the firms. The bulk was absorbed by the tens of thousands of intermediaries who bought the crop on behalf of the firms, including a majority of farmer-traders. The integration of farming and trading and the growing strength of the cooperatives meant that much trade income was in fact retained within the cocoa farming community. Such income had contributed importantly to social differentiation and the rise of an upper stratum of big brokers. The monopoly of the Farmers' Council was a blow to these interests, although a small section of the farmer-trading community had allied itself with the Council and was able to claim a share of the monopoly profits in their capacity as Chief Farmers and committee members. But more important was the massive shift of trade income away from the farming community into the hands of the new bureaucratic trading class. By expanding its own numbers and deciding its own remuneration, the latter left little in the way of monopoly savings or monopoly profits to be returned to the farmers as commissions, raised producer prices or otherwise. Farmers had been promised that once the firms and the old middlemen were gone, the profits of the trade would be used for the development of social amenities in their home villages. The little that was actually spent in the villages, however, was reserved for the construction of cocoa sheds, a useful commercial investment for the Council but not an amenity which could compensate farmers for discontinued commissions and trading profits.

The rise of the new trading class had important consequences for the social organisation of the cocoa economy. It blocked the road for private traders and cooperatives. Growing claims on the surplus were made by the new functionaries, including Secretary-Receivers, Accountants, Storekeepers, Auditors, District Officers and others. For most farmers trading profits and commissions had added little to their farm income in the pre-monopoly days. They were therefore affected more by the rising overall cost of marketing and its consequences for the level of producer prices than by discontinued trade profits and commissions. The new costs included the proliferation of uneconomic buying centres and a growing superstructure of managerial, supervisory and administrative staff. Yearly increments, promotions, upgradings and revisions of salary scales ensured that the costs of this establishment absorbed a growing share of a stagnating or declining national cocoa income. The protection of the trader-bureaucrat earnings contrasted with the serious fall in real producer prices experienced by the farmers in the same period.

Farmers resented the apparent prosperity of the Council officers the more as they suspected, and sometimes knew, that it was not only obtained from official earnings but also from private profiteering. The massive evidence drawn upon in this study, both from the Council's own files and from those of the deGraft-Johnson Committee does not lend itself to quantitative estimates of the overall importance of such unofficial appropriations. Nor is it possible to tell how many farmers were directly affected. Many allegations were no doubt baseless, others exaggerated or distorted for partisan purposes, as in the case of the serialised petitions of the post-coup days. But the Council had a poor reputation in this respect even before the coup. Within a context of strong conflicts over the control of the trade and of cocoa income in general, it is not surprising that rumours and allegations about malpractices were widely circulated and believed. The number of specific reports in the Council's files may by themselves explain the exploitative public image acquired. I have quoted extensivly from such detailed reports of corruption, extortion and other malpractices, not because I consider the evidence to be conclusive but because it offers a concrete picture of the opportunities for private profiteering. Many reports may look petty and trivial, especially in the light of the wider national issues of public accumulation. What did it matter if farmers were cheated of a few pounds now and then, if the basic function of the monopoly was to facilitate the massive transfer of resources from the cocoa economy into public development. It may also be remembered that during the same period producers and government alike were robbed of tens of millions as a result of falling market prices.In the light of these major outflows of value the unofficial profiteering by middlemen or, for that matter, the high official costs of monopoly might seem to be of only marginal importance. But just because the redistribution of cocoa income was so vital for the strategy of national development it was the more important that the political means by which the surplus was appropriated were conducive to the purpose and not self-defeating. It may also be recalled that a main purpose of the strategy was to reduce the dependence on fluctuating world market prices with its debilitating consequences for the country as a whole. But the strategy itself was seriously undermined by the way in which the cocoa trading monopoly was administered. Farmers' opposition to state appropriations was strengthened. There was a build-up of resentment, not only against the petty thieves who occasionally confronted them over the weighing scales and cash counters at the village buying centres, but against a political system which seemed to tolerate, protect and even encourage such a racket. These village centres were the points of direct confrontation between the farmers and the new political and economic order.

Why did the system allow such wide margins for private profiteering? It is important to look beyond individual dishonesty to the sources of power and control within the organisation. These are questions which are discussed in the next chapter. But the examination of the commercial performance of the Farmers' Council permits certain preliminary conclusions.. The official task of the Council was to buy cocoa on behalf of the farmers in order to safeguard their interests and to protect them from exploitation by foreign capital and local middlemen. In the official rhetoric, the Council was the farmers' own organisation and exploitation was therefore eliminated by definition. But the reality was different: a class of bureaucratic middlemen established themselves, exploiting the producers both by claiming an increasing share of the surplus as its own official entitlements and by tolerating widespread private profiteering among its members at the expense of the producers. The performance of the Council stood in sharp contrast to its claim to administer the trade to the benefit of the farmers. But it contrasted also with the claims of the organisation to mobilise the farmers in support of the CPP government, and more specifically, in support of the accumulation of cocoa income for national development.

The character of the Farmers' Council as a farmers' organisation and as an organisation for national development was fundamentally affected by its character as an instrument for the social advancement of a new class of trader-bureaucrats, concerned with monopolising commercial opportunities and using public office as a platform for private entrepreneurship. The class character of the Farmers' Council was determined by the interaction of these factors. Let us examine how the balance was reflected in the political organisation of the Council.

VI. Participation and Political Control

An outspoken attack on the Farmers' Council was made two days before the 1966 military coup by the CPP government itself. It was contained in the *Budget Statement,* read to the National Assembly by the Minister of Finance, Kwasi Amoako-Atta. Part of the criticism referred to the economic administration of the monopoly.[1] The brunt of the charges, however, concerned its political structure and most specifically, its lack of internal democracy:

> One of the main, indeed the basic feature, of a socialist co-operative should be the democratic nature of its organisation. But this is the very feature which is lacking in the UGFCC. The principle of democratic centralism and the democratic election of district farmers and chief farmers by annual conventions of the co-operatives is practically non-existent.[2]

It was maintained that as a result, "power tends to flow from Accra rather than from the individual co-operator or Co-operatives in the villages". The statement recalled that the Council had been entrusted with the leadership of the cooperative movement. It had therefore to ensure that it was run on genuine cooperative lines with a popular democratic base. It was conceded that strong centralisation may have served a useful purpose in the early years of the struggle, "to break the hold of the imperialist buying agents on the farmers" and in the fight against "the former capitalist-oriented marketing co-operatives under reactionary leadership". But the Council had failed "to move gradually towards democratisation of its mass base". This alone could provide "the basis of lasting power in a true co-operative organisation".[3] It was not the first time that the political organisation of the Council was questioned by the government. But it was the first major public exposure: "It seems the time has now come for the Council to put its house in order and save itself from the embarrassment of constant criticism".[4]

How was the Council organised? How were the farmers represented? What were the means of internal and external control? In what way did it reflect the interaction of the commercial and political functions of the Council? How did it reflect the contradictory relations between the three parties most immediately involved: the farmers, the monopoly traders and the national government? The material presented here seeks to answer these questions. This chapter begins with a presentation of the formal structure of the Council: a hierarchy of farmers' representatives alongside a highly centralised bureaucracy under the leadership of the General Secretary. I discuss the way the representatives were elected and subjected to control from below and from above. I document the repeated suppression of elections and the way in which elected officers were effectively incorporated by the bureaucracy, subjected to a centralised system of command on more or less the same line as ordinary employees.

A second section discusses how the system of representation was affected by a variety of cleavages inside the cocoa communities, some related primarily to party politics, others to communal allegiances and socio-economic divisions. Different types of divisions interacted, sometimes reinforcing, sometimes neutralising each other. In the historical chapters above, we saw how the Council developed first in response to, and later, increasingly in opposition to the commercial interests of the big farmers. I discuss what this meant for local representation. The basic contradiction continued to concern the distribution of cocoa income. It influenced the way in which other types of cleavages affected the Council, reinforcing the obstacles to effective farmer participation in the administration of the monopoly.

The centralised structure of the organisation was particularly manifested in the personal powers of the General Secretary. I discuss his position and his role as the vital link between the Council and the party and the government. Nevertheless, there were practical limits to centralisation; regional, district and local officers were also able to exercise a great deal of personal power at their respective levels. The system of bureaucratic self-control—with its accounts, audits and inspections—was elaborate but hardly efficient in preventing the private pursuits of these various officers. I have sought to document the weakness of disciplinary control and the widespread suppression of reports of such malpractices. Fears of reprisals against critics grew with the intensification of political violence and repressive legislation in the country in general. The authoritarian and high-handed style of many Council officers demonstrated how little accountable they were to the farmers. The means of seeking redress for their grievances, however, were not restricted to the official channels. Apart from the ultimate weapon, open resistance, it was possible to make use of the divisions both within the Council itself and in the local political establishment in general. The latter was by no means monolithic and the authority of party leaders, Members of Parliament, government officials and chiefs could sometimes be enlisted in support of grievances.

Finally, this chapter discusses the attempts by the central government to reform the political organisation of the Council. These attempts reflected a growing concern over the weaknesses of the system of centralised command. The spread of reports about the dubious methods and private exploits of Council officers spurred the reformers, who aimed at decentralisation and the introduction of farmer control over local operation on cooperative lines. But the proposals failed to take into account the basic political functions performed by the Council and could therefore successfully be resisted by its leaders. It was only at the very end of the CPP period that the problem of reforming the political structure of the Council was approached as one of combining central political control with effective participation by the farmers at the local level. This was the formula of "democratic centralism" propagated by the Minister of Finance in the 1966 budget speech. Not only farmer participation but also control by the central government had been notably weak. Wide margins were left for the third interested party, the monopoly traders, to pursue their own sectional interests at the expense of the other two.

Electing the farmer leaders

The claim to represent the farmers had been at the heart of the Council's campaign against foreign and local traders. The trade was to be taken over by the farmers through their own organisations, thereby abolishing exploitation and ensuring that profits were reaped by the producers and not by foreigners and middlemen. This was in line with the long-established aspirations of farmer-trader groups and the more recent cooperative ideas of colonial reformers. The ideas were absorbed and adapted by the anti-colonial movement, which added certain elements of anti-capitalism. The commercial involvement of the Farmers' Council was strong from the very start, but its main task when first established was to organise and represent the farmers for general political purposes, as part of the broader nationalist movement. By the time the Council was given direct trading responsibilities a network of representatives, Chief Farmers and committees, had already been developed. The official recognition of the Council as the only organisation entitled to represent the farmers prepared the way for its final claim to market the entire cocoa crop.

How was this representation organised? Let us first look at the formal structure with the help of the 1959 and 1965 constitutions of the Council.[5] The "highest authority" of the Council was the National Delegates' Conference, meeting annually. It consisted of representatives from the districts plus the regional and national executives. The function of the Conference, according to the Constitution, was to elect a National President (National Chief Farmer) a National Treasurer, and a General Secretary, and to "receive and discuss" the General Secretary's Annual Report. It was also to "pass such resolutions as may be put before it".[6] The "highest executive body" was the Supreme Council. Its membership largely overlapped that of the Conference, as all districts were represented by their Chief Farmers. The Supreme Council was entrusted with a number of tasks, the primary one being the control over the commercial activities of the Council. It had wide-ranging disciplinary powers. It could dissolve executive bodies at all levels and remove or expel individual members. The actual administration of the Council was directed by the General Secretary in conjunction with committees appointed by the Supreme Council. Regional executive bodies were headed by Regional Chief Farmers and included representatives from each district. Regional Delegates' Conferences could be called "as and when necessary". Their primary function was to elect regional officers. They met rarely. The controlling and administrative functions of the elected bodies at the regional as well as lower levels were limited, as all field operations were directed from the headquarters.

The District Chief Farmers were the key officers of the representative system. In addition to heading district executive committees and conferences they were members of the corresponding bodies at the regional and national level. The district conferences, by which they were supposed to be elected, consisted of representatives of farmers' committees at the society or branch level. These were the lowest elected bodies of the Council. According to the constitution, they were to be chosen by the farmer members via direct elections in open general meetings.

The combination of highly centralised management and a multi-layer system of indirect elections did not particularly encourage effective farmer control over the cocoa trade. Nevertheless, the constitution provided a formal framework for popular participation. In practice, its formal democratic elements, as well, were largely eliminated.

Tentative observations were made in Chapter 3 about the way in which farmers and farmer leaders were first enrolled by the Council.[7] Formal representation tends to be weak in most organisations during a formative period. Individuals and small groups come together and take on representative roles; outside organisers play an important part in selecting local leaders. Only at a later point, when a local membership is consolidated, can these self-proclaimed or hand-picked leaders be confirmed or challenged in genuine elections.

The Farmers' Council leaders, once selected, were unwilling to submit to such a repeated process of confirmation. The 1953 constitution of the Council contained provisions for elections of all officers every three years. But as early as at the National Delegates' Conference in Tamale in 1955, it was decided that this should be amended, making the officers permanent, elected for life.[8] The amendment was claimed to be in line with traditional constitutional practices governing chieftaincy. Chiefs were "enstooled" for life and could normally only be removed—destooled—if they had seriously neglected their duties over a long period. The traditional destoolment process was complex but essentially a local affair. The element of local control, however, was notably absent in the revised constitution of the Council. It was for the Supreme Council to determine when it was necessary to remove a Chief Farmer.

The abolition of periodical elections met difficulties from unexpected quarters. After Independence, the Farmers' Council, looking for wider recognition, had applied for membership in the International Federation of Agricultural Producers. This organisation found the constitution of the Council unacceptable and proposed the reintroduction of elections at regular intervals.[9] Neither the Working Committee nor the Supreme Council were willing to do so. The minutes of the Working Committee recall that members considered the position of Chief Farmer to be an "ancestral institution". They asked why Ghana should always copy the outside world. As the "foundation members" of the Council they should not have to submit themselves to any further elections. Besides, "there was every possibility of losing their positions if they approved of the amendment".[10]

Not surprisingly, the members of the National Delegates' Conference, which met in early 1959, were similarly unattracted by the prospects of further elections. Their hands were forced, however, by the intervention of the CPP leadership. A senior Cabinet minister read out to them a decision taken by the Central Committee of the party to reintroduce elections in the Council, except in the case of the National Chief Farmer, who would become life-President.[11] The order was explained by the General Secretary in the face of strong opposition from the Conference, led by the Regional Chief Farmer of Ashanti. Resentment was directed not only against the idea of elections but also against the way in which the issue was dictated by the CPP.[12]

The revised 1959 constitution thus once again provided for triennial elections.[13] The following year, the General Secretary noted in his annual report that elections also had been held at the regional and district levels. All the regional representatives were "returned unopposed". The district ones had also been reelected except in five cases. Elections to the Supreme Council were held at the January 1960 conference at Kpandu.[14] Nothing was mentioned about elections at the society level, the only place in the hierarchy where farmers could participate directly.

The Council's commitments at the national and international level reinforced temporarily its faltering allegiance to democratic procedures. An additional incentive came from its aspirations to be accepted as a cooperative organisation, in order to meet the opposition of the "old" cooperatives and their supporters in the government.[15] This may help to explain why the National Delegates' Conference in Bibiani in 1960 decided to change the constitution again and introduce yearly elections at the society level.[16] This was where, according the Cooperative Ordinance, current responsibility for cooperative trading should rest. The Council had no intention, however, of decentralising. On the same occasion, the mandate period of regional and national officers was extended to five years, thereby protecting them against the possible consequences of a higher turnover of officers at the lowest level.

The Council's "merger" with the old cooperatives subjected the constitution to new strains. As mentioned in Chapter 4, the formula of the merger sought to ensure that, at least at the district and higher levels, all the principal elected officers of the Council retained their positions.[17] At the society level, elections would be held in all areas, although they were to be postponed for at least one year in the case of "merged" societies. In the meantime, key Council officers would be in undisturbed control.[18] It may be recalled that, less officially, the Council had declared that the local leaders of the old cooperatives "should never be considered for Executive Posts", because of their suspected political unreliability.[19]

The element of democratic procedure contained in the formula of the merger and in the new constitution did not survive long, once monopoly powers had been consolidated. Less than a year after the merger, a circular signed by the General Secretary suspended all elections of society executives. Only in proved cases of dishonesty or breach of Council regulations could committee members be removed, "unless directed by the National Headquarters". The reason put forward for the renewed suspension of elections was that "some people in authority, other than that of the UGFC have taken upon themselves the undue power to get rid of some of our constitutionally elected executives of our societies".[20] I shall return later to the implied question of political rivalry.

It is not clear whether as early as at the annual conference in Kumasi, in March 1962, it had been decided once again to abandon the idea of regular elections. A district report mentions conference proposals in that direction.[21] The directives issued by the General Secretary were later confirmed by the Supreme Council. Another staff circular gives extracts from the minutes taken on this occasion. Reference was made to recent attempts "to throw off the old District and Society Chief Farmers who have fought for the stability of the Council from the bitter beginning". The General Secretary was quoted as saying that "the Council should have implicit confidence in people who have shown impeachable probity and fidelity and not in new comers".[22] The Supreme Council agreed to amend the constitution. The mandate period of District Chief Farmers was extended to five years. But they should be automatically re-elected unless they had either "traversed the authority of the Government", violated the constitution of the Council, or had been found guilty of "misdemeanour".[23] The conditions were vague and open to varying interpretations. The powers of application rested with the central executive. The new ruling was also applied to the society level. No elections of society officers could be held without the approval of the General Secretary.[24] The latter declared at a regional meeting at Cape Coast that no elected officers were to be "subject to changes in positions, unless they misappropriate some funds or indulge in subversion against the Council, and for that matter, the CPP and the Government".[25]

On the point of elections, the 1959 constitution continued to be suspended until a new one was passed by the National Delegates' Conference in 1965. Regular elections were again to be held. The mandate period was extended to five years for all officers.[26] The right to automatic re-election as laid down by the executive in 1962 was not mentioned. As will be discussed later, the Council was at this time under pressure to reform itself in a more democratic direction. Conditions for the removal of elected officers were spelled out in great detail but remained subject to the overriding power of the central executive. It is not clear to what extent fresh elections were acutally held under the new constitution.[27]

But the scope for interference by senior officers was wide, even when elections were actually permitted to take place. A petition from Agogo society reported a particularly blatant case. When farmers met to elect new society officers, the presiding District Officer had read out a letter from the National Chief Farmer nominating candidates for the most important posts. All the farmers, according to the petition, left the meeting in protest against this unconstitutional procedure.[28] It is probable that the President of the Council only rarely had a direct personal interest in society elections, but there existed a large number of officers from district level upwards who might like a share of local patronage. Junior officers, as well, could enlist the support of senior ones in order to facilitate local acceptance of their own candidates. A sharply worded protest from Atwima Mfensi was lodged against the District Officer for attempts to prevent a certain candidate from being elected:

> He (the D.O.) must not put his nose in the matter which should be the responsibility of the Farmers alone and he should stop threatening the farmers as well as the Secretary-Receiver that if they vote for K.B. he will expell or dismiss that secretary.[29]

The District Officer was warned that he was not wanted at the polling station: "We farmers in the District are electing our chief, not the clerks". In this case, the Secretary-Receiver and the District Officer were apparently pulling in different directions. Chief Farmers, as well, were accused of interfering with the elections at lower levels. One report from Peki district claimed that the local Chief Farmer appointed committee members without bothering about voting.[30] Such cases of open violation of constitutional procedure were probably less important than more subtle forms of manipulations against which the constitution, even when not in suspension, could offer little protection.

The constitution gave the Supreme Council power to remove and reinstate elected officers at all levels. In addition, other executive bodies from the districts upwards could suspend any such officer, subject to ultimate confirmation by the Supreme Council.[31] The conditions laid down in the constitution for the removal of elected officers were not binding on the Supreme Council.[32] The conditions themselves were vague and allowed much freedom of application. What, for example, was "proved insubordination" in the case of an elected officer? A major case of "insubordination" led to the expulsion of Chief Farmers and a large number of committee members in the Konongo and Ejisu districts of Ashanti in 1962. According to the official report, they had shown "gross misconduct" towards the National Chief Farmer by accusing him of dishonesty and corruption. Some of the offenders were disqualified from contesting any future elections.[33] "Subversion" was another flexible cause for removing elected officers.[34] It could be extended to include all sorts of critical or oppositional statements and actions against the Council, the CPP or the Government.

It would be incorrect to imagine that the powers of senior officers in this respect were entirely arbitrary. The system allowed for appeals to higher levels and for investigations and arbitration. A local Chief Farmer, for example, who had been removed by the district executive for "misbehaviour" was reinstated by the regional executive, following investigations by the

National Chief Farmer.[35] Similarly, a society executive member dismissed by the District Chief Farmer was reinstated by the regional executive, no just cause having been found for the removal.[36] The files contain many appeals against unfair removals. One committee member, for example, claimed that he had been judged "in absentia" and had not been "properly informed of the charges" against him.[37]

For the period as a whole, regular elections were a matter of exception rather than the rule. They were suppressed in the interest of both the elected cadres and the bureaucracy. Personal privileges once acquired were in this way collectively protected by the organisation. This aspect was underscored by the General Secretary's speech at Cape Coast. He claimed that those who sought to oust the present elected members were only those who were envious of the benefits attached to office.[38] But by accepting protection from above, the elected officers became captives of the central leadership and vulnerable to pressures from that end. Cash allowances and other privileges were regulated at the top and disciplinary measures could be applied much in the same way as to ordinary employees. The system of representation had little capacity to serve as an independent instrument for popular control. Those who were supposed to represent the farmers were themselves incorporated in the bureaucracy and relegated to a subordinate role.

Elections are no guarantee for democratic procedure nor do they necessarily protect representatives from being subdued by a bureaucracy which they are supposed to control. All systems of representation, even those which are most advanced in a formal sense, face enormous problems in their actual democratic practice. The crude approach of the Farmers' Council, however, left little room for any democratic pretentions whatsoever.

Social cleavages and monopoly representation

The way in which the formal democratic process was manipulated did not by itself necessarily jeopardise the Council's claim to "represent" the farmers. Communities or groups with little internal divisions may accept representation by leaders over whom they have no formal control. The cocoa-farming communities, however, were divided by a variety of social, economic and political cleavages. The manipulation of the democratic process was partly related to the need to supress these cleavages and to conceal the limited local support.

The most apparent division was that of party politics. In the late 1940's and early 1950's the nationalist appeal spread rapidly in the rural areas. The CPP did not face much effective competition to its claim to lead and organise the movement. The early farmer recruits of the Council found little difficulty in combining support for the party and the pursuit of material interests, primarily in the cocoa trade. (See Chapter 3.)

The situation changed radically with the disintegration of the CPP in Ashanti and the rise in 1954 of the National Liberation Movement, which combined a strong regional-separatist appeal with a good portion of "cocoa politics". It placed itself at the head of farmers' opposition to the new tax and price policies of the government. The CPP was reduced to a minority party in most of the important cocoa districts. Much of the Ashanti establishment of the Farmers' Council defected and joined the opposition. Many Chief Farmers and committee members who remained loyal were isolated and surrounded by hostile communities of farmers. Some took refuge in the cities or in the remaining CPP strongholds, as Kumawu.

The organised "cocoa opposition" receded but it is improbable that the CPP ever recovered a majority support in the NLM districts. The basic conflict with the government over the distribution of cocoa income as well as regional allegiances continued to affect the political loyalties of the farmers, even when credible alternative parties had been eliminated. The Council's ban on elections in 1955 and the subsequent resistance to their reintroduction on a regular basis reflected not only the natural desire of individual functionaries to cling to office but also this basic political weakness of the organisation.

A former district organiser in Brong Ahafo recalled that at the time of independence even CPP members felt reluctant to join the Council:

> The handful officers of the Council were hooted and jeered at by hired hooligans wherever they went. Both officials and farmers were intensely terrorised and only the most fearless individuals could openly claim association with the Council.[39]

The picture may be exaggerated but this was certainly not a time when those officers who had remained loyal were anxious to test their support among the farmers through elections or otherwise.

The political struggle in Ashanti drove the Farmers' Council closer to the CPP and affected the recruitment of new cadres. The 1953 platform of the Council had been one of a national farmers' organisation with commercial aspirations. To many of its early members its vague party affiliation was probably of secondary importance. The Council was able to recruit also from groups not committed to the CPP. The 1954 crisis forced members to choose sides. The Council also depended more directly on the local leadership of the CPP when moving into new or abandoned areas. In many cases only the party stalwarts had remained loyal to the Council. Its reconstituted local establishment was often identical with that of the party. Loyalty through these difficult times also raised expectations of rewards once the national victory had been won. Benefits associated with the trading activities of the Council were naturally considered in that context. But the integration of the party and the Council was far from complete. Many Council officers, elected as well as employees, had obtained their positions without passing through the party. Many built up their own network of patronage, which they guarded against all interference from outside.

The Council's claim to represent the farmers was obstructed by the penetration of party rivalries in the cocoa areas. Monopoly over the cocoa trade was chosen as a tool to reinforce this claim. The enforced "merger" with the old cooperatives made it necessary to control such political divisions inside the organisation, as well. The Council sought to achieve this by reserving the important posts for its own functionaries. The decision to prevent destoolments and fresh elections soon after the merger served to make this arrangement permanent and to block the possible influx of politically unreliable elements. People identified as political opponents seem to have been prevented from claiming the minority seats on the new committees to which they were entitled according to the agreement on the merger. A report from Akim Awenade, for example, claimed that members of the former cooperative committee were barred because they had been members of the United Party.[40]

Party discipline and loyalty were stressed when the new rules for removing elected members were introduced in 1962.[41] Allegations about anti-party activities were occasionally raised in attacks on individual officers. Party merits were used to support claims for office or in defence of criticism. But the Council files do not record many cases where disciplinary actions were actually taken on the basis of political charges. There is a reference to a major clean-up operation in Takoradi, where it was discovered that leading functionaries of the rubber-growing division of the Council had voted against the CPP in the elections. This was declared to be a personal disgrace to the General Secretary, who had advised the farmers to cast their votes for the CPP. They were now to "hunt out for all the enemies of the Government and to see that every society member is in possession of a CPP card".[42]

Party politics and party divisions were in their turn affected by traditional political cleavages, local as well as regional, throughout most of the cocoa areas. The cocoa opposition of the National Liberation Movement was closely associated with Ashanti nationalism. The impact was equally felt at the local level, where divisions and rivalries between families and clans penetrated and interacted with party politics. Chieftaincy, although not necessarily involving effective leadership, was at the centre of many of these local or regional conflicts. The status of a particular stool vis-à-vis another reflected historical relations between communities. These were not determined once and for all, but subject to continued efforts to have them reinforced or changed.[43] Similar struggles were fought by families and clans contending for a particular stool.

Local and regional allegiances were mobilised against the CPP and its claims for national political hegemony. But the CPP itself made use of such divisions to break up larger blocs of regional opposition.[44] Thus, traditional Brong resentment of Ashanti domination, for example, could be mobilised, restricting the appeal of the Ashanti-based opposition in these northern cocoa districts. In the south, as well, the not very distant experience of Ashanti expansionism played into the hands of the CPP.

At the local level, the party was similarly involved in disputes and rivalries between and within communities. Government support for one side was traded in exchange for local support for the party. The result was a complex pattern of reinforcing and contradicting relationships between communal and party politics and between chieftaincy and the new leadership. Party divisions which by themselves may have had little bearing on the actual administration of the monopoly were reinforced by communal allegiances. Hostility to the Farmers' Council in many areas became in this way not only a party affair but involved the community and its traditional leaders. An example of how the Council was affected by such solid local opposition was the case of Nkenkasu in Ashanti where it was obliged to remove its district headquarters in 1963, because of persistent opposition and interference by the chief, who was said to have much influence over his subjects and a great dislike for the Farmers' Council. The area was said to be a United Party stronghold.[45]

Chiefs and elders had a standing claim to be recognised as the "natural" representatives of the rural population. As this was predominantly made up of farmers, any new organisation claiming to represent the latter would have to define its position in relation to chieftaincy. The records of the Council indicate a variety of relations with these traditional political functionaries. As in the case of the party, the Council at one time canvassed on a platform which attacked chieftaincy.[46] Individual chiefs were occasionally enrolled. The first Vice-President of the Council, for example, was a chief from Odobeng in Central Region. The Regional Chief Farmers of Ashanti and Western Region were at one time both chiefs. The roll-call of National Delegates' Conferences similarly list a small number of chiefs.[47] On the whole, these seem to have been minor chiefs and in terms of total numbers hardly a significant element in the local representation of the Council. Organisers, however, were advised to seek the cooperation of the chiefs when establishing the Council in a new area.[48] The election of local committee members could also be given some kind of official sanction by the chief when authorising the beating of gongong to call the villagers to the cermony. Friendly chiefs could be consulted on the nomination of committee members.

Occasionally, local chiefs took a strong interest in the way in which Council functionaries, elected as well as appointed ones, were selected. In one case, for example, a chief demanded two seats on the committee as a condition for his cooperation.[49] The Regional Office of the Council in the Western Region complained that the Omanhene of Sefwi Wiawso, an important Paramount Chief, interfered greatly in the affairs of the Council, pushing his own nominees for Council office.[50]

At a more superficial level, efforts were made to enroll chiefs collectively behind the Council. Prominent chiefs were invited to take part as guests of honour in important Council gatherings, such as the annual national conferences.[51] Official backing was also sought from the regional "Houses of Chiefs". The Standing Committee of the Brong Ahafo House cabled its support to the Kpandu conference in 1960: "Chiefs, farmers, and people solidly behind Farmers' Council".[52] The Council's success in enlisting the support of traditional sources of representation for its own purposes was, however, very limited. Chieftaincy served instead primarily as a rival source of leadership which exposed the weaknesses of the Council's own system of representation and its claim to speak for the farming community.

Chieftaincy had not exhausted its capacity for community leadership. Nevertheless, its hold had been greatly weakened by the social forces which were released as a result of colonial penetration. Villagers were reoriented into new productive roles as determined by outside market relations. The peasantry was increasingly differentiated according to the amount of land and capital controlled. Privileged access to land facilitated to a certain extent the transformation of chiefs and members of their families into leading landlords and farm-owners. But there is no simple identity between chiefs and the growing class of big farmers. The latter were recruited much more broadly. Migration, for example, gave rise to large holdings entirely outside the traditional system of land distribution.

The way in which the Council "represented" the farmers must be examined in relation to this growing social differentiation of the cocoa economy and the divisions to which it gave rise. How did the composition of the farmer-cadres of the Council reflect these divisions? Early recruitment was strongly influenced by the Council's links to previous farmer-trader groups and its continued involvement in the cocoa trade. This farmer-trade bias was also largely a big-farmer bias. The farmer-trader groups had appealed mainly to the commercial interests of the big farmers, and so did the joint scheme of the Cocoa Purchasing Company and the Farmers' Council.

There were sound commercial reasons for concentrating on the most "important" farmers in a particular area, apart from their spontaneous interest in the trade. A small number of large individual suppliers involved lower marketing costs than a great number of small ones. This fact was underscored by the highly unequal distribution of control over production. The output of one big farmer would outweight that of scores of petty-producers. But big farmers had also a "natural" claim to leadership, based on personal achievement or on control over large inherited holdings. Big farmers commanding labourers, tenants, and financial resources were able to command the "respect" of their fellow farmers. The latter might also depend on the big farmers for credit and other services. Although the origin of the "chief farmer" institution is obscure, it clearly involved substantial farmers who had demonstrated their success in the farming business.

The right of the big farmers to represent the farming community as a whole seems never to have been questioned by the Farmers' Council. The aspiration of the CPP to stand out as the party of the "commoners" in relation to the established urban middle class and the chiefs did not have any consequences for the Council's approach to the social differentiation of the peasantry. Not only were the labourers left entirely outside the Council's scheme of representation but the wide gap between the fortunes of the mass of small farmers and the small number of big ones was similarly ignored.

This uncritical acceptance of the big-farmer approach to peasant leadership would have caused the Council less difficulties if it had been more successful by these standards when recruiting its farmer cadres. Instead, it often ended up with second- or third-rate candidates. After the 1966 coup, some farmers spoke scornfully about Chief Farmers and committee members with little or no cocoa of their own, but who "lorded it over the real cocoa farmers".[53] This was certainly an exaggeration. It seems as if the Chief Farmers of the Council were normally quite substantial farmers or farm-owners, although not necessarily big enough to impress themselves on their fellow farmers as "natural" leaders in this sense. It should be remembered that throughout the 1950's the Council was a minority group, with powerful rivals in the cooperatives and in groups attached to firms and brokers. Party cleavages, which at times had close bearing on the material interests of the cocoa farmers, also restricted the Council's choice among the big farmers. The latter were likely to be particularly responsive to the propaganda of the "cocoa opposition".

In some areas, the division between tenants and landlords made an impact on the Council's local representation. This was largely a division between communities of stranger farmers and local communal landlords. The migrants were often economically more advanced. It was therefore entirely within the logic of seeking out the "important" farmers that the Council in some areas established its base among the migrant, tenant farmers. But groups of stranger farmers, prosperous or less prosperous, could also provide a foothold for the Council in areas where the host communities were anti-CCP or backing rival trading organisations. The potency of these divisions and the ambivalent position of the Council were fully exposed in the struggle over the "Rent Stabilisation" laws which aimed at maximising the rent paid by tenant farmers on cocoa land.[54]

The institution of chieftaincy was not unaffected by these wider social divisions. Many chiefs were major farm-owners and shared the commercial aspirations of their own class. Many were drawn into the trading ventures of various groups, including cooperatives and those allied with firms or brokers. Relying on communal allegiancies, chiefs could on occasion lead entire villages into the fold of a particular trading organisation. Relations between the Council and individual chiefs therefore involved an element of competition for control over the cocoa trade, as well. Chieftaincy also played an important role in the conflict between stranger farmers and communal landlords. It affected the Council's representation most acutely in the Central Region, where many chiefs, often major landlords in their own right, violently resisted the attempt to enforce the new land-rent regulations. The National Vice President of the Council, an important chief, was expelled from the Supreme Council because of his active involvement.[55] The interests of the chiefs as private landlords were in this case inseparably mixed with their role as guardians of communal land rights against usurpation by strangers.

Party politics, communal allegiances, and other social divisions restricted the reach of the system of representation which had been established by the Farmers' Council. The local foundation of the system was brittle not primarily because of its weak democratic procedures but because it lacked a systematic base among the social forces of the cocoa economy. This produced a fragmented pattern of allegiances, united by a commitment to the CPP and the Council, and by the individual and collective benefits which these organisations were able to provide. The occasional and personal character of the Council's local representation made it vulnerable to challenge and disruption. The manipulation of the representative system reflected the need to prevent these numerous divisions from "interfering" with the commercial and political assignments of the Council. It reflected at the same time the fundamental contradictions between the cocoa farmers, the government and the monopoly traders over the distribution of cocoa income.

Bureaucracy: centralisation and self-control

The leadership of the Council was largely undisturbed by the requirements of representative government. The hierarchic chain of command of the trading organisation permitted little interference from the parallel system of elected committees. A Secretary-Receiver, for example, was responsible only to his senior officer. He could not be removed, suspended or otherwise disciplined by the local committee, whose "secretary" he was supposed to be. Centralisation was strengthened as the monopoly was enforced. The General Purpose Committee noted in 1961 that Chief Farmers in the past had often interfered with the duties of the commercial officers. This practice was to stop: "The penalty for any Chief Farmer who will be guilty of such offence shall be instant expulsion".[56] The official Cadbury adviser emphasised the need to eliminate every element of dual command. There should be "no divided loyalty or authority".[57] Junior officers should be strictly and exclusively responsible to their seniors and not to Chief Farmers and elected committees.

Only at the very top, through the Supreme Council, were elected members entitled to exercise direct control over the bureaucracy. But neither the Supreme Council nor the National Delegates' Conference seem to have been able to give much meaning to this formal supremacy. The General Secretary exercised undisturbed command over the organisation. The weak constitutional restraints were demonstrated when he personally suspended the constitution and cancelled elections in 1962. No reference was made either to the Supreme Council or the National Conference.[58]

Much of the power of the General Secretary stemmed from his exclusive control over appointments. According to the minutes of the Working Committee such control had already been vested in him by 1955, but this may at this point have referred to headquarters staff only.[59] The National Delegates' Conference, however, confirmed in 1960 that his powers in this respect applied to all officers of the Council, thus marking a step in the Council's reorganisation into a major business enterprise.[60] These powers were the normal prerogatives of management. Business management is supposed to be ultimately responsible to the owners. In the Council's case, the assumption that farmers exercised such ultimate control was largely an illusion, especially as the "prerogatives of management" were extended to direct interference with the way these "owners" were represented.

The power of the General Secretary was supported by his special access to the party and the government. The significance of these special relations was enhanced by the poor integration of the Council and the party at other levels.[61] The authority of the General Secretary as a government and party leader may not have been great outside his own domains. His formal position, at any rate was rather obscure. He was awarded the rank of a Cabinet minister in 1960 as "Ambassador Extraordinary and Minister Plenipotentiary".[62] But this did not mean a regular membership in the inner government circle, and seems primarily to have been an honorary title. Likewise, his position in the Central Committee of the party was at first that of a "co-opted" member rather than a "substantial" one.[63] He was made a full member of the committee in 1962.[64] To the functionaries of the Farmers' Council, however, the General Secretary constituted the critical link of transmission for government and party authority.

There were practical as well as political limitations to centralisation, as well as to the personal leadership of the General Secretary. The organisation was spread over a vast area with poor communications. Local functionaries had to be given considerable responsibilities, and their knowledge of field conditions strengthened their position in relation to headquarters. This provided for a strong element of personal rule at lower levels of the organisations, as well, including opportunities for private profiteering. Headquarters, on the other hand, sought to enforce discipline in order to protect corporate profits but also to avoid potentially dangerous conflicts with other centres of power in the Ghanaian political economy.

The problem of bureaucratic self-control was a major one, especially in the absence of any effective channels for supervision based on the participation of the farmers themselves. The monopoly had been born in an atmosphere of public suspicion. Apart from the broader issues of strategy involved, the capability of the Council to manage the entire cocoa trade was widely doubted. The government sought to meet some of the critics by stressing that the Council would be closely supervised by the Auditor-General's Department, just like any government body.[65] Staff was transferred for this purpose from the former Department of Co-operatives. This audit was a major exercise and the heavy costs were charged to the Council.[66] In addition the Council had its own internal audit, also a major operation, engaging at the time of the military coup almost 400 full-time audit officers.[67] Reports by internal and external auditors in the files of the Council show that this heavy investment was not all wasted. Auditors were able to draw attention to numerous irregularities and their mere presence may have prevented others.[68] But the reach of this formal control of accounts and stocks was limited. Only certain types of malpractices could be detected by such methods. The records also tell us of many ways of dodging auditors and manipulating accounts.[69] Nor were auditors immune to corruption.

Auditors could not by themselves enforce discipline even within their own limited field. Their investigations and disclosures had to be followed up by administrative action to have any effect at all. This applied equally well to all other reports about malpractices which reached headquarters from official and unofficial sources. The shortcomings of the system of bureaucratic self-control in this respect were particularly evident: the contrast was great between the number and nature of the disclosures and the evidence of disciplinary action.

The records give many examples of how cases were "settled" without any relevant action being taken against the offenders. Even in serious cases, a marked tendency to keep out the police and the courts could be noted. This was partly a matter of protecting the public image of the Council, but it also involved the cover-up of current rackets involving the interests and positions of other Council officials. An illustration may be given from Ashanti Akim in 1962, where the District Chief Farmer accused other members of the district committee of having misappropriated £1,227 of the farmers' money.[70] The case was investigated by the National Security Officer from Accra, who confirmed the allegations. But counter-charges were made against the Chief Farmer, who also admitted that he had collected a total of £400 from the societies in his district "for a certain purpose". The Ashanti regional executive committee which met to consider this delicate matter arrived at the following conclusion:

> We recommend that no court action should be instituted . . . We further order that from today (the D.C.F. and the Vice-D.C.F.) should not on any circumstances bring up the matter again. Nobody should talk about £20 collection from societies. Every matter is settled, and both parties should retain their positions, as the saying goes 'we do not want to wash our dirty linen outside', so as to give way for unscrupulous persons to speak against our activities and operations. Both parties swore oaths . . .[71]

The only unusual aspect of this particular case was possibly the dutiful way in which the cover-up deal was recorded for the benefit of posterity.

When money was found missing or unjustly appropriated, the case was often settled merely through the refund of the sum involved. The terms of repayment could be quite generous as in the case of a senior officer who had misappropriated over £2,000. He agreed to refund the money in yearly instalments over an eight-year period.[72] His career in the Council was not visibly affected. Reports after the coup similarly tell of the continued confidence the Council showed in its officers after such refunds.[73] On one occasion, court action had been initiated but:

Before the commencement of the case A. replaced the 108 bags of rotten cocoa with good quality cocoa at his own expense. The Council, therefore did not lose any money . . . I suggest that the papers be put away as the case is considered closed.[74]

Auditors expressed concern over the inefficiency of the disciplinary action. Even in cases when culprits were actually dismissed they were often reemployed in other parts of the organisation.[75] Farmers in Nyinahin, for example, reported that when they had found out that the Secretary-Receiver had rigged the scale, they took him straight to the District Officer. "But the D.O. in charge refused to punish him". He only transferred him to another station.[76] The wholesale reposting of staff was also considered an appropriate step to placate the farmers after the military take-over in 1966.[77]

The high tolerance of the "weaknesses" of individual Council officers shown by their seniors may partly reflect a desire to give culprits a second chance if they repented and "made good". It is more important, however, to stress the probable connection between such tolerance and the frequent reports about collusion between junior and senior officers. Personal relations of patronage and dependence, in combination with more crude forms of partnership, including bribes and extortion—all contributed to place obstacles in the way of bureaucratic self-control.[78]

We used to hear that several money for the UGFC Co-operatives were missing but when such cases were sent to courts, we heard nothing about them again—because the officers of the Farmers' Council concerned were interested and had shares in the deals.[79]

Of course, not everyone had powerful friends. Many cases of actual dismissals and prosecutions are documented. The minutes of the Treasury Committee, for example, listed a number of dismissals following a report by the Chief Internal Auditor in June 1964.[80] Five Secretary-Receivers and one Depot Keeper were sacked on this occasion for shortages and fictitious way bills. A District Officer was to show cause why he should not be dismissed for a similar offence.

Many, on the other hand, were protected or could at least count on some assistance from friends or accomplices if exposed. A report from Twifu District claimed that the District Chief Farmer who had been suspended boasted that it was a mere formality to keep his local critics quiet and that he would soon be reinstated.[81] The Chief Farmer in this case was both the district chairman of the party and, according to the petitioners, a "great personal friend" of the District Commissioner. Party merits did no doubt serve as protection in many cases and were also used in appeals against disciplinary actions. One dismissed Chief Farmer claimed that his assistance and sacrifices to the Party were "so enormous that they cannot be enumerated": "I, therefore, crave the indulgence of the Osagyefo [the President] and the Supreme Council to use their good offices to instate me as Chief Farmer again".[82]

But there were also limits to what powerful friends and party merits could buy. Let us look at the case of Mr. A., Secretary-Receiver of the Nsokote Society, dismissed for misappropriating £1,301 and 1,309 empty bags.[83] No lesser person than the Minister of Information intervened unsuccessfully on his behalf.

> I really have sympathy for Mr. A., the petitioner. Because he is the only responsible for his old mother's upkeep and infact Mr. A. has suffered in the Party's struggle and if there is anything you can do to help him to continue in Office I will be grateful to you . . .[84]

The Minister's letter was one of a great number of similar notes received by the General Secretary of the Farmers' Council from ministers, members of the National Assembly and other dignitaries "recommending" their clients, usually for appointments, but also, as in this case, against dismissal.[85] On this occasion, the General Secretary was unable to meet the request. He added, however, that it was the policy of the Council not to report such cases to the police, in spite of their serious nature. The reason stated for this tolerance is worth noting:

> The Farmers' Council, fully conscious of the fact that its employees are party activists has always decided against public criticisms to deal leniently with any officer involved in a stealing or any criminal case by summarily dismissing him.[86]

Faced with audit reports and other clear indictments, the Council may have been obliged to take some action. It was therefore the more important for interested parties to seek to supress reports at an earlier stage, before the damage was done. At the village level, this could take the form of attempts by committee members to prevent farmers from reporting the malpractices to senior officers. A case of this type was recalled in 1966 by a farmer who had discovered that the scale was rigged with a piece of lead:

> At this juncture, I held him (the S-R) by the neck in the presence of the Treasurer, Committee members on duty, my Caretakers and also one farmer. I also dashed the Secretary Receiver on some Cocoa bags, but regret to say, the Treasurer and the members did not come to aid . . . (They) blamed me because they were in favour of the Secretary Receiver . . . (and) called me home and begged me not to forward the case to the District Officer at Ntotroso otherwise the Secretary Receiver will be dismissed.[87]

But senior officers could also choose to ignore reports or to "settle" disputes, hoping to stop the matter from being further pursued by the plaintifs. Cases documented by the Council's own records are of course those where the latter refused to accept such settlements. The petitioners from Nargode Society, for example, took in 1964 their long list of grievances against the Secretary-Receiver to their local committee members. These, however, only "abused and quarrelled" with them. They therefore turned to the District Chief Farmer but with the same poor result. They claimed that he "did not give us ear at all" and treated their charges as "null and void". They had not been allowed to call witnesses and the arbitrators called by the Chief Farmer had been biased against them, as well as "provocative and annoying". The petition was thumbprinted by thirty farmers.[88]

In this case, the farmers insisted and brought their case to the headquarters. There were clearly risks involved in doing so. If they were not successful, their relations with local Council staff would be further strained, and even if they succeeded in removing the Secretary-Receiver, other implicated officers would continue to be around. The capacity of the latter to make life difficult for troublesome farmers at the buying centres would have to be taken into account.

Another reason for the failure of the committees and elected members to provide effective channels for farmers' grievances was their own vulnerable position and dependence on the bureaucracy. If they reported or passed on reports of malpractices involving senior officers or their local agents they too risked dismissal or suspension. The Treasurer in the Asamankese B District, for example, was suspended, "to serve as a deterrent to others", for having passed on allegations against the Regional Vice-Chief Farmer.[89] "The allegations were so serious that if they had been substantiated the R.V.C.F. might have been dismissed". The formal ground for the suspension was that the Treasurer had contravened the regulations by reporting the case direct to headquarters and not to the regional office. Bureaucratic routines could in this way be invoked to block undesirable reports.[90] Temporary suspension, however, was a fairly mild punishment. A village functionary, probably a treasurer, complained in 1965 that he had been sacked by the District Officer for revealing certain malpractices.[91] He was told that "if I wrote any further letter in connection with the topic, he would hand me over to the police to deal with me".

Uncertainties about "the rule of law" probably served as an important deterrent against undesirable reporting. The existence of "special relations" between Council officers and the police may have been more imagined than real. But there may have been good reason to be careful, especially in areas where the political struggle had been particularly bitter and violent. The Preventive Detention Act was regarded with quite legitimate fears. It had been introduced to control political violence directed against the regime, and permitted the detention of suspects for more or less indefinite periods of time without trial. The line between political violence and other forms of opposition, however, was by no means clear and the detention laws were at times used as a general instrument of political repression with wide margins for arbitrary application.[92] Threats of detention were made even in connection with minor expressions of opposition against the government, the party, and its affiliates. A Chief Farmer claimed in 1964 that the District Commissioner had threatened him and others with detention if they refused to allow a special cocoa surtax of two shillings per load for local development projects.[93] Similar allegations were frequent after the military coup. The committee members and the Secretary-Receiver in Akim Awenade, for example, were said to have "pulled their heads together . . . (and) planned and recommended for detention anybody illegally".[94] The committee included not only the CPP chairman but also the two most important local chiefs. The petitioner claimed that he and his sister had been arrested and detained for two months in James Fort Prison because they had "quarrelled with the Committee members over the irregularity of the weighing of cocoa". Farmers from the Volta Region reported similarly that they were threatened with detention if they opposed the manipulations of Council officers. One farmer said that they had "no voice to complain for fear of police arrest and detention".[95] Another claimed that if a farmer "queries a Secretary-Receiver, the only word you will hear is: Take time or otherwise you will be detained".[96] These may have been empty threats and may also have been taken as such by the parties involved. But they nevertheless expressed a repressive political relationship. Many farmers clearly feared that criticism and opposition would be punished in one way or another.

> No farmer dared to say anything against them. If one did say anything one's cocoa would be turned down and one would not have anywhere to sell it. They did not allow criticism.[97]

The means of reprisal were there, although reports of actual cases where cocoa was refused in this way were not common. But there were other, more subtle methods.

The authoritarian style of many administrators and their farmer associates was probably by itself an important obstacle to open criticism even without threats and reprisals. James G. Edusei of Kumasi, submitting his views on the Council to Nkrumah in 1964, stressed that farmers would come forward with their criticism "if UGFCC officers do not address or speak to the farmers from pulpit".[98] The lopsided political relationship was expressed in the highhanded attitude shown by many employees and committee members alike in their contacts with the farmers on whose behalf they were supposed to run the organisation. Farmers claimed that the committee and the Secretary-Receiver treated the society "as if it is their own private property or business organisation" and that they took on "unnecessary airs of superiority and autocratic attitudes".[99] Another petitioner wrote that the society was run entirely as a "family affair" and that only those who were actually members of the committee had a say in the matter.[100] Many reports emphasised the disrespectful way in which farmers were treated. This may look like minor complaints in comparison with others but they were often attributed great importance by the petitioners themselves. In their eyes, rude behaviour, abusive words, and other expressions of disrespect highlighted their humiliation and their lack of power:

> In some cases even labourers at the purchasing station had the impudence to order these farmers about. What a shame and disgrace to a so-called farmers own organisation![101]

A petition from the Juaso district in 1965 sought the removal of the society treasurer because, among other things, he "does not show any good face to the farmers but rather shouts at them and maltreats them as children".[102] Memoranda submitted to the deGraft-Johnson Committee in 1966 complained of insulting language, disregard for old age, and rough treatment in general.[103]

The suppression of reports from below and the stifling of criticism from farmers as well as from within the bureaucracy itself were serious obstacles to bureaucratic self-control. It helps us to understand why farmers were so exposed to the private exploits and official manipulations of the Council functionaries. But the mass of evidence in the Council's own files by which such conclusions may be supported constitutes at the same time evidence that repression was far from total. The numerous petitions and reports demonstrate that channels existed for both criticism and redress. Many allegations were not only reported but also investigated. Officers were dismissed, suspended and transferred on the basis of such reports. The farmers did not passively accept bullying and cheating. Their petitions and complaints may even be said to indicate an element of confidence in the ability of the system to correct its own mistakes and punish offenders within its own ranks, although such confidence may often have been misplaced.

Resistance and redress

The power of the bureaucracy was not monolithic, although largely unhampered by the formalities of representative government. Nor was it total. It was restricted above all by the possibility of organised or spontaneous resistance by the farmers, but also by the need to coexist with other leaders and organisations at the local as well as the national level. Divisions within the local political establishment as well as within the Council itself could be used by the farmers when seeking redress for their grievances.

Farmers appealed on many occasions successfully to higher levels of the bureaucracy against the way they were treated at the village buying centres. Petitions to headquarters or to district and regional authorities backed by a fair number of important local farmers were in many cases sufficient to achieve the most immediate objectives. Such petitions were often strongly worded and carried a weight which could not be easily ignored. In 1963, fifteen petitioners from Kyekyewere society, Offinso, expressed themselves as follows:

> We want absolute peace in this town but we reiterate, that if any disobedient, dishonest and unpatriotic receiver and unscrupulous committee members are forced on us, we will not hesitate to discard their services.[104]

Farmers could be prevented from marketing their cocoa through other channels but they could not be forced to sell it to the Farmers' Council. Withholding cocoa from the market had been used by farmers as a weapon in the fight against foreign firms and their agents. It could also be turned against the Council.

Resistance took sometimes direct and violent forms. A Security Officer of the Cocoa Marketing Board in Akim Swedru reported that farmers had resorted to drastic measures against a committee which refused to render account of its activities and submit to new elections:

> The masses have therefore planned to use positive action on the committee members and that is, they will beat up the old committee members and spoil everything in the society's shed.[105]

The villagers, according to the report, marched to the shed singing party and war songs. The District Officer was attacked but escaped in his car. The arrival of the police prevented the destruction of the shed. The local chief had taken the side of the rebels. There is not much evidence of such violent protest but the possibility of open revolt and "positive action" by outraged farmers set limits to arbitrary treatment and open profiteering.

The effectiveness of protest and resistance depended greatly on the degree of unity among the farmers. In some areas, this may have been easily achieved, whereas in others, they were divided by loyalties to rival sets of leaders, "traditional" as well as "modern". Differentiation on the basis of inequalities in the access to land and wealth as well as between local and "stranger" farmers often mingled with these other divisions. The lack of unity among the farmers widened the scope for manipulation open to the Council and its functionaries. Criticism could be ignored and supressed as partisan or sectional and "unrepresentative". Opposition could be branded anti-CPP or anti-Government.

The prospects of opposition depended also on the ability of the farmers to find allies within the ranks of other local ruling groups. These were above all the CPP leaders, including councillors and members of village and town development committees. The District Commissioners, usually with strong records as party activists, were in a category of their own, as the official representatives of the government. They had wide powers, especially in relation to "national security". The arbitrary rule and private exploits of some of them were the subject of many tales in post-coup Ghana. A further category of local leaders were the Members of Parliament and the important sons and daughters in a position to "represent" a particular area in Accra and influence the central government. Finally, there were the chiefs and elders, with varying influence over the local community. Much overlapping existed between the different types of leaders. Chiefs could be party nominees and even party functionaries. We have mentioned the frequent integration of the local establishments of the party and the Council. Relations of dependence and patronage created in some areas a solid ruling group centred around a leading "big man", for example an M.P. or a District Commissioner. But the Council records show that the unity of the ruling group was far from compact and that farmers were able to use the divisions to protect themselves against the monopoly. The relativly junior position of the Council officiers vis-à-vis other party functionaries facilitated appeals and interventions.

The Council sought to prevent such interference in its affairs. The General Secretary claimed as he suspended elections in 1962 that "some people in authority" encouraged the destoolment of Chief Farmers and other committee members.[106] On another occasion he was even more specific: certain farmers were trying to dispose of the present Council functionaries "through the influence of M.P.s and D.C.s"[107] Many cases of intervention by District Commissioners and other party officers are recorded. Most commonly, such actions were directed against Secretary-Receivers and other junior officers. On one occasion, however, the D.C. in Mampong took action against the Regional Chief Farmer of Ashanti, who was reported to have collected various sums from the societies in the district. He was now requested to refund the money by the D.C., who described the collections as "completely illegal and out of your office".[108] Intervention in this case was clearly in response to local grievances. But issues of rival patronage were also often involved. On one occasion, the District Commissioner in Mampong was accused of having caused the removal of the Chief Farmer of Mampong because he had opposed the D.C.'s candidate for the 1965 National Assembly nominations.[109] A prominent case was the conflict over the installation of the Assesewa District Chief Farmer in 1960. The Council was accused of subversive intentions by local CPP functionaries. The new Chief Farmer had been installed without consulting either the CPP, the District Commissioner or the M.P.[110] The M.P. himself was even more definite:

> It is within my political right and authority that my consent must be sought on all matters of appointment in my constituency; and I am prepared to face anybody who holds a contrary view before the Central Committee of the CPP . . . If it is the policy of the UGFC to operate in some areas as a separate entity from the branch party, then I will request you to understand that I shall never permit it to be so in Manya Krobo constituency.[111]

The question of who should control appointments could of course involve more substantial divisions of local politics. In this particular case, the M.P. claimed that the Council had been influenced by anti-CPP elements—"even worse than those who were recently expelled from the Party".[112] The Farmers' Council was not prepared to accept such extreme claims of supremacy by M.P.'s or other party functionaries. But their effective influence set limits to its freedom of action. It introduced an element of outside control of which farmers were able to take advantage.

Party organisers had good reasons for seeking to prevent the excesses of individual Council officers, as they wished to protect the reputation of the party itself. The executive committee of the CPP in Agogo, for example, emphasised that the support given by the Regional Officer of the Council to dishonest Secretary-Receivers had a disrupting effect on the party:

> Because of the activities of these weighing clerks our farmers are trying to lose confidence in the Party, for the simple reason that the UGFC co-operatives is the sole agent . . . None of the clerks here has the spirit of the Party in him.[113]

Not only individual malpractices but also the Council's general methods of operation were challenged. The Eastern Region Steering Committee of the CPP, for example, protested in 1963 against the way in which farmers' "commissions" were retained by headquarters instead of being made available to individual societies for local development purposes. The Council was accused of not honouring promises made when the monopoly was established.[114] The party could in this way occasionally take on the functions of supervision which the Council's own committees performed so ineffectively. In many areas, however, the close integration of the various ruling groups did not allow the party to play an independent role in this respect.

The position of chiefs and elders was rather different. Although at times members of the ruling group as active CPP supporters or party nominees, they were normally less closely integrated with the party establishment. Many chiefs feared government reprisals and refrained from interfering with the affairs of the various local party organisations. Others, however, felt less restrained, especially when they could fall back on a solid "constitutional" position and wide local support. Chiefs and elders frequently took an active interest in the affairs of the Council. They sometimes put forward their own candidates for office and demanded to be consulted in connection with appointments and elections.[115] But they also intervened to seek the removal or suspension of local officers: in one case, a council of elders suspended the society committee and appointed a new one to take its place. They proclaimed that they were not going to allow the former committee members to receive any money from the Council.[116] In another case, the Secretary-Receiver as well as the Treasurer and Chief Farmer were deposed "for showing gross disobedience to a local chief.[117] Spokesmen of the latter complained that the committee had been established "without properly consulting the chief and the inhabitants".[118] Similarly, the Chief Farmer of Nsaba was suspended by the chief and elders because the former was said to have misappropriated loans intended for the farmers.[119]

As in the case of similar interventions by government and party officials this was at times a matter of securing posts for favoured candidates. Intervention could also be prompted by the failure of the Council to allow chiefs a share in the local surplus extracted from the farmers.[120] But chiefs and elders also acted in response to local grievances and pressures. In Atta Mampong, Juaso District, the chief and the elders demanded the removal of the society treasurer, who was accused of a wide range of malpractices.[121] On other occasions, chiefs and elders joined the farmers in demanding that accounts should be submitted by the Council committees.[122] Again, this may at times have been an attempt to make Council officers personally accountable to the chiefs. But the demand for accounts and public meetings corresponded to widely-felt grievances over the secretive and "private" character of the Council's operations. A report from Akim Swedru may serve as illustration:

> They (the farmers) elected the nine men committee . . . The season is closed and they want the committee to render account . . . The Omanhene has several times summoned the committee members but they refused to attend.[123]

The chances of success were greatest when farmers were able to secure support from a broad alliance of party functionaries and chiefs as in the case of a resolution of non-confidence in the Council's committee in Chiraa, Brong-Ahafo, which was passed jointly by the executive committee of the CPP, the Chiraa Town Development Committee and the Chiraahene and elders, "representing the peasantry". The latter was said to be disgusted by the "horrible mishaps and misdeeds" of the local representatives of the Council.[124]

A close study of the petition files of the Farmers' Council would probably throw much light on politics at the grassroots level during the CPP period. A variety of groupings and leaders were involved, appearing both as rivals and as allies, competing for support and seeking to secure themselves against political risks through organised followings or special relations to key dignitaries at higher levels. It seems particularly important to emphasise the active role of farmers and farmer leaders in this process. Social and political realities of the cocoa communities would be greatly misconstrued if farmers were pictured as passive and defenseless victims of the manipulations of petty-bureaucrats. Farmers were capable of a variety of direct actions from shaking dishonest clerks by the neck to threats of destroying cocoa sheds or withholding cocoa from the market.

Farmers were also able to use and develop means of protest and redress within the framework of local politics, as set by party organisations, government officials, chiefs and elders. The system provided restraints on the operation of its individual components. These were used by farmers as a defence against malpractices and profiteering. We have examined massive evidence indicating the failure of such restraints, but this should not lead us to believe that the latter were absent. The evidence itself (the petitions, appeals, and reports) reveals not only malpractices but also a political process through which these were exposed, and at times rectified.

The ability to enlist external support and to manipulate the divisions of local politics was, however, a poor substitute for effective democratic or bureaucratic restraints within the Farmers' Council itself. While individual malpractices were exposed, the system itself was largely left intact, reproducing the political basis of continued corruption and profiteering.

Cooperative democracy and government control

The government and CPP leaders were not unaware of the weaknesses of the political organisation of the Farmers' Council. Demands for reforms were raised within their own ranks. The *Budget Statement* of February 1966, quoted at the outset of this chapter, was the culmination of official criticism.[125] The chapter will conclude with a discussion of these attempted reforms, leading to the fundamental question of how the political structure of the Council was related to its basic function in the political economy of cocoa.

A principal focus of official criticism was the concept of "society autonomy". In cooperative theory, as introduced and developed in Ghana by colonial officials, a cooperative society was to be an autonomous body, with full financial and managerial responsibilities. Elected executives were accountable to the members: Secretary-Receivers and other employees were hired and controlled by the societies themselves. The basis of their financial autonomy was the share-capital contributed by the members. In practice, powerful district unions and apex bodies assumed much of the financial and managerial functions even in the "old" cooperatives. Centralisation was encouraged by commercial competition and by the dependence on government finance and supervision. An element of local self-management, however, was retained, in contrast to the more highly centralised system of the Farmers' Council.

The Council's critics held that if its "societies" applied true cooperative methods, much of the difficulties and opposition facing the organisation would be eliminated. The centralised buying organisation of the Farmers' Council was a direct continuation of the Cocoa Purchasing Company, although a certain cooperative flavour was added. The official status of the Council's "societies" was highly ambiguous. This reflected not only attempts to provide ideological window-dressing but also genuine divisions over the role of "cooperative methods" in the overall strategy.

In 1956, when the Purchasing Company was on the brink of collapse, the Working Committee of the Council suggested that the village farmer committees, which hitherto had played only a marginal role, should take over the buying of cocoa. Local control over operations would be ensured by requiring employees to submit property deeds as security against misappropriations.[126] This approach was abandoned when the Company was liquidated and the Council took over. But it was not done without opposition. The Brong-Ahafo and Volta Region members of the Board of Directors for the new organisation argued in favour of cooperative methods and independent societies. They also claimed that this line had the support of the National Delegates' Conference. The General Secretary and the Chief Marketing Officer, on the other hand, maintained that marketing should be controlled from the centre through a network of "inspectors". It was not necessary, according to another member of the Board, to "copy the cooperatives to the letter".[127]

There were important advantages to being recognised as a cooperative organisation, including tax exemption, free auditing, and access to central government funds.[128] At first, the Council was refused recognition.[129] When the Registrar of Cooperative Societies later was compelled to register the Council, only the apex body, not individual societies, was accepted as a cooperative. Arrangements were made, according to the Council, to have the "branch societies" registered in the nearest future.[130] A condition, however, was that they were reformed and established on an "autonomous" basis. The Registrar reported that this was the "avowed goal" of the Council.[131] The gap between public ideology and actual policy was wide. According to the minutes of the Commercial Committee, "the idea of creating autonomous societies was discarded".[132] The 1959 annual report, on the other hand, declared that cooperative methods had been proved "beyond any reasonable doubt as the most appropriate and the only safe step" that was required in the struggle for the well-being of the rural population.[133]

In the name of the farmers and under the cloak of a cooperative ideology the Farmers' Council advanced rapidly in the direction of a centralised bureaucratic monopoly. Attempts by the Registrar of Cooperatives to reorganise the local branches into cooperative societies were abandoned as the monopoly plans matured. The Registrar's department and the Ministry of Labour and Cooperatives actively resisted the monopoly plans. Both were abolished as the Council's victory was consolidated.[134]

The illusion that farmers were to exercise direct control over local trading operations was still encouraged by the Council's leaders. The General Secretary painted a rosy picture of cooperative democracy when explaining the new scheme in a radio talk:

> He (the farmer) has officers elected by him to take care of his interest and a secretary controlled by him to weigh his cocoa. The profit that accrues from the selling of his cocoa shall be used in providing for him social amenities recommended by him in conjunction with other farmers in his local society.[135]

In practice, no local control or "dual command" was permitted. The farmers' "profits" were centralised and reserved for the construction of cocoa sheds.[136] "Shares" were sold to the farmers at one point to mark the cooperative nature of the organisation, but their function within a system of centralised financial control was never explained.[137]

A National Cooperative Council had been established in 1959 as a step in the attempt to ensure government and party control over the cooperative movement as a whole, at that time still dominated by the "independent" cocoa-marketing unions and their various apex bodies.[138] The colonially-trained officials of the Department of Cooperatives were not considered suitable as the instrument of this new political leadership. The rivalry between the established civil service "cooperator" and the aspiring party cadres was clearly seen. The Department and the new Cooperative Council struggled over who should inspect and control the cooperatives.[139] The abolition of the Department and the office of the Registrar of Cooperative Societies in 1961 left the National Cooperative Council as the official instrument for guiding the "movement". But as the Farmers' Council absorbed all the cocoa unions, little was left for the NCC. It never succeeded in enforcing its formal leadership over the dominant section of the movement controlled by the UGFC. According to one report, the Farmers' Council withdrew from the central organisation in early 1962.[140]

The problem of how to organise the cooperative sector continued to preoccupy the government. A slight shift towards greater ministerial control took place by early 1962. The National Cooperative Council was abolished and replaced by a Central Cooperative Council with less staff and no direct government subsidy. It depended on voluntary contributions from its member organisations.[141] The office of the Registrar of Cooperative Societies was reestablished, although with much less power and resources. Ministerial responsibility was allocated to the Minister of Labour and Social Welfare.[142] The formal authority of these new organs vis-à-vis the Farmers' Council was not clearly spelt out. In practice, they were much too weak to exercise any real influence on the Council without strong government backing. Nor was the latter forthcoming. The new ministerial organisation nevertheless became a platform for the mounting official criticism directed against the Council.

The first "Progress Report" produced by the new Minister did not mention the Farmers' Council by name. But the criticism was implicit, as also noted by the Council in its comments.[143] The report attacked the tendency to build up central organisations at the expense of the local units. It emphasised the importance of developing "strong autonomous primary societies". The centralisation of control "was unconstitutional and contrary to the democratic principles of the Cooperative Movement".[144] The Minister was asked by the Cabinet to develop his plans for the reorganisation of the cooperative movement in consultation with the Farmers' Council. In a memorandum he outlined a new central organisation, a "Ghana Cooperative Authority", which would include the Farmers' Council. He emphasised in particular the need to allow the societies themselves to use their profits for local development purposes.[145]

The official response to the Minister's proposals was prepared by the Financial Adviser of the Council. He stressed that there was no need for independent capital formation and finance at the society level as the Council "provides its associated Societies with initial capital equipment and with the funds necessary for marketing". This central financial "assistance" was in turn used as the major argument why the Council was "obliged to exercise certain controls over the societies".[146] In a separate memorandum, the spokesman of the Council maintained that the Minister's proposals were derived from "the Western concept of cooperation rather than from that prevailing in a socialistic economy", where the central organisation was said to play a much greater role. He declared that the Council would resist any attempt to impose a new, Ministry-controlled cooperative structure.[147]

The Council was in a strong position to reject or ignore the Minister's criticism. The first year of monopoly buying had been successful. The major disruption in the trade, predicted by many critics at the time of the take-over, had in fact not occurred. Quality had improved and the leadership had used its new monopoly position to "rally" the farmers behind the compulsory savings scheme. The 1962 National Delegates' Conference was inaugurated by a highly appreciative address by Nkrumah, who confessed that he himself had at first had his misgivings and doubts.[148] In May 1962, a well organised demonstration of support for the Farmers' Council was staged in the National Assembly, which passed a motion declaring that "this house views with satisfaction the work of the UGFC".[149] The speeches on this occasion contained abundant praise. Even the lonely spokesman of the opposition United Party joined in and added only a minor remark about the shortage of cocoa bags in some areas.[150]

But an increasing flow of reports and peititions about irregularities and malpractices had begun to reach M.P.s, government and party leaders. During the 1963 budget debate in the National Assembly a number of speakers—mostly CPP backbenchers—attacked the Farmers' Council.[151] The government became more willing to listen to the ideas of the cooperative reformers on society autonomy and cooperative democracy. Growing difficulties in the general field of agricultural policy was an additional reason for a closer look at the way the Council managed the cooperative sector. The country was facing a major crisis: the non-agricultural sector expanded strongly, while local food production as well as the food-importing capacity of the economy stagnated. The Farmers' Council had been entrusted with major responsibilities in the expansion of local production. The number of cooperative "societies" multiplied but their impact on production was hardly noticeable.[152]

The Council's approach to cooperation was increasingly questioned. A special meeting of the Cabinet in April 1964 decided that the cooperative "movement" was in need of a general overhaul. A committee was set up to examine the Farmers' Council and to propose reforms "with a view to the establishment of cooperative societies in the true sense". E.N. Omaboe, the Director of the Central Bureau of Statistics, was appointed chairman.[153]

The report of the Omaboe Committee came out strongly in favour of the approach advocated by the "cooperative establishment" in the ministries. An urgent review was to be undertaken of all cooperatives. The office of the Registrar should be provided with "sufficient powers" for registration, organisation and control. Individuals should be encouraged to purchase shares and management committees should be drawn from these share-holding members. They should prepare their own budgets and bye-laws, although these would have to be approved by the Farmers' Council.[154] The conclusions were supported by a report submitted by the Minister of Agriculture.[155] According to Council sources, the Cabinet proposed that the Council should be reorganised on the lines indicated by the committee. It was the view of the Cabinet, according to Appiah Danquah, that the Council was not a proper cooperative organisation "because of the absence of society autonomy".[156]

The Farmers' Council leaders saw the proposals for the reorganisation of the cooperative movement as a threat to the very existence of the organisation. If its local branches were to become "independent cooperative societies controlled by the Department of Cooperatives" they would be "cut away completely" which would "immediately force the UGFCC out of existence".[157] The Council's defence centred on the need to maintain centralisation as an instrument of political control. The argument was developed in a petition submitted to the President:

> The destruction of the Council shall bring many basic changes to the political and economic structure of the country. The Party will have to find a new method of keeping the country's farmers under control. . . . The autonomous societies will be free to take their own decisions on any issue and the troubles which the old cocoa cooperatives gave to our Party in the past will start again.[158]

The petition claimed that the reason why the Council had decided after independence to enter the cocoa marketing field was that this had been considered to be the best way of organising the farmers politically. To be efficient in this respect, the Council had to be able to play an important part in the economic life of the farmers. As far as "true cooperative methods" were concerned, the Council dodged the critics. Its own methods had been proven to be efficient from the commercial point of view:

If by the use of such methods the UGFCC could not claim to be a properly organised cooperative organisation, then we humbly request that the UGFCC and its activities be taken away from the sphere of cooperation . . .[159]

To pacify the cooperative purists the Council was willing to renounce its claims to cooperative status. It concentrated its defence on the question of political control *over* rather than *by* the farmers. This line of argument had been successful in warding off the major previous threat to the Council's supremacy in 1961.[160] Was it to be equally effective this time?

A round table conference was to be held in August 1964, where the Council would be able to put its case before the President.[161] There are no records of that meeting in the Council's files. For the ultimate version of the government's decision on the reorganisation of the cooperative movement we must rely on the Council's own account in the General Secretary's report to the National Delegates' Conference at Tamale, April 1965.[162] The formula of the Omaboe Committee had been watered down. It no longer seemed to pose a threat to the Council. The report spoke of management by a "Committee of representatives drawn from share-holding members", but only for the "National Apex Society", not for the village units. The Council added its own interpretation, for which it claimed to have the support of the new Ministry of Cooperatives. There would be no financial autonomy in the cocoa societies, as otherwise a majority would have to be closed down. This would not be in the interest of the farmers. The societies were therefore to remain "branches" of the Farmers' Council.[163]

The attack was averted, at least for the time being. It is possible that the Council misreported or misinterpreted the intentions of the government. But it is also quite probable that the Cabinet decision on the Omaboe Report was sufficiently ambiguous to permit different interpretations. The emphasis on "true cooperative practices" may have been thought laudable in itself but not necessarily a matter of high priority to the government. Its relevance as a solution to the concrete problems facing the monopoly was not much elaborated in the report. The official agitation for cooperative methods could even be regarded as another move in a continual tug-of-war between civil servants and party functionaries over who should exercise control over the cooperatives. In that context, the "society autonomy" approach served to enhance the power of the former, while the "branch" approach worked in favour of the latter.

The Council had sought to mobilise support from the party for its own position. The "Farmers' and Cooperative Organisations Committee" of the

CPP met in March 1965 under the chairmanship of Appiah Danquah to discuss and adopt a new "cooperative policy". The document spoke of the need to use cooperative methods "in all walks of life". Most importantly for the Council's purposes, it included branches of apex bodies as well as local societies in its definition of cooperatives. It was accepted that the Ministry of Cooperatives shall "be permitted to deal with all cooperative organisations".[164]

Such a Ministry had been established in February, 1965. The office of the Registrar of Cooperative Societies was strengthened.[165] It was the most concrete outcome of the Omaboe Report and as such a partial victory for the cooperative establishment. The award of separate ministerial status to the former Department of Cooperatives reflected the new importance attached to cooperation by the government. But there is no evidence of any sustained attempt to enforce ministerial authority within the domains of the Farmers' Council. The new Minister seemed primarily concerned with the reorganisation and expansion of the sector which was outside the Council's control. Emphasis was laid on consumer cooperatives as the major field of expansion. No mention was made of "society autonomy" and other controversial issues when the Minister introduced his new cooperative policy in the National Assembly in August, 1965.[166] The elaborate formal organisation suggested for the movement seems to have remained a paper construction. Judging from another speech by the Minister at Winneba, much importance was given to the need to strengthen the Central Cooperative Council as the principal instrument for guiding the movement. It indicated a "Party"—rather than a "civil service"—approach to the question of leadership. But again, the central organisation was primarily to deal with what was not already controlled by the Farmers' Council.[167]

The Minister's new plans and organisational designs were largely ignored by the Council. A new committee was set up, but it is not clear whether it produced any report or even met. Repeated summons to its meetings were sent with little effect not only to Appiah Danquah but also to the Ministers of Finance, Agriculture and Party Propaganda. After three unsuccessful summons it was announced that the Minister of Cooperatives might be compelled to submit recommendations to the Cabinet "without further recourse to the committee".[168] The authority of the new Minister with his government and party collegues seems not to have been particularly great.

But public criticism of the Farmers' Council was mounting. In September 1965, the National Assembly discussed a motion congratulating the cocoa farmers on their decision to accept a drastic cut in the producer price. The Supreme Council of the UGFC had met with the President and agreed to the reduction.[169] No one in the Assembly questioned the cut as such but there was little praise for the Farmers' Council. A number of grievances were raised. The member chosen to second the motion set the tone of the debate by speaking of unscrupulous Secretary-Receivers "trying to get money overnight, no matter how they get it". "The way some of them are treating the farmers is unbearable".[170] The misdeeds of Secretary-Receivers became a major theme of the debate, although hardly in line with the official intentions of the motion. They were accused of being grafty, pompous and disrespectful.[171] They duped and swindled the farmers. Only few were honest party workers. Most looked only to their own selfish ends. "Some of these Secretary-Receivers are even more than our own enemies whom we are trying to eliminate from our society".[172] More specific allegations were raised about false weighing, delays in payments, "horning" of cocoa bags, absence of accounts, and profiteering in connection with the distribution of cutlasses.

The political structure of the Council was attacked. One member maintained that farmers were prevented from voicing their grievances. He mentioned that criticism had been supressed by local dignitaries in his home area, in order to avoid giving the General Secretary a bad impression of the region.[173] The suppression of the formal democratic process was brought up by another speaker:

> I am , therefore, suggesting that the Minister should see that the offices of the UGFCC are made elective. How can we appoint a regional chairman of the UGFCC and say that he is a life-chairman? How can this be when possibly the members of the cooperative concerned might have lost confidence in that person?[174]

Neither the Speaker of the House nor the party leaders present made any major effort to prevent the disgression of the members into matters which easily could have been branded as irrelevant to the motion. The Council was defended by its General Secretary, himself a member of the Assembly, at this time but he received little support from the others. The Leader of the House, in winding up the debate, advised the Council to take note of the various points raised.[175]

The debate in the National Assembly reflected a new concern about the unfavourable reports received on the Farmers' Council from the constituencies. Their numbers may not necessarily have increased but they had acquired a new urgency in the light of the general deterioration of the political situation in the cocoa areas. It is doubtful whether much good will remained after the compulsory savings scheme of 1961 but whatever was left was threatened by the 1965 producer price decision. These successive cuts in combination with shortages and sharply rising consumer prices compelled the government to look more carefully at its representation among the cocoa farmers. It was increasingly important to eliminate such other sources of friction and discontent which might aggravate the government's difficulties with the farmers. The malpractices of the Farmers' Council were in this context a major liability.

It is against this background we may see the government's decision to press for fundamental changes in the political organisation of the monopoly. This time it was no longer a matter for internal reports and committees. The Council was confronted in public. The official silence was broken by the *Budget Statement* of February 1966. The Minister of Finance noted that "for some time now, the UGFCC has enjoyed a position which has been considered almost sacrosant". This was no longer so. The Council, according the Minister, had become a target of constant criticism.[176] The statement has already been quoted in length on the first page of this chapter. It remains to summarise and discuss its position in the context of earlier attempted reforms and the political requirement of the CPP government.

The Minister attacked excessive centralisation, the top-heavy bureaucracy, and the absence of democratic participation by the farmers. He emphasised the need to build a "genuine cooperative movement with popular democratic base". This, however, was not the demand for "true cooperative methods" of earlier reformers. Their proposals had centred around concepts of "society autonomy" and "control by share-holding members". They had their main basis in the "cooperative establishment" of the old Department of Cooperatives, with its background in the colonial administration. Reference to these principles were notably absent in the Minister's statement. The key concept was instead "democratic centralism". The true cooperative methods of the civil servants were primarily a matter of correct economic administration, including separate accounts at society level and individual financial contributions and incentives. The focus was on the commercial aspects of the administration. The emphasis of the *Budget Statement* on democratic centralism and the democratic election of Chief Farmers, on the other hand, concerned problems of power and political control. There was also a difference in the unit of interest. Earlier reformers concentrated their attention on the societies. They were concerned above all with the need to strengthen the separate commercial identity of the latter. The emphasis now shifted to the organisation as a whole and the need to secure popularly based leadership at all levels.

The Farmers' Council was successful in warding off earlier attempted reforms largely because these did not take into account the basic political functions of the Council. "True cooperative practices" may have been useful as a means to improve the commercial administration of the Council. Their relevance for the political mobilisation of the cocoa farmers behind the Government was more obscure, and certainly not spelt out by the reformers. It even seems as if the latter, in their emphasis on share-holding as the basis for finance and management at society level, ignored the political conditions produced by the monopoly position of the Council. It was never explained how such a system was to be made compatible with compulsory membership and the absence of alternative market outlets.

The formula of democratic centralism represented an attempt to combine genuine participation by the farmers in the commercial administration of the monopoly with a strong, popularly based central leadership capable of guiding the farmers at the level of national politics. But could democratic centralism in this sense provide a solution to the fundamental contradiction between the commercial and political functions of the monopoly? Could the Farmers' Council serve simultaneously as an instrument for the farmers' self-management of the trade and as an agent of state appropriation of cocoa income?

The failure of participation and control: concluding remarks

The suspension and manipulation of elections, the incorporation of elected cadres in the bureaucracy and their subordination to the latter were all features which conflicted with the official position of the Council as a farmers' organisation. But they were not mere aberrations caused by excessive "personal rule" and the usurpation of power by the bureaucracy—not uncommon even in organisations with a formal democratic process. The absence of any significant degree of farmer control over the cocoa trade was directly related to the essentially repressive function which the Council performed in relation to the distribution of cocoa income. The Council was designed to ensure the farmers' support for policies which implied a heavy tax on their income for the purposes of national development.

This was indeed a difficult task, as most farmers could be expected to be anxious to protect their own income. The next chapter deals more specifically with the Council's methods of "mobilising" the farmers for this end. Was it possible to create within one single organisation a system of representation which ensured farmers' control over the cocoa trade at the same time as it prevented them from voicing their views on the crucial question of the price received for their products? The need to exercise control *over* the farmers seems to have left little scope for any control *by* the farmers in the solution offered by the Council.

Officially the illusion of the Council as an organisation managed by the farmers was maintained. The question of who the Council leaders really represented in the absence of effective farmer participation was dodged. If pressed on this point, party leaders may have answered, as Kofi Baako did in 1961, that the Council, in line with the other "wing" organisations, was an integral part of the party, that it represented the party and was controlled by the party and that party and state were one.[177] The Council's own records show no evidence of effective external control either by the central government or by the CPP. Attempts to establish some sort of ministerial control were halfhearted and aimed primarily at strengtheing the "autonomy" of trading operations at the local level. There was no current ministerial control of the type exercised over public corporations, such as the Cocoa Marketing Board, where ministries were represented on the Board and where major decisions had to be approved by them. The supervision exercised by the Auditor-General's Department covered only limited aspects of the Council's activities and was no substitute for central political control.

It may be argued that the General Secretary of the Council was himself a minister and that this took care of the need for government control. In the case of public corporations, such supervision and guidance was exercised by separate organs of the state with separate staff and leadership. In the Council's case the Minister and the "Managing Director" were the same person. How was one to supervise the other? Nor did the fact that the Council was the sole licensed buying agent of the CMB provide for any external control from that direction. The Council was represented on the Board, not the other way round, and it was not accountable to the Board.

What about party control? The files containing the correspondence between the Farmers' Council and the CPP do not indicate that the party leaders saw it as their task to supervise and give directives to the Council, although there were cases when the party intervened into internal disputes, especially at lower levels.[178] There was no central party involvement in the overall organisation of the Council. The "integration" with the party was no doubt real in, for example, the overlapping of membership on party and Council committees. Attempts were made in 1961 to formalise this integration. Separate membership cards for the "wing" organisations were abolished.[179] Joint periodical meetings were to be held by senior staff.[180] The regional officers of the Council were designated assistant regional secretaries of the party.[181] But it is hard to discern any tangible consequences of this type of formal integration, at least not in terms of increased party control over the administration of the monopoly. The elaborate network of coordinating committees at the CPP headquarters was largely inoperative.[182] The National Cooperative Council and later, the Central Cooperative Council were conceived as instruments of party control but were never able to extend their leadership to the sector managed by the Farmers' Council. We are faced with the picture of an organisation either left to manage its own affairs or, when occasionally challenged, able to fend off outside interference.

I have explained the weakness of the internal democracy of the Farmers' Council as partly caused by the political function of the trading monopoly as an instrument of the fiscal policy of the government. How do we explain the absence of government and party control? One might have expected that the primacy awarded to this political function should have enhanced the demand for external political control. Instead we find a combination of weak participation from below and weak control from above. Wide margins were created for the emancipation of and private exploitation by the class of bureaucratic middlemen entrusted with the administration of the monopoly. As a result its performance was increasingly brought in conflict with the basic aspirations of the national government, which was concerned with mobilising farmer support or at least to neutralise their opposition to the appropriation of cocoa income by the state. In explaining this apparent contradiction it is not sufficient to point at the self-deception of the CPP leaders and that they deluded themselves about the nature of the Council as an organisation controlled by the farmers. I think it is necessary to seek the explanation in the way in which the class aspirations of the rising bureaucratic petty-bourgeoisie and the national developmental aspirations of the CPP leaders interacted and reinforced each other. We saw this convergence of sectional and national interests in the rise of the Council and in the victory of its monopoly plans. These were victorious because they seemed to solve the primary contradiction between farmer interests and the requirements of national development to the advantage of the latter. But this national solution incorporated and intensified a secondary contradiction, that between farmers and traders, both anxious to protect or expand their respective share of the commercial surplus. This secondary contradiction was solved largely to the benefit of the new middlemen at the expense of the farmers. As a result, the attempt to find a "national" solution to the primary contradiction was seriously compromised. It is time to turn to the accumulation of cocoa income by the state and the role of the Farmers' Council in "mobilising" the farmers for national development.

VII. Cocoa and the Public Economy

The commercial and political monopoly of the United Ghana Farmers' Council must be understood in relation to its role as an instrument of public finance. The CPP state was largely based—directly or indirectly—on the surplus appropriated from the cocoa economy. The government sought to ensure a political basis for these appropriations by controlling the farmers politically through the Farmers' Council. By making the Council the monopoly buyer of Ghana's cocoa it attempted to create a material platform for this effort to organise the farmers behind the government and the party. It was essentially an exercise in political demobilisation: an attempt to neutralise or to block the political activities of groups actually or potentially opposed to government policies. The methods used in this exercise give important information about the character of the regime. The focus in this chapter is no longer on the Farmers' Council itself, but on the producer-price and tax policies of the government. These policies are discussed in the context of the developing public economy and changing market conjunctures. I examine briefly the impact on producer income and, more tentatively, on production and national development. The political contribution of the Farmers' Council is discussed in relation to the massive public appropriations of the 1950's, and the subsequent efforts by the CPP government to protect its financial base in the face of falling market prices. As prices fell, the political costs of maintaining public cocoa income rose.

The first section below examines the conjunctures of the public economy as affected by the development of cocoa production and market prices. During the first post-war decade, extremely favourable conditions were created for the appropriation of cocoa income by the government. The basis was laid for the dramatic expansion of the public sector during the following decade. After independence, however, market conditions deteriorated. The continued expansion or even the maintenance of the public sector were threatened. I discuss below why the surplus available for appropriation radically declined, despite the strong increase in production. This "overexpansion" of production was largely encouraged by the government itself. A belated attempt to unite the world's cocoa producing nations in order to protect prices failed miserably.

Since the introduction of the marketing-board system, the decisions on the cocoa producer price are among the most important recurrent political decisions taken by the Ghanaian government.[1] These decisions, in combination with the development of market prices and production, determine the distribution of cocoa income between the producers and the state. They have important consequences for the level and direction of economic activity in the economy as a whole. Much of this chapter is an account of the cocoa tax and producer-price policies of the government. Four sections each cover one period or phase in the development of these policies. One covers the pre-independence period and, in particular, the Cocoa Ordinance of 1954, which sought to peg the producer price for a longer period, while at the same time raising the level of public appropriations. Another section examines the early post-independence years. It focuses on the "voluntary contributions" collected from the cocoa farmers under the Second Development Plan which was launched in 1959. The following section deals with the controversial 1961 budget and the attempt to step up public development in spite of growing financial difficulties. Cocoa farmers were asked to make a major contribution to the "compulsory savings" scheme introduced in that budget. These "savings" were later replaced with a new cocoa tax. The last section of this review of producer-price and tax policies brings us to the acute crisis of the public economy (and the cocoa economy) in 1964/65, greatly aggravated by the bumper cocoa harvest. The producer price was drastically reduced in 1965.

The Farmers' Council participated directly in most of these price (and tax) decisions, as the official representatives of the farmers. The Council canvassed in support of the government line. It offered the "voluntary contributions" on behalf of the farmers in 1959, and the "compulsory savings" in 1961/62. On other occasions, new producer prices were negotiated between the Council and the government. The Council's ability to represent the farmers and take decisions on their behalf was, however, widely disputed. An increasing number of CPP backbenchers also joined the critics. The methods used by the Council to rally the farmers behind the price and tax decisions carried little conviction. The Council's system of representation and the contradictions between the farmers and the new middlemen added to the picture of fictitious mobilisation. The growing contradictions were also reflected in the widening gap between the farmers' income and the salaries of state employees.

In the final section I discuss tentatively the wider consequences of public cocoa appropriations for national development. I summarise the picture of an emerging, dominant public sector economy. What were the implications for cocoa production and agricultural development in general? In what ways were the foreign capital interests in the cocoa economy affected? What did it mean in terms of economic independence, the key objective of CPP strategy? I discuss the redistribution of economic opportunities and the development of class forces brought about by the state appropriations.

Statistics on the development of cocoa income and the public economy are included in Appendix A. It gives the basic figures on which much of this chapter is based. The tables are partly summarised in the diagrams included in the text but on the whole I have avoided breaking up the presentation with frequent references to the Appendix. The readers are therefore advised to take an early look at how the tables are organised in order to facilitate such references which they themselves may wish to make in the course of reading.

One concept used in these tables and frequently in the text, as well, may require a special note. I speak of *public cocoa income* as comprising not only government cocoa revenue but also the net surplus of the Cocoa Marketing Board. The reason is that the CMB reserves, as we shall see, were largely used to finance the central government, although originally intended to stabilise producer prices. The inclusion of the fresh additions to these reserves therefore gives a better measure of the overall resources made available to the public economy than the actual payments to the government in a particular year.

The conjunctures of public cocoa income[2]

Exceptional market conditions made it possible for the Gold Coast colonial state to appropriate cocoa income on a massive scale in the decade following the Second World War. Foundations were laid for rapid expansion during the following decade of a state sector under nationalist control. Expansion was brought to an halt with the collapse of the CPP government in 1966, but the capacity of the cocoa economy to sustain the high rates of advance were preempted well before that date.

The stagnation and decline of luxury consumption in Europe and America in the 1930's and during the Second World War caused a major depression in the cocoa-producing economies of West Africa. Little fresh investment, poor maintenance of existing farms, and the spread of the swollen-shoot disease combined to hold back production. In Ghana, it declined from an average of some 270,000 tons in the late 1930's to 225,000 in the second half of the 40's.[3] Production remained at this level for most of the following decade. (Appendix, Table 1.) This depressed production was entirely out of tune with the rapid recovery of demand in Europe and America, once the war and the period of enforced austerity was over. There was sharp competition between manufacturers for the meagre supply of beans. The result was a strong upward pressure on prices and extreme instability, with manufacturers' stocks forced to bottom levels. Heavy increases in the cost of imports reduced the real value of the augmented export earnings but the average selling price obtained by Ghana in the first half of the 1950's was still in real terms more than two and a half times what was received in the late 30's.[4] The development of production and market prices during the CPP period is summarised in the diagram, Fig. 1.

The sharp increase in market prices, not the growth of Ghana's cocoa earnings as such, created the favourable conditions for raising public income. The state share grew dramatically, while allowing simultaneously for increases in producer prices. During the peak years of state accumulation producers were paid prices which were two or three times those received

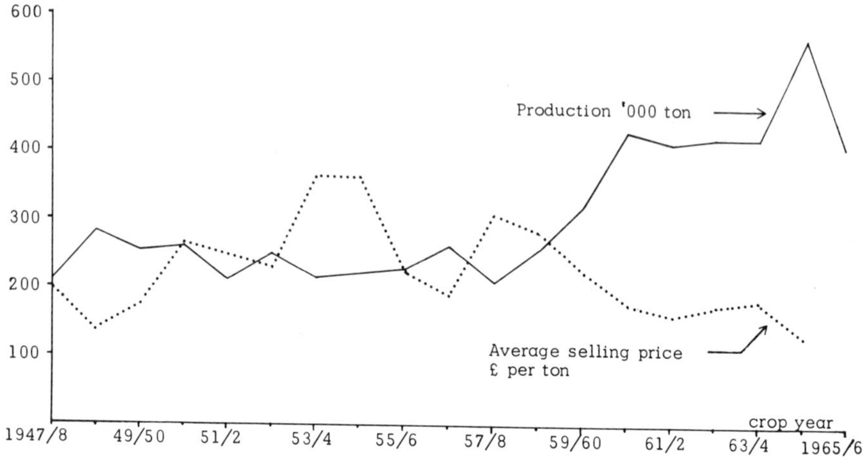

FIGURE 1. GHANA: COCOA PRODUCTION AND MARKET PRICES
(Source: See Appendix A, Table 1.)

before the war, if the Accra retail price index is used as a measure of changes in real value.[5] Violent swings in market prices facilitated the expansion of public income. They were used to justify a policy of stabilising producer prices at "safe" levels.[6]

The marketing-board system was a powerful instrument for socialising cocoa income. Government revenue from cocoa during the inter-war period was only equal to five per cent of sales.[7] Between 1947, the first year of the new Board, and 1965, government collected by means of export duties and other levies almost one-third of the total value. When the net surplus of the Marketing Board is included the public share was as over 40 per cent. For the development of sales and public cocoa income, see Fig. 2, below, and Tables 1 and 2 in the Appendix.

State control over Ghana's cocoa earnings began during the war and the Cocoa Marketing Board inherited a surplus of £14 million. In the period before the nationalists were allowed to participate in government in 1951, the Board had already accumulated reserves of as much as £71 million.[8] During these first four years the public share was already over 40 per cent of the market value. The CMB reserves accounted for the bulk of these collections while direct government revenue was still quite modest. A major part of the reserves had been invested in British government securities, contributing to the post-war reconstruction of the metropolitan economy.[9] But the direct government share grew fast, from a mere one per cent in 1947/48 to almost one-fifth of the value of the 1950/51 crop. (Appendix, Table 2)

As the nationalists were permitted to join the government in 1951, the accumulation of reserves slowed down. The peak war reached in 1954/55, when they stood at £87 million.[10] The shift towards direct payments to the

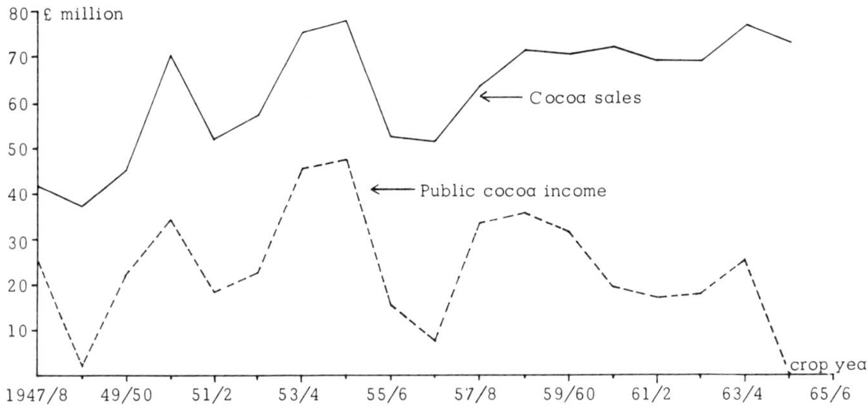

FIGURE 2. COCOA SALES AND PUBLIC COCOA INCOME
(Source: See Appendix A, Tables 1 & 2.)

government accelerated. In the peak year such receipts equalled exactly half the total sales value of Ghana's cocoa. Together with the surplus of the Board the public share at this point reached 60 per cent.

High market prices during these years and the large share retained by the government and the Marketing Board provided the foundations for the dramatic expansion of the public sector associated with the CPP regime. As early as the second half of the 1940's government expenditure had increased by some 50 per cent in real terms above the level of the late 30's.[11] Public expansion accelerated in the following decade as can be seen from Fig. 3, below. (See also Appendix, Table 3.) In real terms, average government expenditure in the 1950–54 period was probably three and a half times that of the late 30's and almost two and a half times that of the immediately preceding five years. But government revenue, which can be studied in the same graph, grew even faster. By the first half of the 1950's it was four times the 1935–39 figures and almost three times that of the late 40's.[12] A large budgetary surplus was piling up. The average surplus of the first half of the 1950's was equal to one-third of total government expenditure. But more than half of the total surplus was obtained in one single year, 1954, the year of the controversial Cocoa Ordinance and the rise of the "cocoa opposition". That year government in fact spent only 58 per cent of its revenue.[13] That this public sector boom was essentially a cocoa boom stands out clearly when it is considered that total government receipts in the 1950–54 period, that is, current tax revenue and grants as well as capital transfers from the Marketing Board, were equal to 80 per cent of government expenditure.[14]

The cocoa price boom of the first post-war decade was largely a result of the disrupting impact of the war. But it was also influenced by the dynamics of

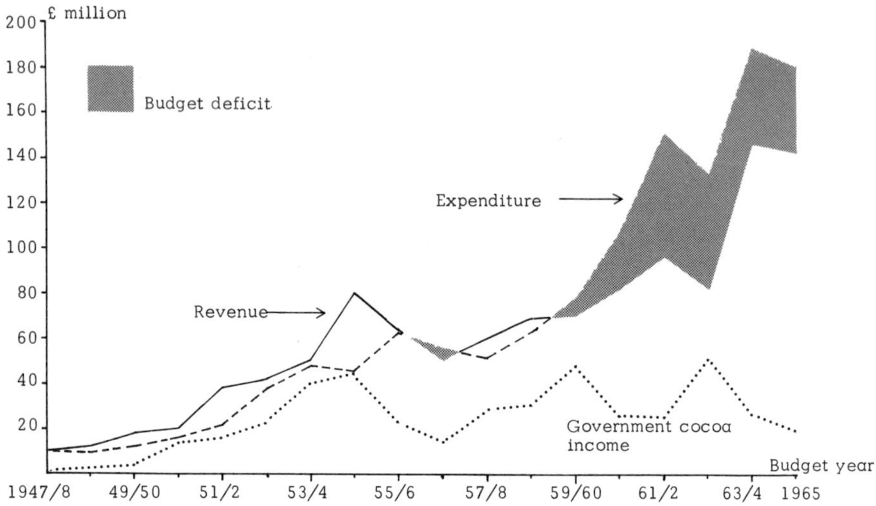

FIGURE 3. GOVERNMENT REVENUE, EXPENDITURE AND COCOA INCOME
(Source: See Appendix A, Table 3.)

tree crop production with its low short-term "elasticity of supply", which meant that it took a long time for output to expand in response to favourable prices. It takes five to ten years after planting before the cocoa tree yields a significant harvest and a full yield may take much longer. A survey from the early 1950's suggested that full capacity was on an average only reached about the 18th year after planting.[15] (New varieties have subsequently shortened the period of maturation.) Low elasticity in this sense also meant that once fresh planting had been undertaken capacity continued to expand even when the initial price incentives were no longer at hand.

This is also what happened in Ghana. The response of production to the good prices of the first post-war decade caused not only the termination of the price boom but also its reversal into the deep depression of the 1960's. New planting in combination with heavy public investment in the control of crop diseases and other services to the farmers had a dramatic impact on output around 1960.[16] Production jumped to a new level of about 400,000 tons. Market prices, on the other hand, had already begun to drop, and became even further depressed under the heavy weight of Ghana's new production. Average market prices in the second half of the 1950's were down by 17 per cent from those of the first half, partly as a result of manufacturers accommodating to the slow growth of primary production, partly because of production increases elsewhere.[17] In the 1959/60 season, Ghana's production expanded by 24 per cent, and in the following season by another 36 per cent. Prices obtained at the same time fell by 20 and 22 per cent, respectively. They declined by another 10 per cent in the following season. For four seasons production stabilised at the new level until a new 36 per cent rise was achieved in 1964/65. The value of the increase in output was eliminated by the accompanying fall in prices.

During the first half of the 1950's, an average crop of 230,000 tons was sold by the Marketing Board for £66 million. A decade later, during the first half of the 60's, an average crop of 451,000 tons, that is, almost twice as large, brought in £72 million, an increase of less than 10 per cent. Comparing the extremes, we find that Ghana was paid £78 million for 220,000 tons in 1954/55 but only £71 million for the 572,000 tons produced ten years later.

Production doubled and prices were cut by half. Although total national cocoa earnings were largely the same, conditions for maintaining public cocoa income had radically changed. The cocoa boom of the early 1950's put large resources to the disposal of the nationalist government. Accumulated funds continued to sustain the expansion of the public sector for some time but its ultimate basis in the cocoa economy was undermined. Public expenditure in the second half of the 1950's was 64 per cent above that of the first half or even 80 per cent, if the revised estimates presented by G. B. Kay are used.[18] But the share of cocoa receipts, current revenue as well as capital transfers, in meeting this expenditure had already declined sharply.[19] Expenditure had caught up on revenue and there was an average budgetary deficit of about 10 per cent according to the revised estimates.[20] These deficits grew as expenditure continued to rise in the early 60's and revenue stagnated or declined. Average deficits for the first half of the decade including the 1965 budget year equalled more than one quarter of total expenditure, or almost £40 million. Total cocoa receipts, although roughly of the same magnitude as in the previous decade, were now only equal to one-fifth of government expenditure. The major new source of revenue, import taxes, was nevertheless also closely tied to the fortunes of the export economy. The growing deficits were at first largely financed from the government's own external reserves but increasingly by way of monetary expansion and short-term foreign borrowing.[21]

The government sought to protect its share of cocoa income by drawing heavily on the CMB reserves and by successive cuts in the prices paid to the producers. It succeeded in this way in stabilising its current revenue from cocoa at about £25 million per annum until this strategy was finally wrecked by the 1964/65 bumper crop. The share of these receipts in the total value of cocoa exports remained in fact remarkably stable, only declining from 35 per cent in the 1950–54 period to 33 per cent for the first half of the 1960's. But the Marketing Board had difficulties in finding liquid funds, and by 1965, they were virtually exhausted.[22]

The maintenance of current tax payments by the Board concealed the sharp fall in the public share of cocoa income. This was reflected in the large deficits in the Board's current accounts. Public cocoa income, net of these internal transfers, fell from an average of £33 million per annum in the first half of the 50's to £24 million in the second half and to £15 million in the first half of the

60's. While import prices had remained fairly stable throughout the period, the internal purchasing power of these receipts were undermined by domestic price inflation in the 60's.[23] The public share of cocoa income declined from over half to about one fifth, looking at period averages. *In 1964/65 the net contribution of cocoa to the public economy was even negative!* The deficit of the Marketing Board in that year was larger than its payments to the government.[24] Payments to the producers and marketing costs were sufficient to consume current export receipts. Only a transfer of public funds already collected could be effectuated. But this was a paper transaction which did not place any fresh resources at the disposal of the government. The Board's current tax payments were merely cancelled against reductions in its holdings of government bonds. The transaction illustrates well the depth of the financial crisis.

The Marketing Board reserves, £87 million in 1954, stood at £14 milliion by the end of the 1964/65 season.[25] The cumulative surplus for the full 1947–65 period was £128 million, some 40 per cent of which was absorbed by current deficits. Almost one quarter was spent on the control of crop diseases and rehabilitation. Other major expenditures against the reserves included development grants and scholarships.[26] But the reserves served above all as a source of credit to the government. At first, the British metropolitan government was the principal beneficiary but after 1951 these resources were gradually redirected to the local government. Loans were granted for specific projects, such as harbours and railways or as general support for public development, such as the £16 million loan for the Second Plan. The bulk of these capital flows occurred during the late 50's and early 60's. In 1962/63 most of the Board's holdings of U.K. securities were exchanged for Ghana government bonds.[27] The reserves were in this way placed at the disposal of the government. Their ability to support current tax payments was nevertheless consequently exhausted.

Why did public cocoa income decline so strongly while national cocoa income in fact merely stagnated? Considering the new ambitious development plans and the accelerated growth of public expenditure one might have expected an opposite trend, especially in view of the greater political control established after independence. In order to maintain the level of cocoa income achieved in the first half of the 1950's it would have been necessary to let producer prices follow export prices all the way down. This was not done, although producer prices were subjected to a number of successive cuts. In 1957, after independence had been secured, a first ten per cent cut brought the price down from 80 to 72 shillings per load (one ton = 37.3 loads; one load = 60 lbs.). (Appendix, Table 4.) In 1959, the farmers' "voluntary contribution" reduced the effective producer price by 17 per cent to 60 shillings per load. In 1961, the compulsory savings scheme took another 10 per cent off the price, leaving it at 54 shillings for the subsequent four years until a final 26 per cent cut in 1965 forced it down to 40 shillings. Only this last cut was of a magnitude equal to the drop in export prices. Up to that point, the share of the producers in total income had grown from an average of 47 per cent in the first half of the 1950's to 56 per cent in the second and to 65 per cent in the first half of the 60's.

In the context of expanding production this relative "improvement" in total producers' income brought little satisfaction to individual farmers. They were concerned with the return for the labour, land and capital which they individually had engaged in production, not with the gross receipts of the cocoa community as a whole. The increase in production meant that new resources had been committed and remuneration was expected. The cost of producing 400,000 tons of cocoa was not the same as for producing 200,000. The use of insecticides, largely subsidised by the government, contributed to the growth in output and to a reduction in average costs of production. But other factors worked in the opposite direction, especially the growing cost of labour, rising food prices, and price inflation generally. In addition, the total to be divided between the government and the producers was reduced by increases in the costs of marketing. These rose from an average of six per cent of the value of sales in the first half of the 1950's to some 14–15 per cent in the early 60's. (Appendix, Table 1.) In 1964/65, the bumper crop caused a sharp rise in marketing costs, claiming on this occasion perhaps as much as one-quarter of Ghana's total cocoa earnings.[28]

The cocoa economy as a whole could be said to be "compensated" for the fall in export prices by the increase in production. But the cost of producing and marketing the same *worth* of cocoa rose sharply, reducing the surplus available for appropriation by the government. The decline in public cocoa income reflected this basic change in the conditions set by the market for taxing the farmers. It is also against this background that the political costs of the successive cuts in producer prices must be evaluated.

The dynamics of the international cocoa market were largely beyond the regulative capacity of Ghanaian political institutions, Yet, the export monopoly, the control over producer prices, and the public investment in pest control and other services to the producers provided points of departure for a planned development of production and income. International cooperation between producer countries could also be geared to such purposes. The Ghana government seems to have been largely unprepared for the "over-expansion" of production and its disasterous effects on prices and its own income. Measures taken to restrict production and sales were belated and ineffectual in meeting the crisis which culminated during the 1964–65 season. A single-minded expansionist approach to production seems to have prevailed until the early 1960's, reflected in the first place in massive public support for the capsid spraying scheme of 1958.[29]

In 1960, when output passed the 300,000-tons mark and field reports pointed to a new large increase, the question was raised by the Ministry of Finance as to whether the government should continue its heavy investment in the cocoa industry in the face of falling prices. There was a need to plan future production and a coordinating committee was suggested for this purpose.[30] The Ministry of Trade held a different view, supported by the Ministry of Agriculture. Expenditure on disease control and agricultural extension work should be kept up. "Nothing must be done to damage the enthusiasm and energy with which this healthy evolution is unfolding itself."[31] The Ministry of Finance replied that "the energy and funds expended by the Ministry of Agriculture should not result merely in reduced incomes both for farmers and for the government, through paying insufficient attention to market conditions".[32] But the proponents of the existing policy insisted that production should be encouraged to the "highest possible level".[33] The Marketing Board welcomed the idea of a planning committee but added a fatalist note: "As you know cocoa is a tree crop and once the tree comes into bearing there is nothing one can do."[34]

Against the background of current public investment in the cocoa industry the declared unwillingness of the Ghanaian government to interfere with "the normal interaction of demand and supply" had a somewhat hollow ring.[35] Earlier that same year the government had turned down a proposal by Brazil for the coordination of the marketing operations of the cocoa-producing countries. It declared that Ghana did not want to enter into any formal agreement, as consumers could be expected to react sharply, which "might well be detrimental to our own best interests".[36]

The continued decline of market prices and the growing financial difficulties of the government encouraged rethinking on this point. By 1962, the Cocoa Producers' Alliance, an international organisation comprising the major producer countries, was activated.[37] The demand for an international cocoa agreement with price-stabilising and price-supporting functions was intensified. Preparatory studies and discussions had been going on for many years within the framework of the Cocoa Study Group of the UN Food and Agriculture Organisation.[38] A major UN conference was convened in Geneva in 1963 but the minimum prices suggested under the proposed agreement were unacceptable.[39] The producer countries decided in this situation to take action on their own. A series of meetings during 1963 and 1964, in the face of deteriorating prices, resulted in a decision in October 1964 to withdraw from the market as long as prices remained below an agreed minimum level of £190 per ton.[40] But manufacturers were well stocked as a result of series of large crops. The 1964–65 crop promised to be a record one. They were therefore prepared to wait. After a three-month, ineffective hold-up, the producers were obliged to reenter the market without having achieved any positive results.[41]

Here is not the place to discuss the failure of the Cocoa Producers' Alliance or the limits and possibilities of international actions by producers in the world cocoa economy. Unplanned expansion of production and the belated attempts to restrict and coordinate sales may go far to explain the failure on this occasion.[42] The abortive nature of the action, however, demonstrated how vulnerable the producers were to the "natural" dynamics of the market and how blunt were the weapons at their disposal. It was not possible to prevent these dynamics from precipitating a major crisis in the Ghanaian economy, a crisis which was aggravated by the hold-up itself.[43]

Producer prices and public development: the pre-independence experience

As the Cocoa Marketing Board was established on a permanent basis after the war, many farmers feared that it would be used to suppress prices in the interest of the manufacturers. The price ring of the foreign cocoa-buying firms was a fresh memory.[44] Farmers had been denied the fruits of the victory over the firms in the 1937–38 cocoa hold-up. The temporary restrictions on competitive marketing accepted by farmer leaders in the 1938 "truce" were continued in a more rigid form under the war-time marketing scheme.[45] The conflict was frozen. The collapse of the international market greatly reduced the bargaining strength of the farmers. As the end of the war and a recovery of the market was in sight agitation revived. The immediate target was the low price offered by the temporary marketing authorities but the proposed permanent scheme was also attacked. In September 1945, a cocoa-farmers' delegation went to London to register their protests.[46] In October the same year, representatives of various farmers' organisations met in Suhum to decide whether a new hold-up was to be organised in opposition to the producer price announced for the coming season. After intervention by the Standing Committee of the Joint Provincial Council of Chiefs the new price was accepted.[47]

But the market situation changed in favour of the producers. The high export prices made it possible for the marketing authorities to raise the producer price in real terms two or three times above post-war and war-time levels during the first few years after the war.[48] In this situation, the farmers' resistance to a continued public export monopoly seems to have subsided. Their resentment, as those of other cash earners at the time, was instead largely directed against shortages and inflation. The rapid spread and devastating effects of the swollen shoot and the intense controversy over the appropriate methods of controlling it contributed to drawing attention away from the new marketing scheme.[49]

In the volatile market situation of post war years it was difficult for producers and other local interests to obtain a clear picture of the much-advertised price-stabilising functions of the new marketing board. Much suspicion had already been aroused by the delay in transferring the surplus accumulated during the war from London to Accra.[50] Suspicion increased as the funds retained by the Board continued to grow at a spectacular rate.[51] It is possible that these very large reserves and their investment in British government securities were thought to be necessary for the purposes of long-term price stabilisation. It has been pointed out, on the other hand, that this policy was eminently suited to meet the financial and hard-currency requirements of the metropolitan economy.[52] Nevertheless, the local colonial government also began to make claims on the growing surplus. Rates of taxation were revised and by 1950/51 the Accra government collected one-fifth of the cocoa export earnings in duties.

Farmer opposition to the accumulation of cocoa earnings by the marketing board became part of the national grievances and aspirations which brought the CPP shared ministerial responsibilities in 1951. In Chapter 3, the attacks on the marketing-board system by the Ghana Farmers' Congress, the Council's predecessor, were noted.[53] The 1951 election manifesto of the CPP promised that the farmers would be given control over the funds of the Board. But the manifesto also outlined the massive development programme which the party wanted to launch if brought to power.[54] It required a high level of public income and fresh claims were likely to be made on funds already saved. Double track railways and branch lines were to be extended on a massive scale; pipe-borne water, electric lights, good housing, education and other social and cultural amenities were to be brought to towns and villages. The manifesto envisaged free and compulsory elementary, secondary and technical education up to the age of 16. Numerous high institutions for liberal, scientific, technical, and agricultural education were to be established as well as a network of recreational activities, stadiums, theatres, concert and dance halls, cinemas, community centres, holiday resorts, parks, swimming pools, playgrounds, libraries, museums, art galleries, and exhibition halls. There would be free medical service, family assistance, and public housing estates. It was a catalogue, in fact, of much of the public services offered by the advanced industrial states. "At no very distant date it should be possible to see water on the tap in most homes in the country." Mail would be delivered to the door.[55]

The party had also an economic plan. Its cornerstones were the modernisation of agriculture and the electrification and industrialisation of the country. The Volta hydro-electric scheme was attributed prime importance. Agriculture was to be progressivly mechanised. A countrywide network of silos and cold storage facilities and a modern transport system would ensure efficient marketing. Large scale irrigation projects were to be undertaken in arid areas. Industrialisation was a principal objective. A number of factories for consumer goods were envisaged, including canned food, electrical goods, building materials, machinery, cutlery, crockery, provisions, hardwares and textiles. Little was mentioned about the social and political organisation of production but the nationalist state was expected to play a leading role in promoting and financing development.

Theoretically, there was no contradiction between this huge public expenditure programme and the promise of farmers' control over the funds of the marketing board, to be used "for the benefit of the farmers primarily, and the country in general", at least not as long as the farmers were willing to accept the national goals as outlined in the manifesto as being in line with their own long-term interests.[56] The immediate concern of the farmers, however, was less with the administration of public capital resources than with the price received for their cocoa. The 1951 election was fought at a time when farmers for the second season in a row were receiving less than half the market value of their cocoa. They clearly expected more. In spite of strong professional advice pointing at the inflationary dangers, the new government decided to meet the farmers' demand for a larger share of the country's booming cocoa earnings. The 1951/52 cocoa price was raised to 80 shillings per load, a record level over three times immediately pre-war prices in real terms and about five times what had been received during the war.[57] The country escaped the feared inflationary consequences of the high producer price only because cocoa production dropped by almost a quarter in the 1951/52 season. Had the crop been of the same size as in the previous year, another £5 million would have been added to the purchasing power of the producers in an already "overheated" domestic market. The political expediency of the price increase was apparent. Once more firmly established, the new CPP ministers were more willing to take "professional advice" and the producer price was reduced to the pre-election level. The political difficulties of justifying this more cautious approach to the farmers were reflected in the problems met by the CPP in its attempt to develop the Ghana Farmers' Congress into a reliable party organisation.[58] Continued suspicion against the Marketing Board and its price policies combined with commercial rivalries to paralyse the Congress.

The Farmers' Council, established in 1953 primarily on the basis of the pro-CPP elements of the Congress, ran immediately into far greater difficulties, as world market prices began to soar to extraordinary heights while the CPP government, confirmed in a second general election, set out to "stabilise" producer prices at the existing level. The Cocoa Ordinance of 1954 and the rise of an organised cocoa opposition were important factors in shaping the early history of the Farmers' Council. The main features have already been outlined in Chapter 3.[59] It remains to examine the events within the context of fiscal policy and the related problems of political mobilisation.

The bulk of the pre-1951 state appropriations had been absorbed by the reserves of the Cocoa Marketing Board. Only in 1950/51 did a substantial part go directly to the government as export duties. In the following years these proportions shifted radically. Most public cocoa income now went to the government, and only little to the Board. As long as the Board was the main recipient it was possible to argue that the wide gap between producer prices and market prices primarily served the purpose of price stabilisation in expectation of future bad days. The amount spent by the Board on "development" in this early period was also primarily justified in terms of the specific requirements of the cocoa industry. It included money for swollen shoot rehabilitation, cocoa research, scholarships for farmers' children, and agricultural education.[60] The essentially fiscal or public-revenue-supporting function of the reserves became apparent only in the late 50's and early 60's, when they were locked up in public-development schemes or used to finance current tax payments.

The shift from reserves to government revenue was effectuated through a steeply graduated export duty. Although not part of the marketing board system itself, such a tax would have been hard to enforce without a public monopoly of exports. It was only in 1948/49 that a duty based on value replaced a flat piece rate.[61] The new duty increased progressively with the market prices obtained by the Board. The tremendous potentiality of the new system revealed itself in the 1950/51 season when an increase in average market prices of about 50 per cent in combination with a modest crop increase raised government tax receipts almost three times. It is unlikely that the tax schedule had been designed for such price levels. The new, partly-elected government of 1951, was nevertheless intent upon defending and expanding these gains. The rate of graduation was raised further by a new tax schedule: half the market price per ton in excess of £100 was to be paid as export duty.[62] The revision ensured in the subsequent two seasons that the government managed to raise its own direct receipts by half to almost one-third of the total value of the crop, in spite of lower export prices. But more was to come. The abolitition of rationing in Britain, depleted manufacturers' stocks and a poor new crop sent market prices rocketing to new record levels in the 1953/54 season.[63] It allowed the government with the help of the 1951 tax schedule to claim almost half the total value. Cocoa revenue was more than doubled. But also the reserves of the Marketing Board were further augmented. The share received by producers was reduced below 40 per cent.

In the middle of this extraordinary price boom, the CPP government, confirmed in office by the June 1954 elections, now with full "internal self-government", decided to introduce the *Cocoa Duty and Development Funds (Amendment) Bill* of August 1954. The producer price, widely expected to be raised, was to be pegged at its present level for four years. A further revision of the export duty sought to ensure that a larger part of the export earnings went straight to the government, rather than accumulated by the Board. If market prices remained above £260 per ton, the full difference between the actual price and £180 was to be paid as export duty.[64] Only half the value of such increments had so far gone to the government. The Bill also contained provisions for the allocation of these public super-profits between different development funds. One-fifth was assigned to general revenue.[65]

The government justified the scheme primarily by reference to the threat of inflation and to the requirements of national development. The current producer price was said to be the maximum which the country could pay without the risk of "grave economic harm". "The Government is determined to prevent if at all possible an increase in the cost of living, not only because of the hardship which it brings to individuals but also because it would reduce the actual amount of development which could be obtained from the funds that are available."[66] The inflationary argument was not entirely implausible. The 1953/54 crop was very small. A larger one, for example of the size harvested in 1950/51 and 1952/53, in combination with even a modest rise in producer prices, e.g. back to the 1951/52 level, would have increased the purchasing power of the cocoa producers by some 30 per cent. In the latter season, only "good luck" in the shape of a small crop had prevented a similar effect. On the other hand, consumer prices had been remarkably stable for four years.[67] Earlier restraints on import capacity had been largely eliminated. Imports expanded rapidly.[68] It could be argued that the country now had resources to allow a substantial expansion of consumption without malignant inflationary effects. The question, however, was not whether more purchasing power could be released but which sector of the economy was to be given priority: should the new resources be used to meet the private requirements of the cocoa producers or those of the rapidly expanding public-service sector?

The primacy actually awarded to the public sector stood out clearly. Total payments to the cocoa producers not only stagnated but declined after the nationalists entered the government in 1951. Government expenditure, on the other hand, had grown by an average of over 40 per cent per annum during the same period. It was equal to only half the income of the cocoa producers in 1950/51, but exeeded it by 70 per cent in 1953/54! The explanatory statement issued with the Ordinance declared that "the Government considers that the taxation of cocoa should be so devised as to use the present prosperity of this crop to build up the resources of this country". "In that way the country will be freed from its dependence on the vagaries of one crop ..."[69] The exact meaning of this far-reaching goal, which was to remain at the centre of CPP strategic thinking, was by no means clear. Judging from allocations under the 1951 Development Plan, emphasis was heavily on infrastructure and social services, with special stress on education.[70] The way in which this public-service economy was to reduce the dependence on cocoa may not have been spelt out but the necessity, on the other hand, for massive public development was widely accepted. Nor was there much difference between the priorities of the nationalist leaders and the colonial administrators at this particular stage. A comparison between the CPP plan of 1951 and its "fully" colonial predecessor of 1950 reveal only minor changes of emphasis.[71]

The wisdom of freezing the producer price at its present level was clearly more controversial, even within the government. The new tax schedule was designed in such a way that the producer price in fact could not be increased substantially unless the Board were prepared to use its reserves.[72] The farmers were offered a guarantee for four consecutive seasons that the price would not fall below the existing level regardless of what happened to market prices. In the booming market conditions of 1954 the need for such guarantee can hardly have impressed the producers as particularly urgent. By May 1954, the Cocoa Marketing Board had already asked the government to consider an increase in the producer price. Such an increase, it was argued, would "barely touch the Board's profits". It would be regarded as "a kind gesture on the part of the government to assist the farmers".[73] But the Cabinet had already decided against an increase.[74] In August the same year, the Board met the day after the government had announced its four-year "stabilisation" plan. Board members were worried: "present world prices justified an increase". "Farmers were aware of the world price and were expecting an increase."[75] The danger of intensified smuggling was emphasised. Prices offered in the neighbouring "French" territories were reported to be more than twice those paid in the Gold Coast. It was also reported that manufacturers in the principal consumer countries stressed the need to raise producer prices in order to stimulate production. The Board made a last-minute attempt to change the mind of the government but it was already firmly committed.[76]

The political repercussions of the Cocoa Ordinance of 1954 were greatly underestimated. As noted in Chapter 3, the cocoa price and tax issue provided a platform for a powerful opposition movement, based primarily in Ashanti, but with allies further north and in the Volta region.[77] The Farmers' Council, at this time only one year old and with little time for consolidation, broke up, as most of its Ashanti establishment joined the cocoa opposition and the National Liberation Movement. The Council had without success petitioned the government for an increase in the price and had great difficulty in supressing continued agitation in this direction among its own functionaries.[78] Meetings of loyal Chief Farmers were called to declare support for the government. One was held in Agona Swedru in September 1954. Emphasis was laid on the threat of inflation but also on the benefits that would accrue to the farmers from public expenditure. The conference declared:

> We have carefully studied the Cocoa Duty and Development Ordinance and have seen the wisdom in it as priority action will be given in developing our areas that have been neglected by the old government.[79]

The immediate benefits of the loans scheme of the Cocoa Purchasing Company were also stressed.[80] Leading officers of the Coouncil canvassed the cocoa areas to explain and defend the Ordinance. The General Secretary recalled in 1963:

> This was by no means an easy job. The political opponents of the Convention People's Party used this issue to stir up some of the farmers of the country to mutiny and rage. The situation was most critical, yet the officers of the Council fearlessly toured all the cocoa growing areas and explained to the satisfaction of the farmers the need for the government's action.[81]

The Cocoa Marketing Board, as well, was alarmed by the agitation that was "brewing throughout the country" and decided that special propaganda officers should be employed for the purpose of putting the government's views across to the farmers.[82] The impact of these attempts to mobilise opinion in favour of the Ordinance is hard to evaluate. The cocoa-price issue, however, was soon overshadowed in the campaign strategy of the opposition groups by the demand for a federal type of government as a means to break up CPP domination. This was an issue better adapted for the purpose of holding together the alliance of regional groupings constituting the new opposition bloc.[83] The decline in the market price of cocoa during 1955 in combination with an increase in the producer price reduced the effectiveness of the price issue as a basis for continued agitation. From the point of view of Ashanti, the federal issue and the cocoa price were of course closely related. Greater regional autonomy would serve to redirect a greater portion of cocoa revenue to its origin. Although growing about half the country's cocoa, Ashanti (including Brong-Ahafo) was clearly disadvantaged in terms of public development. There was legitimate fear that the increasing emphasis on the latter at the expense of private consumption and investment would further enhance developments in the south and in the capital in particular. Not only vicinity to this growth centre but also historical advantages in education and skills seemed to ensure that the people of the coast would be able to claim a larger share in government-derived business than those of the interior. As far as the expenditure of the Cocoa Marketing Board was concerned, the southern cocoa areas had greater interests at stake. Ashanti was only marginally affected by the swollen shoot and most of the heavy payments for rehabilitation and compensation were collected by the southern farmers.[84] In addition, Ashanti dependence on cocoa was far greater than the southern regions, where the urban economy was more advanced. Historical and cultural divisions also contributed to bring the federal issue into the focus of the struggle and to restrict the appeal of the "cocoa opposition" in the south.

By mid-1955, the Farmers' Council used the opposition shift to federalism to revive its own demand for a higher producer price, a demand which it had been forced to discontinue in order not to be identified with enemies of the government. The General Secretary argued "the necessity of approaching the government constitutionally for a review of the cocoa price".[85] He noted that it was a very delicate issue. A demand in this direction could be misconstrued as being anti-government. But "the appetite for an increase in the cocoa price is still alive in every farmer". He thought that the new emphasis on federalism on the part of the National Liberation Movement created a "golden opportunity" for the Council to act.

The producer price for the 1955/56 season was increased from 72 to 80 shillings per load. Facing a major threat to its own survival, the government was anxious to remove the issue of the cocoa price from the immediate struggle. The general position on the need for price stability taken in defence of the 1954 Cocoa Ordinance was abandoned. The average market price obtained in the 1954/55 season remained at the extreme heights of the preceding season. Half the value was collected by the government in the form of cocoa duties. Another £8 million were added to the accumulated surplus of the Marketing Board. A refusal to raise producer prices in this situation was increasingly difficult to justify. The decision to raise the price was taken, however, at a point when market prices were on their way down. Average selling price in the 1955/56 season was reduced by almost 40 per cent, matched only by a slight increase in production. As a result the public share of cocoa income was halved, from 60 to less than 30 per cent. As the total value of sales fell by one-third, actual public receipts were even further diminished. Marketing Board payments to the government fell from £38 to £15 million and the Board itself barely broke even. The change in the market may have confirmed the wisdom of the stabilising attempt of 1954, but the government was at this stage determined to keep the producer price out of the political struggle, and especially out of the 1956 general election. Addressing the Working Committee of the Farmers' Council in early 1956, Gbedemah, the Minister of Finance and a leading campaign strategist, expressed the conviction that the Cocoa Ordinance of 1954 "was the sinews of all the political tension in the country".[86] In spite of the continued deterioration of the market, the government decided to maintain the 80 shillings per load producer price for the 1956/57 season, as well. The General Manager of the Marketing Board noted that "it was felt desirable, since the general elections were pending, to remove the vexed question of cocoa price out of politics". But there was also a need, according to him, to demonstrate that reserves were actually used to support the producer price, especially in view of "adverse criticisms overseas".[87]

The cost to the government of depolitisising the cocoa-price issue was high. As market prices continued to fall and production recovered, the public share of cocoa income was reduced to a mere 13 per cent, which may be compared to an average of about 50 per cent for the first half of the decade. Government receipts were further depressed and the Board recorded its first major deficit. The share of the producers on the other hand was raised to almost 80 per cent.

Independence: Farmers' voluntary contributions

The concessions to the "cocoa season politicians" were only temporary. In March 1957, having successfully survived the 1954–56 political struggle, the CPP was in charge of an independent state, capable not only of exercising greater control over political opponents but also of redirecting attention to problems of national development. Frequent elections (including the Togoland plebiscite) and constitutional controversies had absorbed much of the time and energy of the government. Political tensions remained and the "political kingdom" required consolidation. With independence achieved, however, the government was clearly anxious to reestablish a healthy financial basis for continued public expansion. The first two years of independence were treated as a preparatory phase for the Second Development Plan, to be launched in 1959.[88] Protracted negotiations over the Volta River Project, to which such importance was attributed, formed one reason for waiting. This huge hydro-electric scheme was expected to provide the basis for the diversification and industrialisation of the economy. Continued dependence on cocoa was considered incompatible with the goal of economic independence. The realisation of the scheme was delayed by protracted negotiations with foreign aluminium interests, the major prospective customers of Ghana's cheap electricity, and on attempts to secure external finance from other sources. The people of Ghana were also expected to make "very real sacrifices".[89]

In the meantime, the government acknowledged its dependence on cocoa and set out to restore the level of public income, which had been permitted to decline during the pre-independence struggle. The producer price was returned to its 1954 level of 72 shillings per load, in spite of the reluctance of the Cocoa Marketing Board. A majority of the directors were in favour of supporting the price at its present level of 80 shillings, "because they felt that the idea behind the Board's reserves was to cushion the producer price". The Managing Director warned that it might lead to another £7 million reduction in the reserves, "but members felt that these features should be offset against the political repercussions of a lower price and the economic needs of the farmers".[90] Against the background of the depressed export prices of the preceding season, the government was obviously less worried about the political consequences of a cut and the Cabinet overruled the Board.[91] Judging from the National Assembly debates, there was some protest from the section of the parliamentary opposition which in the past had acted as spokesmen for the cocoa farmers' interest. It was maintained that the reserves were large enough to support the producer price even if market prices remained low.[92] The Togoland National Farmers Union, a regional opposition group, deplored the decision and asked for a return to the old price.[93] Similar appeals were probably submitted from other parts.[94] There is no evidence at this point, however, of a revival of a militant cocoa opposition on the 1954 model.

The regional distribution of cocoa earnings was brought to the fore of the public debate when the government decided to remove certain legal restrictions on its access to the funds of the Board. Existing laws assumed that the funds were administered for the benefit of the cocoa-farming communities.[95] The main purpose was price stabilisation but development expenditure to meet their needs was also sanctioned. The Board was supposed to act as trustee for the farmers. While the destination of some of the early expenditure of the Board, such as the grants to the University College, was clearly national rather than sectional, most payments up to that point may have been compatible with the letter of the law. Much had been spent on the rehabilitation of cocoa farms, scholarships to farmers' children, health centres in cocoa areas, etc.

The government wanted to use the accumulated funds of the Board for its development programme, without such regional or sectional restrictions. Legislation to that intent was presented to the National Assembly in July 1957.[96] The Minister introducing the Bill declared that the cocoa funds "should properly be regarded as being held in trust for all the people of Ghana". The main argument used in support of this thesis was that the people of the Northern Region, where cocoa was not grown, had provided the main labour force of the cocoa industry. Their home areas should therefore equally be entitled to a share of the funds:

> . . . it is due to the pioneering work and effort of the ordinary Northern Region farm labourer, tilling the farms now spread all over the cocoa producing areas of Ghana, that the cocoa industry has been established.[97]

Spokesmen for the parliamentary opposition, mainly a coalition between groups based in the Northern Region and in Ashanti, explained that they had nothing against CMB funds being used in the North but insisted that it was the cocoa farmers' money and that the farmers had to be consulted.[98]

Speakers, mostly from cocoa areas, none from the North, attacked the way in which the government used the Board for its own purposes. Control over the Board should be vested in the farmers themselves. Farmers were said to be treated as slaves working for the government.[99] One speaker thought that the new Bill savoured of communism: "It is only communist Governments which do not respect the opinions of the people who work to amass the wealth of the country."[100] The Marketing Board was compared with the Trade Union Congress. The government could not "freely dip their hands into the coffers of the T.U.C. and use the money to develop all parts of the country without consulting and having the consent of the T.U.C. and the workers". Why, then, should it use the farmers' money in this way? The government was warned that the farmers would not take such actions lying down.[101]

It is unlikely that many farmers harboured illusions about the Marketing Board being *their* organisation, as distinct from one of the government. The Board's trusteeship for the welfare of the farmers was little more specific than that of the government itself. The allocation of funds between price stabilisation and development projects was a matter of government discretion. The new law may have added little to the powers of the government in this respect. The division of public cocoa income between the government and the Board was also regulated by tax laws without the participation of the Board. Nothing prevented the government from using funds made available in this way for the development of non-cocoa areas. Compared with such expenditure, the direct undertakings of the Board were marginal. The revision of the statutes may, on the other hand, have served a broader purpose of demonstrating "the spirit of uniting the whole of Ghana together".[102] At the same time, it served to drive another wedge into the shaky political partnership of prosperous Ashanti and the poor North by underlining the unity of interest between the latter and the central government in the redistribution of cocoa income.

The recovery of market prices during the 1957/58 season, induced largely by the poor West African crop, increased the political pressure for a higher producer price. The new market situation, in combination with the price cut of 1957, had swiftly restored a rate of public appropriation equal to that of the first half of the 1950's. Public cocoa income was again almost 50 per cent of the value of sales. The Marketing Board, having more than recovered the "loss" incurred in the previous season, recommended a restoration of the producer price to 80 shillings per load. Members made reference to the deplorable living conditions of the farmers and the rising cost of labour.[103] A higher price was not approved by the government.[104] In defending its position in the National Assembly, the Minister of Finance pointed to the insecure market situation—("It is not unlikely that the price can drop considerably during the season")—and the need to prevent new "losses" on the 1956/57 scale. Farmers, he claimed, were better served by a "stable income" than by repeated fluctuations caused by changes in the producer price.[105] Spokesmen of the opposition questioned the "wisdom and equity of saddling only a small fraction of the community—the cocoa farmers of the country—with the entire burden of the development of the country". Farmers were said to be "pressed to the crushing point", especially as the cost of production was rising sharply. A twofold increase in the producer price was even suggested.[106] The question of price was also raised repeatedly in the committee debates on the Ministry of Agriculture estimates.[107]

Another opportunity to question the distribution of the cocoa export earnings arose when a Bill consolidating various cocoa-based development funds was introduced in the National Assembly in August 1958.[108] The Bill itself may have been merely "technical" but new pleas for higher cocoa prices were made. The cocoa communities were said to receive little in return for their huge contribution to public finance: "We see that all development is centred mostly in the municipalities of this country where no cocoa is grown."[109] At the same time as opposition speakers from the South argued that the burden of financing national development should be distributed more evenly, members from the North were more anxious to increase the share of their own areas in public cocoa income.[110] The regional division of interests within the ranks of the opposition was again demonstrated. The alliance of regional groups, now reconstituted in the United Party, was unable to provide an effective platform for challenging the cocoa policies of the CPP government.

The *Second Development Plan* was launched in March 1959.[111] It envisaged a radical increase in public development expenditure. The First and the "Consolidation" Plans, covering nine years (1951-59), had absorbed some £118 million. The second proposed to spend £350 million in a mere five years, including £100 million reserved for the Volta River Project. The realisation of the full plan depended on a high rate of foreign finance, private investment and loans. To take account of the uncertainties, a minimum plan—a list of projects "for immediate implementation"—was outlined at a planned cost of £132 million, excluding the Volta scheme. The latter was kept outside the main plan as the negotiations for external finance were still unconcluded.[112] The plan emphasised productive investment in agriculture, industry and trade. "Specially high priority will be given within the next five years to promoting the establishment of no less than 600 factories of varying size producing a range of over 100 different products."[113] Most of the factories were to be in the private sector. Foreign capital was expected to play a major role. The connection between the financial commitments of the government and the expected growth of manufacturing was rather obscure. Rubber growing in the Western Region and modern cattle ranching in the North were major projects in agriculture but most of the major investments were not included in the minimum plan.[114] The reorientation of the strategy towards more directly productive investments was less marked in the "small plan" than in the full one. Public services still dominated although emphasis had shifted away from communications (mainly roads, harbours, and railways) and education to health, water, public administration and national security.

A discussion of problems of finance was notably absent both in the plan and in the introductory statements made in the National Assembly. The amount allocated for the "small plan" was obviously what was considered realistically available.[115] But no details were given about the way in which the government intended to raise even this smaller amount. The Prime Minister emphasised that it was now a task for the various political and other organisations of the country to ensure that the plan was realised. He called in particular on the Farmers' Council, the Trade Union Congress and the "Co-operative Movement" to support the development effort. The intention to hold consultations with these and other organisations was announced.[116]

At least as far as the Farmers' Council was concerned these consultations led to quick and very material results. Two weeks after the Second Development Plan had been introduced in the Assembly, the Prime Minister was able to read out a statement by the Executive Committee of the Farmers' Council declaring, "of our own free will" and on behalf of the farmers of Ghana, that they were prepared to accept a reduction in the cocoa price from 72 to 60 shillings per load, a cut of 17 per cent, in support of the plan. The 12 shillings difference was to be the farmers' "*voluntary contribution*" to the intensified development effort.[117] The offer was made "on behalf of all the farmers in Ghana in the belief that it will receive their full approbation". The Council also expressed its belief that "the success of the second five-year development plan will mean more benefits for us than probably for any other section of the working masses of Ghana". The estimated contribution of the farmers for the five-year period was some £25 million, in addition to the amount collected as cocoa duty and the grants and loans received from the Marketing Board. The government undertook in return to maintain the producer price for five years. The Prime Minister appealed to other sections of the nation to take note of the patriotic action of the farmers.[118]

A motion of appreciation was passed in the National Assembly in which the patriotism and spirit of sacrifice demonstrated by the farmers were acclaimed. No opposition member participated in the debate. They were unlikely to accept the right of the Council, a CPP organisation, to make an offer of this kind on behalf of the farmers. Even CPP members of the Assembly, however, felt uneasy about the rather hollow pretence of this cermony. W. A. Wiafe of Kwahu South suggested that it would perhaps have been better if the Farmers' Council "had taken the trouble to go round the whole country to explain to the farmers the benefits they would derive from the Development Plan before passing the resolution".[119] Another CPP member, M. O. Kwatia of New Juaben was even more outspoken. He suggested that as the farmers had not been consulted, the declaration made by the executive of the Farmers' Council could not be taken as a declaration by the country's farmers. The Minister of Justice found this suggestion "completely out of order", as the task of the debate was "either to appreciate what the farmers have done or not to appreciate". The Minister of Agriculture informed his misguided party collegues that the leaders of the

Council were the "accredited representatives of the farmers".[120] Most speakers, however, found no difficulty in expressing their full appreciation of the action taken by the farmers, "an unmistakable sign of the economic wisdom of our peasantry".[121]

It was suggested in the debate that the government should respond to the generous offer by assisting the farmers to take over the local cocoa trade.[122] In fact, it was only one month after its "voluntary contribution" that the Farmers' Council presented the bill: the Marketing Board was asked to withdraw the cocoa-buying licenses of all the foreign trading companies. This marked the beginning of the Council's struggle for supremacy over the local cocoa trade. The timing of the two actions underlined the close connection between the Council's commercial aspirations and its willingness to shoulder political responsibilities in connection with the fiscal policies of the government. A £500,000 government loan to the Council in October 1959 was one major token of the government's appreciation, which greatly facilitatied the expansion of the Council's share of the trade.[123] The imposing multi-story headquarters, inaugurated that same month was another. The £100,000 annual subsidy granted a year earlier should also be included in the picture, to emphasise the element of reciprocity in the transactions between the government and the Council.[124]

The ability of the Council to convince farmers about the soundness of the "voluntary contribution" may be doubted. A press release signed by the Administrative Secretary painted a rosy picture of the simple and virtuous life of the farmers, close to nature and with little interest in "big cars, radiograms and luxurious living". Their bold manifestation of support for the Development Plan would not go unrewarded. They had "given away with the right hand but are receiving back with the left what they themselves gave away". The government, however, having learned from earlier mistakes, could not give the money directly to individual farmers "to play ducks and drakes with". Instead, the funds which they had contributed would be invested in agricultural development projects in which the Farmers' Council, on behalf of all of them, would take close part.[125] Leading Council officers toured the cocoa areas to explain the voluntary contribution.

One petition maintained that the farmers were "the victims of the wild decisions of a clique of few officials within the U.G.F.C.". The petitioners, the National Agricultural and Plantation Workers Union, although mainly organising workers on government agricultural stations, also claimed to speak for the cocoa farm labourers in general. The First Annual Delegates' Conference of this union, meeting in early May 1959, declared that the offer made by the Farmers' Council "should be termed null and void". The monopoly of the Council as the only "recognised" spokesman for the farmers was challenged. The conference demanded that all major farmers' and farm workers' organisations should be consulted, including the Cooperative Marketing Association (the "old" cooperatives) and the National Farmers' Union, the Ashanti based opposition group.[126] The former had issued a statement on behalf of its "more than 51,000" members that they did not support the deduction.[127] The latter organisation had only a year earlier demanded that the producer price should be raised to as much as 100 shillings per load.[128]

The parliamentary opposition to the new cocoa tax was half-hearted. The annual budget debate in July 1959 would have provided an opportunity for a major attack on government's fiscal policy.[129] Most speakers seemed, however, to be more concerned about the way public money was spent than how it was obtained. A major opposition spokesman acknowledged that farmers may not have been properly consulted but as "the farmers have made their contribution", it was now up to the government to make sure that the money was properly used to develop the rural areas.[130] Other speakers dwelled on the hardship and indebtedness caused by the reduced price. But this was only one among many large and small complaints and there was no organised demand that the old price should be maintained or be replaced by a higher one.[131]

The limited parliamentary opposition to the farmers' "voluntary contribution" may be contrasted to the intense reaction to the government's decision to cut the salaries of Members of Parliament by ten per cent, "as a sacrifice to the nation to assist in the implementation of the Second Development Plan".[132] Speakers were outraged. The new, legally enforced "contribution" was seen as a measure "calculated to fool the toiling masses".[133] Members challenged the constitutionality of the deduction, as property, according to the constitution, could not be confiscated without adequate compensation.[134] Others warned that similar contributions could also be enforced on workers and civil servants.[135] Repeated references were made to the cocoa farmers' contributions but without the same objections to their fictitious voluntary character.[136] The ten per cent deduction of the parliamentarians was intended to demonstrate to the nation that the cocoa farmers were not the only "patriots" willing to assist the development effort. But no other major groups were made to come forward.[137]

The farmers' voluntary contribution was not the only way in which cocoa was assigned a special role under the new plan. £25 million were to be lent to the government by the Cocoa Marketing Board, an amount equal in size to the extra revenue expected from the voluntary contributions. £16 million were in fact paid out during the first year of the plan.[138] It was a soft loan, carrying an interest of $2^1/_2$ per cent and it was repayable over a period of 20 years. The liquid resources of the Board were drastically reduced. Most of the funds used to finance current trading were transferred to the central government.[139] The Minister of Finance explained in his Budget Statement that these funds were unnecessarily tied up in trading operations and could be better used for national development purposes. The current trade was instead to be financed from bank overdrafts.[140] The ability of the Board to "stabilise" farmer or government income was greatly reduced as the major part of its funds were now locked up in long-term government securities.

The large-scale transfer of funds to the government in addition to the removal of regional restrictions on the Board's own expenditure served as an ultimate confirmation of the transformation of the Board into an instrument of public finance rather than a vehicle for the stabilisation of producer prices. It also meant an end to another, less important illusion: that the Board somehow played an independent role in fixing the producer price. Overriding importance had infact been attributed to the impact of the producer price on the national economy at an early point, but this illusion had still marked formal procedures. In practice, the Board had little to say, for instance, in pegging the price in 1954 and in its reduction in 1957. In 1959, however, it was apparent that the price-fixing function of the Board had been rendered entirely inoperative, at least for a five year period. The price was pegged for the duration of Second Plan through an agreement between the government and the Farmers' Council, without consulting the Board.[141]

Against the background of the 1958/59 successful trading operations, the decision to make a major cut in the producer price, the second since independence, must have been difficult for the Farmers' Council or any other agency of the CPP government to justify to the farmers. In spite of the recovery of production from its very low 1957/58 level, average market prices had remained high, resulting in a substantial increase in overall earnings, and allowing the government and the Marketing Board to retain jointly about 50 per cent of cocoa income, a rate so far only associated with small crops and extreme market prices. Total payments to the producers also improved but were still well below the level of the independence season of 1956/57. There was little in the market performance itself which could justify another price cut. In this respect, the conditions of the voluntary contribution of 1959 differed markedly from previous controversial price decisions. Price pegging in 1954 was argued, although not too convincingly, in terms of possible inflationary consequences of a price increase. In 1957, the price cut could be defended with reference to the "losses" made by the Board and their possible long-term effects for price stability. On both occasions, the arguments served to make the essentially fiscal purposes of the price decisions more compatible with the official function of the marketing board system. In 1959, the market did not permit this type of justification. A further cut in the producer price at that point could only be argued in terms of the need for a further redistribution of national cocoa income, at the expense, at least in the shortrun, of the cocoa producers.

The market situation in which the voluntary contribution would have been able to augment the resources available to the government above the level achieved in 1958/59 never materialised. On the contrary, the following years saw a sharp decline in public cocoa income, only temporarily prevented from affecting government revenue through a run on the reserves of the Marketing Board. The "over-expansion" of production and the related decline in market prices have been discussed above. In two seasons production jumped from a level which for over a decade had averaged well below 250,000 tons to one of over 400,000. This sudden change was clearly unforeseen by those who launched the voluntary contribution simultaneously as they transferred much of the liquid resources of the Marketing Board to the central government. The Board soon ran into difficulties as a result of these double financial obligations to the government while retaining a producer price of 60 shillings per load as promised under the 1959 agreement.

The expansion of public expenditure, however, accelerated. By 1960/61 government expenditure had been raised to a level twice that of the mid-1950's. A major budgetary deficit was incurred in 1950/60. It was trebled the following year. The CMB loan ensured a high degree of cocoa backing for the 1959/60 budget but this could not be easily repeated. The urge for public expansion was nevertheless undeterred by the deterioration of its financial basis in the cocoa economy. Negotiations over the Volta River Project were brought to a successful conclusion, making major claims on the resources of the government.[142] The government was determined to procede with the industrialisation programme, in spite of the difficulties of finding the private foreign investment which had been assigned a major role in this respect. A reorientation of public strategy in a more radical direction was accompanied by heavy investment in a variety of public enterprises. The nature of this reorientation was discussed briefly in Chapter 1. The reorientation was encouraged by the new international contacts of the government. Economic and political links were established with the Soviet Union and other socialist countries. By 1961, much of the economic philosophy which had marked the "liberal" 1959-plan had been replaced by a professed belief in the overriding importance of direct state participation in production and comprehensive state planning.

It is probable that the frustrating experience of the cocoa industry during the first years of the Second Development Plan contributed importantly to this reorientation of strategy, and to a confirmed belief that only a massive redirection of resources into "modern" industry and agriculture could break the country's debilitating dependence on the world cocoa market. In 1958/59 season, Ghana received some £70 million for a total crop of 255,000 tons. In the following two seasons—the first two years of the plan—production was raised to 317,000 and 432,000 tons respectively, a remarkable national achievement, but without adding anything to the level of earnings already reached.

The sharp fall in market prices meant that the basis of the 1959 producer price agreement, as seen from the government's point of view, was jeopardised. Public cocoa income, that is, government revenue plus or minus the surplus or deficit of the CMB, was reduced by over one-third in 1960/61. The Board incurred a £6 million deficit. The joint share of the government and the Board in national cocoa income was down to one-quarter, which may be compared with an average of 43 per cent for the 1950's as a whole. The amount of money in the hands of the cocoa producers increased at the same time by one-third, adding importantly to the internal and external economic pressures released by the accelerating of public expenditure. A liberal foreign-trade policy ensured that the increase in local purchasing power was transformed into a strong rise in imports at the same time as foreign-exchange earnings stagnated. During the first half of 1961, imports were up 20 per cent compared with the corresponding figures for 1960.[143] The balance of payments which had begun to show deficits on current account in the second half of the 1950's, deteriorated rapidly. In 1960 the deficit trebled and reached a level equal to half the current value of the cocoa crop. In 1961 it reached a record level of £53 million. The deficits could be financed for the time being from the large foreign reserves which had been built up during the 1940's and the first half of the 50's. In 1955, the reserves reached a peak of £208 million. By the time the Second Plan was launched they were still about £170 million. At the end of 1961, however, they had been reduced to £73 million, or less than required to finance two deficits on the 1961 scale.[144]

The 1961 Budget and Compulsory Savings

Radical measures were required to reduce the large gap between foreign receipts and payments as well as between government revenue and expenditure. The government was intent on doing so without putting a brake on the national development effort, as represented primarily by the expansion of the public sector. It looked for new sources of external and internal finance at the same time as it sought to redirect national expenditure from private consumption to public investment. The July 1961 "austerity" budget contained a number of fiscal reforms and increased taxes. Import and excise duties were raised and a new purchase tax was imposed on a variety of durable consumer goods. Much controversy was also caused by a new property tax. The coverage of income taxation was broadened.[145] Foreign-exchange controls were extended in order to preserve the country's dwindling hard currency resources. Comprehensive import licensing covering almost all commodities was introduced later in the year as it turned out that the new high duties on imports were insufficient to achieve a balance on that account.[146] A number of foreign-trade and payment agreements, especially with East European countries, were expected to boost the resources available to the country.[147]

The most important part of the 1961 budget as far as the cocoa farmers were concerned was the compulsory savings scheme. A compulsory deduction was to be made in exchange for National Development Bonds which were redeemable after ten years, together with accumulated interest. In the case of wage and salary earners the deduction was five per cent but for companies and self-employed it was to be ten. Cocoa farmers were taxed according to the higher rate, or six shillings in the current price of 60s. per load. They were the only section of the farming community covered by the scheme.[148]

The 1961 budget is commonly considered a major turning point in the fortunes of the CPP government. Up to this time the government is believed to have had a fairly broad popular support. The "Tighten your belt" policy of 1961 alienated many groups which had been close to the party, including workers, junior public servants and small businessmen. The demoralising impact of the budget on the party was reinforced by internal divisions over strategy and leadership which had marred the preceding year, involving cabinet changes and open controversies between CPP members of the National Assembly and the party press. The budget was followed by a serious outbreak of revolt in the important railway-workers' union, which threatened to develop into a general strike. After some vacillation it was suppressed by the government and a number of real and alleged leaders were imprisoned under the preventive detention laws.[149] Workers opposed the compulsory savings scheme, which reached down to income groups previously exempted from taxes, but they were also particularly concerned over the steep rise in taxes on imports, which constituted an important element in the urban consumption pattern.

Although certainly also affected by rising consumer prices, farmers were hit more directly by the ten per cent cut in the producer price. As in the case of the earlier "voluntary contributions", government was anxious to emphasise that the new reduction had been agreed in consultation with the farmers. Again the Farmers' Council was the mediating agency. The President declared when first presenting the scheme that the government planned to exempt all farmers from income and property taxes but that he felt "that our farmers will be just as anxious to participate in this great national savings scheme as the rest of us".[150] A government spokesman declared that it was really left to the farmers to say yes or no to this invitation.[151] Only two days later, the National Assembly was informed that the Supreme Council of the Farmers' Council had met and approved of the scheme on behalf of the cocoa farmers.[152]

The Farmers were praised and congratulated for again taking the lead in support of the national development effort.[153] The benefits to the farmers of the scheme were also elaborated. They would be helped to save money for the education of their children and prevented from unnecessary spending, such as the heavy expenses on funeral rites.[154] Opposition speakers felt that the farmers had already contributed sufficiently to national development and should be left alone. The procedure was also challenged. "How can we say that these farmers of their own free will have agreed to contribute six shillings a load?"[155] One member spoke contemptuously of the Farmers' Council leaders as "well-groomed and well attired city dwelling farmers of Accra, riding in big cars with bulging brief cases", making agreements with the government.[156] Members of the governing party also expressed uneasiness about the method of consultation. One Assembly member quoted Central Region farmers as telling him, "eh Mr. Quaidoo, you people are going to kill us . . . we were not consulted on these things. The whole thing has been done so hurriedly. When we go to meeting they ask us so quickly whether we agree, and we have to say yes before we have had time to think about the whole thing". The member added that "if you were to hold a secret ballot on this thing you would be surprised to see how many would say no".[157]

It is unlikely that either the CPP government or the Farmers' Council leaders held any illusions about the popularity of the new cut in the producer price. The effect of the reduction was reinforced by the economic conditions of the time, with growing domestic price inflation, shortages of local food supplies, and import restrictions. The new large cocoa crops contributed importantly to these difficulties and the purpose of the compulsory savings scheme was also explicitly to reduce the purchasing power of the farmers. Farmers were warned as early as in November 1960, when the size of the new crop could be safely estimated, that their profits might be eaten up by inflated prices and that the government therefore had asked the Farmers' Council to suggest a system whereby a share of the proceeds could be saved "for the individual farmer himself".[158]

The political impact of the 1961 budget and the compulsory savings scheme in the cocoa farming communities was further magnified by the drastic reorganisation of the local cocoa trade which had just begun. The cocoa purchasing monopoly of the Farmers' Council, preceded by much public controversy, was put into operation with the mid-crop of 1961. Apart from open hostility from sections of the old cooperative movement, private traders, and anti-CPP groups, there was wide-spread suspicion against the new marketing scheme and the ability of the Farmers' Council to ensure a smooth flow of money and non-partisan services.[159] The first season of the monopoly was also no doubt marked by much confusion and genuine difficulties. The task of rallying farmers behind the fiscal policies of the government must have been particularly hard in this situation. Nor did the 1961 budget make the farmers more favourably disposed to the commercial monopoly of the Council, so clearly identified as a tool of the central government.

It was no coincidence, however, that the Council managed to secure its commercial supremacy at the same time as the government prepared another cut in the producer price of cocoa. The need to enforce the latter against probable opposition enhanced the dependence of the CPP government on whatever channels of political control and influence that were available in the cocoa-farming areas. The network of local representation established by the Farmers' Council was bound to play an important role in this respect. The loyalty of the local Chief Farmers and committee members had to be ensured. The government may or may not have been convinced that the purchasing monopoly of the Council was a good way of strengthening its political representation in the cocoa areas. Its dependence on the established network of the Farmers' Council to "control" opposition made it difficult to resist the commercial ambitions of its leaders. The latter, on the other hand, were equally dependent on government support in order to realise their own plans. They had almost been robbed of their victory in the campaign against the foreign cocoa-buying firms as a result of the last-minute hesitation of the government. Not only opposition politicians but also high CPP members continued to press for a revision of the decision to grant monopoly rights to the Council.[160] Its leaders could hardly afford to be unenthusiastic about the compulsory savings scheme. While not necessarily a matter of explicit bargaining, the obligations involved were still quite clear.

The Council's campaign to explain and defend the compulsory savings scheme started the very day after President Nkrumah's sessional address to the National Assembly. The scheme was introduced at a meeting of the General Purpose Committee of the Council. Members were asked not to falter in their faith but campaign for other farmers "to accept the Policy without any grudge" because the the Council was "the bulwark of the Government and because of the implicit confidence the President has in it". The decline in world market prices was given as the main factor necessitating the savings:

A scan through the past shows clearly that there has been great attempts to dwindle the resources of the country by the Imperialists. Cocoa Prices in the U.K. soared very high in those Imperialistic days. But there has been a sharp drop in the price since the attainment of Independence.[161]

It was emphasised that the behaviour of the market was in itself sufficient to justify a reduced producer price. The fact that the price was still "maintained" was attributed to the President's personal love for the Farmers' Council. Great stress was laid on the fact that farmers had been exempted from paying income tax. This gesture on the part of the government was seen as a response to the farmers' willingness to contribute compulsory savings.[162] The value of the exemption may have been less apparent to the farmers: few, if any, had been charged income tax in the past. A press release praised the farmers for the lead they had taken in supporting the economic reconstruction of the country. In this respect they had surpassed all other groups of workers.[163] Senior regional officers of the Council were briefed at the Ideological Institute at Winneba before they toured the districts to explain the budget to the farmers. Rallies were held at the district level and resolutions were passed in support of the government's fiscal policy and the compulsory savings scheme.[164]

Only half a year after the 1961 budget was introduced the compulsory savings scheme took an entirely new turn, at least as far as the farmers were concerned.[165] Contributors to the scheme were encouraged to forgo their savings and turn them into a donation to the government for the national development effort. The farmers were the first to be made to respond to this new call for national sacrifice. Regional meetings of the Farmers' Council met in the early month of 1962 and took unanimous decisions to abstain from reclaiming the farmers' savings at the end of the scheduled ten-year period. An Emergency Regional Delegates' Conference of the Eastern Region branch of the UGFC, for example, resolved in early February with "the greatest pride to make a further heroic sacrifice" to the nation.[166] Other regions followed suit.[167] In an attempt to underpin the illusion of a spontaneous movement from below in support of the new fiscal deal, the National Conference congratulated the regional branches for their sense of civic duty.[168] The Conference was inaugurated by President Nkrumah, who expressed the gratitude of the government and the party for this new example of the "courageous and generous sacrifice and patriotism which we hope many will be persuaded to emulate".[169] The sum of money involved was considerable. If an average crop of 400,000 tons is assumed, the donation volunteered by the Council would amount to some £45 million for the ten-year period. The gratitude of the government and the party was further elaborated in a special motion of praise for the achievements of the Council which was passed in the National Assembly in May 1962. This was a carefully staged manifestation covering the entire range of the Council's activities, with but few discordant notes. It was emphasised that it was "through the instrumentality of the Farmers' Council" that the farmers had again freely contributed to national development. The Council had made it possible by explaining to the farmers the good uses to which their money would be put by the government.[170]

As in the case of the voluntary contribution to the Second Development Plan only few were willing to "emulate the example" of the cocoa farmers, or obliged to do so, in spite of public exhortations. Only a small number of individuals apart from MPs and the President himself seem to have responded to the call.[171] It is particularly telling that it was not until early 1965, three years after the farmers' "courageous sacrifice", that the Ministry of Finance was notified that the administrative staff of the Farmers' Council itself had decided to forgo their compulsory savings.[172]

By this time the scheme as a whole had been discontinued. It was maintained for only two years. £23 million had been collected of which some £9 million derived from the cocoa farmers.[173] The abolition of compulsory savings in 1963 was for wage-earners a "concession" meant to facilitate the acceptance of another tough, "Tighten your belt" budget. For most of them, increases in direct income tax offset the immediate effect of this concession and unlike the compulsory savings the new charges were of course not refundable.[174] In the case of the cocoa farmers, however, this change from savings to tax was even more straight-forward and without any element of relief. Their voluntary donated compulsory savings were replaced by a farmers' income tax charged at a flat rate of six shillings per load, that is, the same as the previous deduction. The exemption from income tax, although not previously paid, had been a major argument in justifying the farmers' contribution in the first place. In two years, the transformation of the scheme was complete. In this brief period, the farmers' contribution had turned from loan to donation and now finally to a plain tax. The National Assembly was informed by the Minister of Finance that the deduction of 6 shillings per load was to continue in spite of the abolition of the savings scheme because the farmers had "announced their wish to regard their contributions as additional grants to Government".[175]

The 1963 budget was heavily criticised in the Assembly. The few remaining opposition members were joined by CPP backbenchers. A variety of new taxes or tax increases had been proposed. The government sought to raise £22 million by additional taxation, the equivalent of an increase by 20 per cent on the expected yield of existing taxes.[176] To many of the members this was "too much a burden to bear".[177] Speakers from the cocoa areas were particularly disturbed by the continued deductions on cocoa although the savings scheme itself had been abolished. "What happens when we talk with our farmers? They say that we are using them completely in the interest of the country, and when they say this, it is very difficult to argue with them."[178] While some members demanded that the farmers should receive the unreduced producer price, others were more concerned with the method used by the government in handling the issue. If it was not possible to pay farmers more than 54 shillings per load, why not fix such a price, instead of pretending that voluntary deductions had been made by the farmers? "It is a wrong sort of propaganda for anyone to say that farmers have agreed to continue with the Compulsory Savings Scheme. It is absolutely unrealistic; no farmer would like to continue to pay the 6s."[179] The procedure by which the farmers had been "consulted" through the Farmers' Council was rejected:

> I am at loss to understand who gave this authority to the U.G.F.C.C. to deduct 6s. from the farmers' cocoa. I know that this is not true, because when I go to my village or my constituency the farmers themselves come to me and ask; "What are you doing? You are taking all our money." The farmers do not know anything about these arrangements . . . Whenever there is conference, the U.G.F.C.C. call only the officials to meet and the Council ask these officials to convey any decision they take at the meeting to the Farmers.[180]

A spokesman for the government retorted that he could not see why "Members should make it their business to destroy the patriotic moves which the farmers themselves have voluntarily made" by critisising the Farmers' Council.[181] Another speaker maintained similarly that there was no reason to doubt the many authoritative assurances about the genuine nature of the farmers' sacrifice: "Let us applaud and thank the farmers most heartily and let us not put funny ideas into their heads."[182]

The precise form of Farmers' Council involvement in the voluntary income tax decision of 1963 is not clear from the records available to me. References were made in the Assembly debates to the 10th National Delegates' Conference, which met in Accra one month before the budget. It was staged as a huge anniversary celebration involving, according to official reports, some 10,000 delegates, including society representatives, not normally present at the annual conferences.[183] It is probable that they were asked on this occasion to continue to support the compulsory savings scheme, but there is no evidence that the conference was told what would happen if the scheme were discontinued. In practice it made little difference, as the Council already earlier had decided to forgo the savings. Criticism in the Assembly was met with renewed demonstrations of support from Farmers' Council organisations. The regional executive in the Eastern Region, for example, put out a statement denying allegations that farmers had demanded their six shillings back.[184] One report from Akim-Oda tells of a District Chief Farmer who was "jeered, booed at and had nearly been beaten" by infuriated farmers because he had agreed to donate the farmers' compulsory savings to the government "without first consulting the contributors".[185]

The confusion and strong feelings aroused by the farmers' voluntary income tax added greatly to the growing hostility to the Farmers' Council among government supporters, as well. This was evidenced by the increasingly vocal CPP backbenchers in the National Assembly. The general praise showered on the Council in connection with the special motion of May 1962 had been replaced by a more critical mood. The debate on the estimates of the Ministry of Agriculture turned into a major demonstration of opposition to the Council both as an instrument of fiscal policy, its way of administering the cocoa-purchasing monopoly, and its growing involvement in food crop production.[186] Particular criticism was directed against a decision, confirmed at the Accra conference, to deduct "surplus weight" for feeder roads and other existing or planned Council expenditure. The non-payment of bonuses and the scanty off-season advances were other issues brought up in the debate.[187] Much of the allegations against the Council discussed in Chapter 5 were in fact touched upon this occasion. The Assembly was warned by one member that if the present system of cocoa buying was not changed "we shall be here one day . . . when the farmers will come here with their cutlasses to trouble us".[188] Farmers were said to be heavily burdened by government taxes and the official and unofficial charges of the Council were added on top of them. The Council's claim to represent the farmers was contrasted to the way in which they were exploited by the officials of the organisation, as in the words of the United Party spokesman:

> It is treacherous for the Farmers' Council to impress upon the Government that the farmers have agreed to contribute, while in fact they have not. It is surprising to know that Secretary/Receivers who are paid between £15 and £20 a month ride in cars, while the farmers go on foot.[189]

Speakers defending the Council, on the other hand, pointed at the overall success of its trading operations and stressed its important contribution to national development. The farmers did not give their voluntary contributions and donations to the government individually, "but collectively through the U.G.F.C.C." The Council should therefore be congratulated for "the political and economic help they have given to our Government and the people of this country", instead of being criticised.[190]

The Assembly debates in connection with the 1963 budget showed that a growing number of CPP members felt that cocoa farmers were not treated fairly. This was less because of the continued low producer price than because of the way in which the whole issue of prices, deductions and donations was manipulated by the Farmers' Council on behalf of the government. The relationship between the Council and the farmers was questioned, as complaints about maltreatment, injustices, corruption, and self-enrichment accumulated. Confidence in the Council was strained as the gap widened between these fiscal and commercial manipulations and the glossy public image paraded at the 10th anniversary conference.

The Seven-year Plan and the 1965 cocoa crisis

The financial position of the government continued to be critical. Import restrictions and foreign exchange control were partly successful in reducing the deficit on current account in the country's balance of payments but these measures simultaneously retarded the growth of government revenue from import duties, a major source of public expansion in the immediate pre-1961 period. The 1961 budget did not provide a sufficient basis for continued growth. In real terms government revenue declined in 1961/62 and 1962/63. The gap between stagnating revenue and growing expenditure continued to widen.[191] The level of current cocoa payments to the government was at first kept up by voluntary contributions and compulsory savings, which cut deeply into the reserves of the Cocoa Marketing Board. In 1961–63, the deficits of the Board amounted to over one-third of the payments due to the government. The reserves, however, were no longer capable of sustaining current payments, as they were frozen in loans to the government. The Board had great difficulties in meeting its commitments. It was unable to pay the farmers' voluntary contribution both in 1961/62 and 1962/63.[192] Its liquid balances had been lent to the government during the first years of the Second Development Plan. Most of its remaining foreign holdings were exchanged for Ghana government securities during the 1962/63 season.[193] The Managing Director suggested in view of this situation that the Board should in fact be directly incorporated with the government as far as its finances were concerned. Any profits or losses should be automatically passed on to the latter.[194] The Managing Director observed later that it was only with greatest difficulty that the Board was able to convince the government that the voluntary contributions should be discontinued altogether from 1963/64 season, as there were no liquid reserves from which payments could be effectuated.[195] It was agreed that the official producer price should be fixed at 60 shillings per load, less the 6 shillings "income tax". To producers it of course meant no change, but it served to reduce the deficits of the Board and, as suggested by the Managing Director, to make government's financial estimates "more realistic".[196]

The problem of assessing accurately the resources available was indeed a crucial one as the government went ahead with the *Seven-year Development Plan.* The Second (five-year) Plan of 1959 had for all practical purposes been abandoned by 1961, as a result of the reorientation of economic and political thinking, including growing contacts with the centrally planned economies of Eastern Europe. The strategic role of the state productive sector was emphasised. Efforts were made to integrate the private sector more closely within the framework of a comprehensive national plan. A Planning Commission was set up in October 1961 and a draft plan was circulated in 1963. The final document was published in early 1964. The new plan was launched by the President in the National Assembly in March that year.[197] The general political aspirations underlying the plan had been outlined in the CPP Programme, *"For Work and Happiness"*, adopted by the party congress in Kumasi in 1962. According to this document, the heritage of imperialism and colonialism had left Ghana without an independent capitalist class capable of developing the country. Socialism was therefore necessary. A "rapid change of the socio-economic structure" was seen as the precondition for socialism.[198] Accelerated economic growth, rapid development of the state and cooperative sectors, and the eradication of the colonial structure of the economy were the three principal objectives of the plan.[199] Excessive dependence on cocoa was of course the major feature of this still-lingering colonial pattern, but the success of the plan itself depended largely on the cocoa market. This was also explicitly acknowledged by its authors. The calculation of the financial resources available to the plan was "overshadowed by uncertainties regarding the future course of the world cocoa price". The chances of reaching the targets of the plan were said to depend on the extent to which the price of cocoa could be maintained at a level near the post-war average of £240 per ton for the duration of the plan period.[200] The price during the past three seasons had in fact been well below £200. The authors, however, saw signs of recovery from "the deepest depression of the post-war period".

The bottom remained to be reached and the Seven-year Plan did not receive much of a chance. The government's inability to impose "financial discipline" on itself within a framework of stagnating real resources may doubtless have been a major cause of the public financial crises which preceded the overthrow of the regime. The plan postulated strict priorities in public expenditure, "contrary to the trend of Government's operation in the recent past when high rates of expansion have been allowed in practically all lines of Government activity".[201] But the collapse of cocoa prices in the second half of 1964 in the face of the West African bumper crop ensured that calculations were thoroughly upset from the very start. The impact was enhanced by the unsuccessful countermeasures attempted by the Cocoa Producers' Alliance. The depressed prices and the hold-up of sales for several months aggravated the balance-of-payments difficulties and added to the inflationary upsurge.[202] The farmers had already been paid the agreed prices for a much increased crop. Public cocoa income collapsed. There was in fact nothing left to be paid to the government after the purchasing and marketing costs of the Cocoa Marketing Board and the Farmers' Council had been covered. Most of the reserves and liquid resources of the Board had already been transferred to the government and there were therefore no liquid funds left from which current cocoa duties could be paid. "Payments" were instead mainly effectuated by cancelling government debts to the Board.

The January 1965 Budget was introduced while the hold up of sales by the Cocoa Producers' Alliance was still in force. The critical financial position of the Marketing Board and its implications for government finances were reviewed by the Minister of Finance.[203] He declared that the present state of affairs could not continue much longer. As far as the producer price was concerned a reduction of four shillings or seven per cent had been volunteered by the Farmers' Council to finance a planned Agricultural Credit and

Cooperative Bank. It was heralded as another example of the "public spirit and foresight" of the farmers, but was hardly a solution to the financial difficulties of the government.[204] Part of the scheme was also a ten-year moratorium on cocoa farmers' debts to private money lenders. During that period no mortgaged farms could be sold by the creditors. Interest rates would be frozen.[205] According to the General Secretary, the Farmers' Council would take over the mortgaged farms and ensure their redemption. The Council would serve as a "bridge between the Bank and the farmers".[206]

The farmers' special contribution to the Agricultural Bank was only paid for the mid-crop 1965. It was replaced by a more drastic cut in the producer price when the new season approached. By mid-1965 the full impact of the catastrophic cocoa season of 1964/65 was making itself felt on the national and public economy. Ghana had been forced to sell most of the large crop at greatly depressed prices and even had difficulties in disposing of the old crop before the new one arrived. The national consumer price index rose by 35 per cent between October 1964 and July 1965.[207] The external payments crisis was acute, as short-term trade and suppliers' credits were falling due. The government saw itself compelled to call in the International Monetary Fund to ask for help in securing the renegotiation of some of these debts.[208] A preliminary report by an IMF mission recommended a number of measures to reestablish the economy of Ghana on a "sound" footing.[209] On one point the government was able to declare that it had already implemented the recommendations. This was the suggestion that the producer price of cocoa should be cut as part of a strategy to stop inflation. It had been possible "thanks to the realism and patriotism of our farmers who on their own initiative volunteered for a cut".[210]

The Supreme Council had met with the President at the Castle on 17 July 1965 in the presence of a number of ministers and high officials of the Cocoa Marketing Board.[211] The "accredited heads" of the Farmers' Council agreed on this occasion that the producer price should be reduced to 40 shillings, or ten shillings lower than what had already been agreed earlier in the year and 14 shillings, or 26 per cent, lower than the price paid during the past season. The Supreme Councillors agreed also that government subsidies on insecticides and spraying machines should be discontinued and that farmers should impose voluntary restraints on further cocoa planting. In return the government offered to cancel the four-shilling contribution to the Agricultural Bank and to exempt farmers from income tax. Local Councils and other bodies would also be barred from imposing special levies (surtax) on the cocoa farmers "without their consent and approval".[212]

The agreement at the Castle had been preceded by discussions and meetings involving Council leaders, the Cocoa Marketing Board, the Cabinet, the Ministry of Finance, Bank of Ghana and other government bodies. The formal basis was a request from the Marketing Board for a reduction in the producer price in view of its "grave financial crisis". The chairman of the Board declared that "no good businessman can afford to make losses at this rate" and expressed the hope that "it is not your wish that the Board should sink".[213] A full report was submitted to the State Enterprises Secretariat, the Board's new principal. Market prices would probably remain depressed, as it would take time before world cocoa consumption caught up with the higher level of production.[214] It was emphasised at a meeting of the Board that there should be no more savings, deductions, and special contributions but that the farmer should be paid "what his cocoa was worth". The gap between the "official" producer price and what the farmers actually were paid should be closed. For the special contribution to the Agricultural Bank it was proposed that the Board itself should provide the initial capital rather than pretending that the farmers' collectively through their own organisations "invested" in the Bank.[215] The Board's report formed the basis of a Cabinet Memorandum prepared by the State Enterprise Secretariat. Discussions were held in the "Special Finance and Foreign Exchange Committee", which had been set up under the chairmanship of the Minister of Finance to deal with the national emergency.[216] The committee endorsed the recommendations submitted by the CMB.[217]

The General Secretary of the Farmers' Council reported to the committee that his organisation had "attempted but failed to put the farmers in the proper frame of mind to receive such an important announcement".[218] The National Delegates' Conference meeting at Tamale in April 1965 had passed a resolution where it appealed to the Cocoa Marketing Board to maintain the producer price "by cushioning the world price with the reserves which have been accumulated in the past for this purpose".[219] One reason for the failure, according to the General Secretary, was the indifference of some of the District Commissioners, the local government and party officials, on this vital issue. It was essential that "the Party Propaganda machinery should be used to the maximum in preparing the minds of the cocoa farmers".[220] The Council, in its own campaign to convince the farmers, argued that market prices were intentionally suppressed by the forces of imperialism and neo-colonialism in order to crush Ghana's independent strategy. Farmers were urged to rush to the aid of the government in the present crisis and help it to hold out against this blackmail:

> Ghana being a thorn in their flesh as far as the fight against the forces of imperialism and neo-colonialism is concerned, they feel that by constantly lowering the world price they will succeed in strangulating her economically, thus forcing her to turn to them, cap in hand, for aids with strings attached.[221]

The agreement with the Farmers' Council on 17 July was acknowledged by the President in his Sessional Address when opening the reconstituted National Assembly in August, 1965:

> The readiness of our farmers to accept a voluntary reduction in the cocoa price paid to them has given us greater strength and encouragement to deal with this national emergency. It has given us a first-class lesson in the value of socialist morality and socialist responsibility. We congratulate the farmers. This is how a socialist society should function.[222]

The manifestation of public gratitude was followed up in a special motion passed in the Assembly "congratulating the cocoa farmers on their bold decision". Their "ususal generosity" and their "shining example of sacrifice" were acclaimed. The farmers had demonstrated their sense of "socialist responsibilities", "knowing very well that they belong to a Socialist Society in which the interest of the State is paramount".[223]

No voice was raised against the sharp reduction in the producer price in the National Assembly debate on the motion. Instead, the debate developed into a general attack on the Farmers' Council as a monopoly buyer of Ghana's cocoa. As mentioned in previous chapters, members took this opportunity to bring forward a number of specific and general grievances about the malpractices of Secretary-Receivers and other Council officers.[224] Members had to be reminded by the Speaker of the actual text of the motion, which congratulated the farmers (= the Farmers' Council) for their noble sacrifice.[225] Here we need scarcely recapitulate the nature of the reported grievances, as they have been fully discussed in Chapter 5. We may ask, however, why they were brought up on this particular occasion, as their direct relevance to the cocoa-price issue may seem slight. The way in which commercial administration and political representation were intertwined in the organisation of the Council, however, made the grievances in fact acutely relevant. The sharp cut in the producer price was itself difficult to attack in the general atmosphere of national emergency which had developed by mid-1965. But the virtual collapse of the public cocoa economy led to a growing concern with the quality of the relations between the cocoa farmers and the Farmers' Council. The Council's performance as a cocoa trader had direct bearing on its capacity to act as a cocoa tax broker and to represent the government as well as the farmers. The 1965 price agreement, the most critical point in the Council's career in this respect, contributed to the self-examination of the regime. Its vulnerable political basis in the cocoa areas was highlighted by the collapse of market prices. It was hardly a coincidence that it was the Minister of Finance who launched the belated attempt in early 1966 to reform the Farmers' Council in a more democratic and representative direction.[226] Being responsible for fiscal policy, he had particular reason to be worried about the government's political relations with the cocoa farmers and the fictitious nature of the agreements struck on their behalf by the Farmers' Council.

The decline in producer income and the crisis of production

The 26 per cent cut in the producer price in 1965 had most serious consequences for the incomes of farmers, sharecroppers, and the cocoa-growing communities as a whole. In real terms the cut was much deeper, as the value of the new crop was further reduced by inflation. The average national consumer price index for the first six months of the crop year (October to March) was almost 30 per cent above the corresponding figure for the preceding season.[227] The bumper crop itself had greatly contributed to this inflationary upsurge. But the impact on the cocoa communities was aggravated by the simultaneous sharp fall in production from the peak of 572,000 tons in 1964/65 to just slightly over 400,000 in the new season. The latter figure was of course more typical of the period as a whole, but the combination of the reduced crop and the low producer prices was disasterous. Total producers' income was almost halved in nominal terms, dropping from £58 to £30 million. When the 30 per cent inflation was added, the full extent of the catastrophe may be imagined. The rate of inflation in this case is calculated on the national average. If we turn to rural price indices and especially to areas a bit off the main market routes we find that local consumer prices had risen even faster.[228] For many farmers the 1964/65 crop may have brought some windfall profits. The cost of harvesting such a large crop was certainly raised, but it is still probable that the increase in total receipts was sufficient in many cases to cover both higher costs and price inflation. In 1965/66, there was no crop increase to compensate for price reduction and inflation. On the contrary, the drop in output greatly magnified the damage.

There is evidence that the Ghanaian cocoa industry by the middle of the decade had in fact entered into a serious crisis. The fall in production in 1964/65 may to an important degree have reflected a natural cycle where an unusually large crop is followed by a "recoil" before production resumes its momentum.[229] But there was no easy resumption this time. Production continued to decline and the average crop for the second half of the decade was below 400,000 tons.[230] The critical effects of falling producer prices and rising costs have been widely acknowledged. Other factors, such as the shortage of insecticides may be added. High costs were associated with the shortage of labour, related in its turn to the fall in producer prices. The prevalence of sharecrop contracts ensured that price cuts were automatically transferred to the most important group of farmworkers. Farmers complained about the exodus of labourers to the cities or to the neighbouring Ivory Coast.[231] The Techiman Traditional Council in Brong Ahafo reported to the deGraft-Johnson Committee in May 1966 that not all cocoa could be harvested, because of shortage of labour.[232] A farmer from Kenyasi No. 1, another Brong Ahafo village, claimed that he could have plucked 50 loads but as no caretakers were available much of it was spoilt. He was left with only 20.[233] Another farmer reported that his output had dropped from 160 till 100 loads "because no caretakers, no cutlasses and low price".[234] A farmer in Abodom, Bekwai, declared that many farms had been deserted and turned into bush.[235] Or in the worlds of an "absentee" farmer in Accra:

> At present many farms are in ruins because caretakers and labourers have abandoned their posts and left unceremoniously. The farmers have lost hope on their farms.[236]

The petition files of the deGraft-Johnson Committee contain many similar lamentations.[237] Some of the reports are certainly exaggerated: they were part of an effort to convince the government of the urgent need to raise producer prices. Still, the essence of these pieces of personal evidence is firmly supported by the hard facts about the deterioration of producers' income. Most of the petitions were written from the perspective of the big farm-owners and employers of labour, who naturally formed the most vocal section of the community. The mass of small producers working their farms alone or on a family basis were of course also affected. But while these small producers dominated in terms of numbers, the labour-hiring holdings contributed the major part of total output.[238] It is therefore correct to see the acute crisis of the Ghanaian cocoa industry in the mid-60's as primarily a crisis of the labour-hiring economy.

Increased schooling and the pull of the urban economy taxed at the same time the pool of family labour and raised the demand for hired labour even from farmers, who previously had been self-supporting in this respect. Labour shortage and rising costs affected current output on existing farms as standards of maintenance and harvesting were lowered. Marginal farms, those with low yield or those which were difficult to maintain or to reach, were the first to be hit. The steep decline in the return for capital invested in cocoa had also long-term consequences for production, as fresh planting to replace aging trees was discouraged.

There is some evidence that, as a result of the labour shortage, farm workers were able to improve their contracts.[239] On the whole, however, these revisions could not shield the predominantly sharecrop workers from the impact of the steep fall in producer prices. It is worth noting that some of the fiscal measures discriminated directly aginst the farm labourers. This was the case of the compulsory savings scheme, where, originally, the ten per cent deductions were supposed to be paid back to the farmers—with interest—after a ten-year period. The chances of the sharecroppers being able to claim back their share of these savings were slim. The Regional Commissioner's Office in Sunyani reported in 1961 that a series of petitions had been received from various "caretakers' unions"—that is, organisations of sharecrop workers—demanding that their shares should not be affected by the deductions. A petition from the Farm Caretakers' Committee, Kenyasi No. 1 and 2, was enclosed as a specimen. It was signed by the "chiefs" of the ethnic groups among the workers.[240] In the end, of course, these savings were changed into a "donation" and later, to a special farmers' income tax.

The workers' protests underline the fact that the efforts to ensure the producers' acceptance of fiscal policies through the Farmers' Council did not take account of the social organisation of production and the key role of the sharecrop contract. The various contributions agreed by the Council on behalf of the farmers automatically affected the workers. No attempt was made to "consult" or "mobilise" the latter. The Council was the farmers' organisation, despite its claims to represent the agricultural community as a whole.

The sharp decline in income for the cocoa producers may be compared with developments in the urban wage sector, which to such a great extent had been financed from cocoa. The table below shows the decline in producer prices from independence to 1966 and the development of the statutory minimum wage, and the "average" earnings in the "recorded" wage sector. These figures are taken as rough indicators of the relative fortunes of the cocoa growers and of the employees in the expanding urban economy. It does not, of course, give an adequate picture of the development of effective income. A full comparison would have to take into account both changes in costs of production and the different types of cocoa income, profits, wages and rent. It would also be necessary to make allowance for the various ways in which cocoa production is integrated in a food-growing economy. The figures have been deflated by the index of retail prices for Accra. There is no comprehensive rural price series available before 1963. The Accra index is of course not a good measure of changes in rural costs of living. Post-1963 figures indicate that rural prices rose faster than urban ones.[241] In this respect the Accra index underestimates the relative decline in the real value of the cocoa producers' income. Costs of production, on the other hand, are likely to have increased more slowly than production, which means that the fall in producer prices in that respect overestimates the decline in the producers' net earnings.

The table shows how the purchasing power of a load of cocoa was, over the ten-year period, reduced to one-third its former value. It was an unbroken decline. Already by 1959 the real producer price had been cut by one-quarter. By 1964 it was worth roughly half of its immediate pre-independence value. The sharpest decline took place in the last two years. The development of urban wages showed a rather different pattern. Both the government-controlled minimum wage and the average earnings of employees in reporting establishments rose during the early part of the period. A peak was reached in 1960. Subsequently a decline set in, somewhat faster for the minimum wage than for average earnings. Only in 1964–65, however, did the real minimum wage drop below the pre-independence level. In the case of average earnings, this only happened under the impact of the exceptional price increases of 1965. For the majority of employees in the urban sector it is probable that the minimum wage was more representative. The index of average earnings was raised and maintained at a higher level primarily due to the increases at the senior and middle range. The multiplication of such more advanced positions was an important concomitant of the expansion of the public sector. Employees at these levels were in a better position to secure promotions, upgradings, and increments, thus, at least partially, compensating themselves for price inflation. This was not the case for the daily rated or unskilled workers. The wage and salary developments for the staff of the Farmers' Council itself, as discussed in Chapter 5, may be taken as an illustration. The performance of the Council in this respect was likely to emphasise in the eyes of the cocoa producers the contrast between their own declining fortunes and the prosperity of the growing class of state and party

Cocoa producer prices and urban wages 1956–65: Indices of real value[1]

	Cocoa producer prices[2]	Minimum wage, Accra[3]	"Average" earnings, Accra[4]
1956	100	100	100
1957	89	100	106
1958	89	107	109
1959	73	104	110
1960	72	120	130
1961	61	113	129
1962	59	104	123
1963	55	100	121
1964	49	88	106
1965	33	69	86

Sources: CMB *Annual Reports, Economic Surveys* and Rourke and Sakyi-Gyinae 1972, pp. 4–5. See my Appendix, Table 4.
Notes: [1] All values deflated by average monthly index numbers of retail prices for Accra (June 1954 = 100). [2] Main crop season. [3] Statutory rate. [4] Average monthly earnings of African employees in establishments reporting to the Central Bureau of Statistics.

functionaries living off the surplus of the cocoa economy. A general upward revision of the salary scales of the Farmers' Council was undertaken in 1965.[242] Although inflation may have nullified the actual gains for most employees, the decision to raise salaries contrasted to the decision taken at about the same time to reduce drastically the price paid to the producers. The contradictory development of these different types of cocoa-based income may also be illustrated by the decision taken by the Cocoa Marketing Company, the sales subsidiary of the Marketing Board, to grant to all its employees a bonus of one month's extra salary "in consideration of a year's successful operations".[243] A disaster for the producers was turned into a "success" for the new middlemen.

Redistribution of cocoa income: some wider implications

The appropriation of surplus from the cocoa economy was of fundamental importance for the development of the CPP state and the Ghanaian political economy. The growth of productive forces, economic opportunities and class relations were affected. The appropriations were on a large scale, averaging over forty per cent of the market value during the CPP period as a whole (1951–1966). Public cocoa income, including the net surplus of the Cocoa

Marketing Board, accounted for much of the real financial resources available to the government. Taxes on imports, an important item in later years, were indirectly derived from the earnings of the cocoa dominated export economy. Foreign borrowing similarly represented claims on future export earnings, to which cocoa was likely to remain the major contributor for the foreseeable future. It is therefore no exaggeration to say that the CPP state was built on the cocoa economy.

Income was withheld from the cocoa economy on a massive scale. At first, this did not seem to impose any major restraints on production which expanded very fast in the late 1950's and early 60's. The extremely favourable market situation of the first post-war decade permitted producer prices greatly above those received during the previous two decades, in spite of public appropriations. The impact of these price incentives on output was delayed, this being a natural consequence of the time-lag involved in tree-crop production. If the appropriations actually served to restrain this rapid growth, it was a beneficial but, as it seems, unintended side-effect. Little was gained from this growth: production doubled but income was halved. Fresh resources of land, labour, and capital were committed to the established colonial export sector, adding to the distorted and dependent character of the economy. There is no evidence, however, that the government was aware of these dangers or consciously sought to restrict growth by keeping producer prices down. On the contrary; public investment in insecticides added importantly to overproduction.

As market prices fell, the impact of state appropriations on production was intensified. It was no longer a matter of the state merely sharing "super-profits" with the farmers. The viability of production was threatened by the attempts to protect public income. Farmers were squeezed between falling prices and rising current costs. The latter were partly caused by the expansion of cocoa production and the greater demand for labour which it involved. But the cocoa-based growth of the public economy bore a major responsibility for rising labour costs: the public sector had become the most important source of wage employment and government wage rates exercised upward pressures on wages in the economy as a whole. The massive expansion of public education added to the labour shortage by withholding child labour from production. It also reoriented a growing portion of the labour force towards types of employment where education seemed to matter. The number of people withheld from agriculture was far in excess of the number who actually found employment in the new wage-earning sector. In this way, the rise of the public economy undermined both directly and indirectly the system of production which provided its financial base. Both the peasant capitalist system of production and the state were threatened by the cost-price squeeze. This led to the combined crisis of the state and cocoa economies in the mid-1960's, aggravated by the refusal of the government to adapt public spending to the decline in real resources at its disposal. The prospects for public investment to yield income which could replace the threatened cocoa earnings were at best very long-term.

The reallocation of cocoa income to the public economy had important consequences for the development of agriculture outside the cocoa sector, as well. The expansion of export crops had not been accompanied by the development of food production capable of supplying the growing local market. An important part of export earnings had instead been absorbed by imports of foreign food. The process was self-reinforcing as these imports created new patterns of consumption which could be less easily satisfied from local sources. Moreover, even food which could be produced locally was imported. Price relations were heavily weighted in favour of the export crops. These price differences, however, were to a large extent a result of differences in access to the market. Foreign capital and the colonial state had developed market channels which suited the requirements of the export sector. Public transport, for example, favoured the exporting areas, while regions with potential for producing food for domestic consumption remained isolated from their markets.

The cocoa-based public economy created new conditions for the development of food production. The price difference between cocoa and food crops was reduced both as a result of high cocoa taxes and the expansion of the local market. During the 1950's the market continued to be restricted by a liberal imports policy but after 1960, food imports were cut back in favour of the more advanced requirements of the public economy. Investment in public transport helped to integrate local food production into the market economy. But the reorientation of productive forces was no simple process. Cocoa had been encouraged to absorb much of the fertile land close to existing settlements and roads. Price incentives would have to be sharply turned against cocoa in order to justify the destruction of bearing cocoa trees. Distant areas suitable for food crops, like the northern savannah, remained disadvantaged in terms of communications. Commercial food production was also affected by the growing shortage of agricultural labour. Food shortage continued to be a major factor in the inflation of the 1960's.

The redirection of cocoa income was reflected in the distribution of economic opportunities. Wage employment, formerly a marginal source of income, rose rapidly in importance. The urban economy expanded. A multitude of private economic activities grew up around the centres of state power. The cocoa economy was outdistanced by the public sector as the major field for personal advancement. Average earnings in the public sector rose steadily during the CPP period, except in the last two years. The income of the cocoa growers on the other hand, declined. Fortunes were not uniform. Public employees at the bottom of the scale, for example, were less fortunate than those higher up. Some farmers, similarly, were able to compensate themselves for falling prices by increased output. But most were not: the overall tendency was for income in the cocoa economy and in the public sector to develop in opposite directions.

The forms in which national production was consumed shifted from private commodities to public services. Free or heavily subsidised services like education, health services and roads became important items in the new pattern. Many public services were widely diffused and enjoyed on an individual basis. Others, like defence and diplomatic missions, were either "consumed" collectively or by a narrow section of the population.

There were important regional dimensions to this redistribution. Urban economic opportunities expanded rapidly, partly at the direct expense of rural ones. The rise of the cities, and the capital in particular, was the most conspicuous expression of the growth of the nationalist state. But there was also a certain redistribution from the "privileged" export-producing areas to those with a peripheral position in the colonial economy. The savannah regions had received little "development" in the past in return for their contribution of migrant labour to the export sectors of the south. They benefited from the redistribution of cocoa income. The commercial value of the savannah soils was upgraded and public services were more widely diffused. The value of their principal exportable commodity, labour power, also increased. This may have been a more dubious benefit as it may be argued that the pull of the southern labour market increased the underdevelopment of the north by obstructing the growth of local productive forces.[244]

The redistribution of resources and opportunities was manifested in changing class relations and in the growth of new social forces. The most conspicuous feature once again was the rise of the state bureaucracy, which contained a top stratum made up of the managers and senior executives of the public economy; this was a prospective "state bourgeosie" in charge of a growing state-enterprise economy, but in no way capable as yet of securing its maintenance and advancement from the profits of the latter. Below, there was a broad middle stratum of public-service employees with limited managerial and supervisory functions which distinguished them from the mass base of the bureaucracy: the clerks, typists, messengers, drivers, watchmen, and other workers. An important section of the new working class was composed of the growing number of employees in state enterprises. The public economy was serviced by a wide collection of groups, including traders, contractors, artisans, and other small entrepreneurs. At the bottom, we find a broad stratum of marginally employed, with odd jobs in the periphery of the public economy, often subsisting on the income of others.

Class developments in the cocoa economy were less manifest and also more ambiguous. Certain tendencies can be discerned but they require further exploration and documentation. The combination of falling prices and rising costs struck, as we have seen, in particular at the labour-hiring farmers, who dominated in terms of total output but not in numbers. The predominance of the share-crop system served at the same time to pass on the impact of low prices to the labourers who found themselves in a position similar to that of small tenant farmers. There were growing difficulties in recruiting share-crop workers. The more flexible wage-labour contracts seem to have spread. The working-class element in the cocoa labour force therefore probably increased at the expense of the semi-peasant element. Large and efficient holdings may have been able to hire labour on the same terms as, for example, state farms and the Workers' Brigade. Low-yielding, awkwardly-located, or otherwise "marginal" farms were in a more difficult position. The price-cost squeeze led to the "marginalisation" of a large number of previously profitable holdings.

What was the impact of public cocoa appropriations on the foreign-capital interests in the Ghanaian economy? In a first phase, these appropriations were firmly under official colonial control. They served, as discussed in Chapter 2, colonial interests both by reserving hard-currency earnings for metropolitan use and by "stabilising" the overheated post-war conjunctures of the Gold Coast economy. Cocoa manufacturers had a special interest in public investment in the rehabilitation of primary production, damaged by war-time neglect and swollen shoot. The local government seemed to give high priority to the expansion of production. Price incentives to the producers were high, in spite of public appropriations. The idea of long-term price stabilisation had also a definite attraction in this particular conjuncture. Market prices were far in excess of long-term prospects. More moderate producer prices, supported over a longer period, could be expected to encourage a stable growth of production.

Towards the end of the 1950's, it became increasingly clear that the accumulated funds of the Board were absorbed in supporting the public economy and not for price stabilisation. At this point, manufacturers expressed worries about the impact on production and the need to raise or maintain producer prices.[245] But as output began its steep rise at about the same time, the immediate cause of these worries was removed. Manufacturers' stocks were replenished and market prices were low. The position of the manufacturers was strengthened. The viability of production at the new level of prices and output may have been uncertain, but neither stagnant nor moderately declining supplies and higher prices were likely to upset this comfortable position for some time to come. Most importantly, the heavy dependence of the nationalist state on the surplus of the cocoa economy tied it to the continued prosperity of the latter. In the context of international competition this was likely to mean continued expansion of production as well as competitive restraints on the development of production costs, including such additional "costs" imposed by the nationalist state itself.

Cocoa based manufacturers had of course, a special interest in the continued viability of Ghana's cocoa industry, but metropolitan capitalism had also a more general stake in the growth and strength of the Ghanaian public economy. As a customer and as a field for investment, Ghana became considerably more attractive as a result of the reallocation of resources from cocoa to the public sector. Cocoa farmers were poor customers from the point of view of advanced industrial capitalism. Their purchases had a low content of capital and technology. The new Ghanaian state and the "modern" types of consumers it bred offered better prospects. The expansion of the cocoa-financed public-service sector required imports of capital goods, machinery and equipment. The imported needs of the new urban consumers were both greater and more sophisticated than those of the peasants. The expansion of the managerial and middle strata of the bureaucracy, for example, broadened the market for advanced consumer "capital" goods such as private cars, refrigerators and air conditioners. But did not the strategy of import substitution threaten this new and more profitable pattern of colonial trading? This may have been the case to the extent that poorly planned and implemented strategies in this direction held back the overall growth of the economy.[246] But import-substitution itself created an attractive market for the metropolitan capital-goods industry. Technology and management were important items in this new lucrative trade. Foreign capital played an important role in financing the import-substituting industries. The ability of the state to secure foreign credit on a large scale added to its superior attraction as a customer of metropolitan capital, as compared with the peasant producers. One crux, of course, was the growing insolvency of the CPP state. This, rather than the strategy of import-substitution, was the real threat to the interests of foreign capital. Dependent industrialisation, on the other hand, seemed to lead to a closer integration of Ghana in a world economy, still firmly controlled by metropolitan capitalism.

In an early phase, the role of the CPP state was primarily to redistribute the surplus of the colonial economy. This was not a matter of drastically changing the direction of the development of productive forces, although "industrialisation" was a long-term objective. The appropriated surplus was used for the general development of public services, which, in the absence of any serious attempt to redirect productive efforts, served primarily to support the existing pattern of production. High hopes were attached to the ability of an advanced economic and social infrastructure to attract foreign investment. In a second phase, here represented by the Second Development Plan of 1959, the government gave a new emphasis to the need to industrialise. A number of projects were listed, but foreign investment was still seen as the principal source of finance. The accumulated funds and the current income of the government continued to be primarily allocated to the expansion of public services. While large funds were committed to the Volta hydro-electric scheme, its role in the development of national productive resources continued to be obscure. It still depended heavily on expectations of private foreign investment. In the third phase, from 1961 onwards, the state assumed the role of leading entrepreneur for the industrial programme. Large resources were committed to a great variety of projects. The financial source, however, was primarily foreign credits and local monetary expansion. The continued growth of the public-service sector claimed not only all current revenue but also most of the savings of the state. It was in fact a heavy overcommitment of future earnings which would have to cover the current costs of the industrial programme plus the maintenance of public services at a level which so far had been supported by savings. The industrial projects depended on current imports of raw materials, spare parts, technology and management. Together with initial capital costs, this meant that these projects were unlikely to make any net savings of foreign exchange, not to mention net contributions to public finance, for many years to come.

The dependence of the CPP state on the surplus of the export economy increased. It became more vulnerable to the fluctuations of the world cocoa market, as demonstrated by the price depression and financial crisis of 1965. This increased dependence was a long-term problem which continued to affect its successors. A large service-oriented public sector had emerged on the basis of the colonial export economy. It claimed for its maintenance whatever current surplus could be extracted from this source. The state had not been able to create the conditions for its own "reproduction" outside the inherited colonial system of production. The fortunes of the state, its functionaries and dependents, continued to be closely tied to the fortunes of cocoa in the world market.

It may seem inevitable that increased state control over the surplus of the colonial export economy would also make the state more dependent on this sector. Yet, the contradiction could be bridged if these resources were used to develop new local productive forces capable of overcoming the limitations imposed by the colonial economy. In the case of the CPP, however, most of the surplus was absorbed by public services, while investments in production were financed to a large extent by foreign borrowing. A double burden was imposed on the colonial export sector. It was expected both to maintain the inflated public economy and to service the foreign debts.

The failure of the CPP was no doubt partly due to insufficient understanding of the practical difficulties involved, but there were also fundamental weaknesses in strategic conception and basic priorities in the allocation of the resources. It is argued in the concluding chapter that these weaknesses were closely related to the political organisation of the CPP state and the balance of class forces which manifested themselves in the party and in government. It is suggested that the close scrutiny of the political and commercial monopoly of the Farmers' Council undertaken in this study helps us to understand the nature of the limitations of the CPP and the contradictions on which its policies and performance were based. The Farmers' Council experience highlights, as I see it, a fundamental contradiction between national and sectional aspirations. At one level these reinforced each other in generating particular policies and institutions, but national objectives were simultaneously distorted and obstructed by the sectional interests which succeeded in entrenching themselves in the party and state apparatus. Let us sum up the key elements in this experience and examine their relevance for our understanding of the CPP and the political economy of the post-independence state.

VIII. The Farmers' Council and the CPP: Summary and Conclusions

This book has examined how the CPP organised the farmers and the cocoa trade through the United Ghana Farmers' Council. I have sought to explain the origins and functions of the Council's combined commercial and political monopoly within the context of the political economy of cocoa. It remains to summarise the contribution of this experience to the political history of the CPP period and to our understanding of the economic and political basis of the CPP state. The first section below focuses on the Farmers' Council itself: the quest for national control over the cocoa trade and its relation to the appropriation of cocoa income by the state. A second section discusses the rural political basis of the CPP state and the impact of cocoa on political development. In a final note I return to certain general questions about the class character of the CPP state raised in Chapter 1.

The Farmers' Council: the politics of monopoly

The United Ghana Farmers' Council will be remembered in the first place as the monopoly buyer of cocoa in the years 1961 to 1966. The cocoa trade engaged the bulk of its employees and it was as a cocoa buyer that the Council made its direct impact on the social organisation of the economy. The historical basis of the monopoly, however, and its significance for overall national development lay in the political role of the Council as the farmers' wing of the CPP. This was a role closely linked to the appropriation of cocoa income by the CPP state. Before I summarise this wider contribution let me first return to the impact of the Council within the context of the cocoa trade.

The cocoa-buying monopoly was a national solution, replacing a trading structure which had been dominated by foreign capital. It was in line with a general tendency of the CPP to seek national monopoly solutions in the commercial sector. The Ghana National Trading Corporation, for example, took over important sections of the import trade hitherto handled by foreign firms. The Ghana Commercial Bank similarly received monopoly protection at the expense of the foreign-owned commercial banks. The position of foreign capital in the cocoa trade had already been circumscribed by the introduction of the state export monopoly in the 1940's. The firms acted as licensed agents of the Cocoa Marketing Board at a fixed commission and with capital costs partly covered by advance payments from the Board. Much of the actual buying was undertaken, as before, by the local agents of these firms. Some were employees paid on a commission basis. Most were semi-independent entrepreneurs receiving capital advances from the firms. The contribution of the latter was primarily managerial. Foreign control, however, held back the growth of independent local trading organisations. Discrimination by the government and the Marketing Board reinforced the obstacles to the development of local capitalist forces in this sector. Farmer cooperatives, on the other hand, had succeeded in making inroads in the foreign-dominated trade with the administrative and financial help of the colonial government. The cooperatives were seen as a constitutional outlet for the activities of the farmer-traders who had threatened the flow of colonial trade on several occasions, most successfully in the hold-up of 1937/38.

The CPP intervened in the trade at an early point by sponsoring the Cocoa Purchasing Company (1952–57). The Farmers' Council, established in 1953, performed certain local organising functions on behalf of the Company as well as the party. The Council took over the cocoa business of the Company when the latter was liquidated as a result of financial irregularities. After independence, the Farmers' Council organised a campaign against the foreign firms, demanding the withdrawal of their buying licences. The United Africa Company, the major foreign buyer, announced its decision to leave the trade soon after the campaign had commenced. In pushing its own total control over the trade, the Farmers' Council met strong resistance from the cooperatives and their supporters in the government and the party. After some vacillation, a "merger" was enforced. The Council incorporated its rivals.

The new monopoly scheme was developed in cooperation with the second major foreign buyer, Cadbury & Fry, the local representatives of one of Ghana's most important overseas customers. The withdrawal of foreign capital from upcountry produce trading was in line with the general reallocation of resources into more advanced forms of commercial and industrial activity. In the cocoa trade, this process had already commenced with the introduction of the marketing-board system where foreign capital had a protected but transitional role. The Farmers' Council accelerated the transition. In the case of the dominant United Africa Company, the withdrawal was part of a growing, not a decreasing, engagement in Ghana's economy. For Cadbury, local trading profits were of secondary importance. Its primary interest was in ensuring expanding and high-quality supplies. Participation in the local trade had been one way of influencing production, as well as giving access to market intelligence. Close and friendly contacts with the local government and its agricultural establishment were other means for the same purpose. A national monopoly was compatible with the continued presence of the company in such official or unofficial advisory positions. As long as the new trading organisation was reasonably efficient, there were no important objections from this end. The company also emphasised the positive achievement of the monopoly when asked to give its views by the post-CPP government.

There is no evidence that the cocoa purchasing monopoly significantly obstructed foreign capitalist interests in Ghana. On the other hand, it is clear that the growth of local capitalist forces was seriously impeded. The cocoa trade was an important field for the rural petty-bourgeoisie. Many cocoa brokers were substantial businessmen with private capital engaged in addition to what was supplied by their foreign principals. Cocoa trading was a natural field of investment for big farmers. Many carted cocoa on their own lorries. Cooperatives as well, allowed considerable room for private entrepreneurship and profits. Many farmer-traders became cooperators in order to gain access to the trading capital and buying licences provided by the government. Opposition in the CPP against the liquidation of the "old" cooperatives was therefore also an attempt to protect private entrepreneurship in general against encroachment by the state and party bureaucracy. Although usually under the control of big farmers and farmer-traders, the cooperatives contained an important element of farmers' self-management. Opportunities in this direction were also blocked by the centralised monopoly of the Council.

The organisation introduced as a result of the monopoly differed greatly from the prevailing pattern. The village cocoa trade was incorporated into a bureaucratic system. Officials and clerks replaced traders and businessmen. An informal and overlapping organisation, engaging a large number of seasonal traders, many of whom were farmers or farm-owners, was replaced by a formal one, staffed by full-time employees. An important section of the rural commercial petty-bourgeosie was suppressed in favour of a new bureaucratic stratum. The village buying centres of the Farmers' Council became the advance posts of the state and party bureaucracy in its penetration of rural society.

The basic method of surplus appropriation within the trade did not change. The amount absorbed was decided, as before, primarily by the buying allowance granted by the Marketing Board. This was revised upwards but cost items beyond the Council's control may have accounted for much of the increase. On the other hand, no monopoly savings seem to have been achieved. Potential savings from the elimination of overlapping and competition were cancelled by the high cost of full-time clerks and officials. The new costs of bureaucracy included an elaborate accounting and auditing system and a heavy administrative superstructure.

Farmers were also paid official producer prices as before. Certain additional appropriations were imposed by the Council. The payment of bonuses or commissions, for example, was discontinued. These had been paid in the first place by the old cooperatives but also by other buyers, especially in order to tie up more important customers. The amount reserved in the official buying allowance for this purpose was used by the Council partly for the construction of cocoa sheds. Most of it was absorbed in financing its general business. Minor official deductions were made for a variety of purposes, including allowances for committee members, purchases of landrovers, and the construction of feeder roads. In addition, unauthorised deductions and collections were made by individual officials for private purposes. Cheating, manipulation of scales, weights, and quality grades had always been practiced by traders. A competitive market had certain built-in restraints in this respect. Monopoly trading required a different type of control system. The extent of the complaints against the Council, both before and after its downfall, indicate that the political and bureaucratic restraints on the monopoly were not particularly effective. Other monopoly functions entrusted to the Council, like the sale of agricultural implements and insecticides, added to the scope for private profiteering.

Farmers had reasons to be unhappy about the monopoly. Those who formerly had a direct stake in the cocoa trade had more reasons than others. Former cooperators, farmer-traders, and big farmers with special advantages under the old system may have been a minority among the farmers, but theirs was an influential minority. Their role as leaders and patrons in the cocoa community helped to consolidate opposition to the Council. The mass of small farmers were less directly deprived and might even have drawn some satisfaction from being treated on more equal terms with the big ones. But the Council did not use this to extend its political support. New privileged groups replaced the old ones. The Chief Farmers and committee members of the Council were given a share in the appropriations. They continued to be recruited from the minority of big and commercially oriented farmers who were considered to be "representative" of the farming community as a whole. For most farmers, big and small alike, the advantages of monopoly were probably not significant enough to compensate for the loss of freedom to sell cocoa to the men whom you "trusted" in a competitive market. Relations of mutual patronage between farmers and traders under the old system were certainly no protection against exploitation. The methods of private exploitation encouraged by the monopoly, however, were not only more easily visible but also more obviously illegitimate, as they contrasted against the formal rules and professed ethics of the bureaucracy.

Farmers' opposition to the monopoly had more fundamental causes than the petty-exploitation by its functionaries. The central purpose of the monopoly was not to liberate the farmers from exploitative middlemen but to enforce the political control of the CPP government, primarily for the appropriation of cocoa income by the state. Monopoly control over the cocoa trade was important to the CPP because private trading—foreign or national—seemed to stand in the way of full political control in the cocoa areas. Traders had played an important role in organising farmers in the past. On several occasions this had led to open confrontation with foreign firms and colonial authorities. Commercial aspirations were important in giving birth to the many farmers' groups which emerged in the 1930's and 40's. They followed the development of the new marketing-board system with suspicion. Agitation against the appropriations of the Board was part of nationalist attempts to enlist support among the farmers. Farmer-trader groups dominated the CPP-supported Ghana Farmers' Congress. This body split largely as a result of attempts to enforce party control. Yet even the pro-CPP elements, regrouped in the Farmers' Council, were faced with disruption as a result of the break-away of the Ashanti leaders only a year after the foundation of the Council.

Cocoa trading served as a vehicle for political competition. The cooperative movement offered a platform and a financial base for anti-CPP politicians. Foreign firms were rumoured to assist anti-CPP groups to buy over farmers from the CPP camp. The Cocoa Purchasing Company and the Farmers' Council worked closely together in the effort to use trading as a platform for the CPP. The merger of the two organisations after independence was a step in the consolidation of this strategy. The joint commercial and political monopoly of the Council was its logical conclusion. The purchasing monopoly ensured the "presence" of the CPP all over the cocoa areas. Other trading organisations which might provide protection, support and lines of communication for opposition politics were supressed.

General considerations of political security were reinforced by the critical role of cocoa in financing the CPP state. The organisation of the cocoa trade was manipulated to meet the political needs of the state. But these needs, in turn, were manipulated by groups aspiring to promote their commercial interests in the cocoa economy. These groups used the political monopoly claims of the party as a platform for outmanoeuvering commercial rivals, local as well as foreign, in order to achieve full control over the trade. Political and commercial aspirations reinforced each other. Political as well as commercial rivals were suppressed. In this way, monopoly not only served the appropriations of the state but also the advancement of sectional interests in the cocoa trade itself. The class character of the monopoly will have to be understood in terms of the interaction of these "national" and "sectional" aspirations.

What were the sectional forces which manifested themselves in the cocoa buying-monopoly? The cadres of the Council belonged to a commercial, petty-bourgeois stratum, which sought to advance itself by help of the financial and political support of the state. Their methods were primarily bureaucratic and not private-capitalist, although strongly marked by private entrepreneurship and closely associated with entirely private undertakings in related lines of business. The leading cadres had a background in the farming community but distinguished themselves from the common farmers by some form of education which made them look for business or clerical, rather than farming careers.

We traced the origin of the Farmers' Council to the attempts by CPP leaders to organise existing farmer-trader groups behind the party during the early phase of the national struggle.[32] The most important of the groups opted out of this shaky alliance when it was realised that the party, once in power, was unwilling to finance private trading organisations and to reduce the high rates of state appropriation. The disintegration of farmers' support was accelerated by the subsequent open confrontation over producer prices and cocoa taxes. An important section of the farming community became directly involved in the new anti-CPP front. The genuine farmer cadres of the Council were scattered, and depended for their survival on very concrete forms of patronage, such as credits and trading advances. The Cocoa Purchasing Company, the principal vehicle of this patronage, was in no way a farmers' organisation. It was staffed and managed by CPP businessmen and clerks, recruited from the primarily urban cadres of the party. The Council, on the other hand, was supposed to organise the customers and clients of the Company. When the Company was liquidated, and the Council took over, the centralised bureacratic methods of operation were retained. The farmer cadres continued to have little effective influence in spite of their official supervisory functions as Chief Farmers and committee members. Their role was largely symbolical as the "officially recognised" representatives of the farmers. In practice, they were subservient and dependent on the non-farming officials, who ensured their allowances and other privilegies and protected them against threats of "destoolment" from below. They were cut off from their presumptive electorates and incorporated as junior partners in the bureaucracy.

Why did the petty-bourgeois cadres of the Farmers' Council and the CPP opt for a bureaucratic monopoly solution to their personal and class aspirations? An important part of the answer lies in the structure of the colonial economy, where the power and resources of foreign capital, as represented by firms like the United Africa Company, contrasted with the extreme weakness of local private capitalist forces. To stand a chance against the superior power of foreign capital, the local petty-bourgeoisie sought to use the nationalist state as an instrument for advancing its own positions. Various roads were open. Private entrepreneurs could seek state support and protection, for example, through legislation which barred foreign capital from certain sectors of the economy. Alternatively, they could choose to become functionaries of the state support and protection, for example, through legislation which barred foreign capital from certain sectors of the economy. Alternatively, they could choose to become functionaries of the state or party and engage in the building of "public" enterprises with direct access to state funds and protection. These two roads could be combined in various ways.

The Farmers' Council represented the second road, but also the first had its supporters in the party. We have seen, for example, how the former chairman of the Cocoa Purchasing Company, a Kumasi businessman, received a big loan from the Cocoa Marketing Board to set up his own trading company. State support for the cooperative movement represented another alternative which appealed in particular to the big farmers but which also created opportunities for bureaucratic entrepreneurship. The "State Agency scheme" of 1961 was an attempt to create an umbrella under which various national buying organisations could operate. The ultimate victory of the Farmers' Council solution may be explained in terms of the advantages it offered as a vehicle of political control. But it was also a victory for that section of the petty-bourgoisie which had most successfully entrenched itself in the party apparatus and which a "public" solution therefore placed in an advantagous position.

The declining fortunes of the private capitalist forces in the party were also a consequence of the growth in state power. A "state capitalist" solution, as represented by the Farmers' Council monopoly, was in this perspective the logical outcome of the underdevelopment of private capitalism and the "overdevelopment" of the state. As such, it was part of the colonial legacy. The colonial export economy had created favourable conditions for the growth of an advanced state apparatus, but not for the growth of a national capitalist class.

At the ideological level, a "non-capitalist" solution could draw support from the radicalisation of a section of the party leadership. The shift was encouraged by the intensification of the anti-colonial struggle on the continental level (the Congo crisis). It meant a reassertion of the anti-capitalist element in the tradition of the nationalist movement. The deterioration of the cocoa market and the new urgency in changing the colonial structure of the economy highlighted the inadequacy of local capitalism and the need for active state participation in the development of new productive forces. The desire to protect the cocoa income of the state and to enforce political control in the cocoa areas were enhanced by the same factors.

National and sectional factors thus reinforced each other in a variety of ways. The cadres of the Farmers' Council, with the help of commercial capital supplied by the state, set themselves up as political brokers between the cocoa farmers and the state. A farmer "following" was enrolled. This following could in turn be marshalled in support of claims for more funds and privileges. The financial and political support of the state became simultaneously the basis of surplus extraction from the cocoa trade by the cadres of the organisation. The role as political brokers served in this way to ensure them a profitable position as middlemen in the relation between the producers and the world market. The state, on the other hand, made use of this combined commercial and political brokerage to pursue its own interests in the field of political security and public finance.

The rise and performance of the Farmers' Council monopoly can be explained in terms of these reinforcing factors. The Council was an organisation for controlling the surplus of the cocoa economy in the interest of the state and of its own cadres. It was not, as it was made out to be, an organisation for increasing the farmers' share in the profits of the cocoa trade. It served instead to raise or protect the share of the state and of those who assisted in this undertaking. The Council had a double repressive function: it prevented farmers from challenging both public and private appropriations.

This explains the failure of the repeated attempts by certain government leaders to enforce decentralisation and "true cooperative practices". These attempts were based on the illusion, actively propagated by the Council, that the monopoly served to ensure the farmers' self-management of the cocoa trade.

What did state protection and monopoly rights mean for the advancement of the sectional interests of the new middlemen? The monopoly was a large and prosperous enterprise. The trade was made to carry an imposing superstructure of functionaries, including Chief Farmers, committee members, and a hierarchy of marketing officers, accountants, auditors, storekeepers, and other clerks. Annual increments, promotions and revisions of salary scales ensured steady and sometimes quite dramatic improvements in official wages. This was in rather strong contrast to the declining fortunes of the cocoa farmers. In addition to the buying allowance paid by the Marketing Board, the Council could draw on special grants for its political work plus certain minor collections from the farmers. For most functionaries the official allowances and grants were the main source of income. Private profiteering, extortion, and corruption were widely spread but contributed probably only in a marginal way to total earnings. Individual officers, on the other hand, may have been highly successful in this respect, as well. The variety of methods involved are widely documented both in the Council's own records and in the files of the deGraft-Johnson Committee of 1966.[41] Real and artificial shortages and irregularities in the flow of resources controlled by the Council created favourable conditions for the privatisation of monopoly profits. Unofficial "prices" were attached to the various services performed. In some cases, for example, when there was an acute shortage of cutlasses, such monopoly prices permitted high private profit margins.

The essentially private and unofficial character of this profiteering should be stressed. There were examples of appropriations which seem to have been carried out with the support of the highest authorities, as in the notorious case of the "horning" of sealed cocoa bags. But on the whole, private profiteering was not sanctioned by the leadership. Proved cases usually led to dismissals or at least to the reposting of the culprits. But binding proofs were not readily forthcoming, especially not when those who were supposed to exercise supervision as senior officers or as committee members were themselves involved. Sanctions were often surprisingly mild. If possible, cases were not reported to the police or brought to court, in order not to damage the public image of the Council.

Just as in the pre-monopoly days, the resources of the monopoly served as a platform for a wider range of private economic activities more or less directly connected with the cocoa trade. They included transport, moneylending, and general trading. This was partly a matter of engaging funds handled by the monopoly for private purposes, partly of administering contracts and patronage. The carting of cocoa and the building of cocoa sheds, for example, involved opportunities of both sorts. The performance of Council officers in this respect supports the picture of private entrepreneurs seeking to promote their interests through the monopoly institutions of the bureaucracy.

The benefits and opportunities obtained and administered by the monopoly traders were substantial. What did the state get in return? What was the quality of the political organisation and control offered by the Council? Did it increase the political security of the state and strengthen the political basis for its appropriation of cocoa income? In answering these questions it is necessary to leave the reinforcing aspects of the interaction between national and sectional aspirations and turn to its basic contradictions. It is argued here that the character of the Council as an instrument of commercial class forces was a basic obstacle to the successful performance of its role as a political organisation. The cadres of the Council were imposed as new middlemen between the farmers and the market, concerned with appropriating surplus for their own purposes. Their class position was opposed to that of the farmers, anxious to protect themselves against middlemen of any kind. But these were also the cadres of the CPP, backed by the full authority of the state. It placed the party and the state in the same position as middlemen in the cocoa market.

The taxes imposed by the CPP state were unlikely to be popular with the farmers. The government, on the other hand, could give weighty reasons why it was necessary to use the surplus of the cocoa economy for purposes of national development. These arguments may not have been sufficient to satisfy the farmers, but they could still serve to reduce opposition and give moral backing to the claims of the state. The combined commercial and political monopoly of the Farmers' Council obstructed the efforts to legitimise the appropriations of the state in the eyes of the farmers by firmly identifying such attempts with the petty-exploitation and self-enrichment of its cadres. The potentially antagonistic character of the relations between the farmers and the state was enhanced by the natural contradiction between traders and producers.

The usefulness of the Farmers' Council as an organisation for political mobilisation and control was based on the assumption that it represented the pro-CPP farmers. It was by strengthening and extending this following that the Council was expected to make a positive contribution to the political and financial security of the state. The commercial monopoly, in turn, was thought to improve the conditions for mobilising a broader section of the farmers for the party and to help in demobilising potential opposition elements. By placing this delicate political exercise in the hands of a section of the community concerned with pursuing their own class aspirations at the expense of the producers, the operation defeated its own purpose. The Council had little credibility as a genuine farmers' organisation and its pretensions in this direction intensified the political alienation of the farmers. The pattern of political control and the private and corporate appropriations of the new middlemen demonstrated the fictitious nature of the claim that the Council represented the farmers.

The political credibility of the Council was further reduced by the crude methods by which the "support" of the farmers was simulated. The pretentions that the farmers had volunteered major reductions in producer prices in support of national development increased their natural unhappiness about these government decisions. The launching of the Second Development Plan in 1959 was the first major occasion for such voluntary contributions. In 1961 the Council sanctioned a compulsory savings scheme, only to organise rallies in the following year to turn these savings into voluntary donations to the government. When the scheme as a whole was discontinued in 1963, farmers again "volunteered" to continue payments, now in the form of a special "income tax". The large cut in the producer price in 1965 was surrounded by "consultations" and "agreements" of a similarly fictitious character.

By accepting the illusion of the Council as a representative farmers' organisation, the state renounced at the same time its own direct responsibility for the administration of the monopoly and its political consequences. Many government and party leaders were disturbed by the discrepancy between the Council's posture as a farmers' organisation and the lack of effective participation by farmers in the administration of the monopoly. Some looked for a solution to this dilemma in cooperative forms of self-management. But the suggested reforms failed to reconcile the commercial and political functions of the organisation and were therefore discarded. A last-minute attempt was made in the Budget Statement of February 1966 to suggest a different type of solution, one which sought to combine genuine participation at the local level with central political control.

The question as to whether these suggestions offered an answer to the Council's problems must remain a matter for speculation. It is important to stress, however, that the problems cannot be treated as peculiar to this organisation and distinct from those of the CPP state itself. The Farmers' Council was a creation of the party and the government and continued to have official backing and support. The origin of its political weaknesses and the contradictions on which they were based must be sought in the class character of the CPP state itself and in the nature of the social forces which brought it into existence and developed under its protection. The contradictions of the Farmers' Council reflected the limited organised support of the CPP among the rural masses and its predominant base in certain narrow petty-bourgeois strata. The fictitious mobilisation of the cocoa farmers became a substitute for any sustained attempt to reach and organise the mass of rural producers who in theory were represented by the Council as the agricultural wing of the CPP. While the CPP state expanded at a high rate, its political mass base remained undeveloped or was further underdeveloped as a result of this expansion. Let us look at the way in which this political underdevelopment was rooted in the dynamics of the colonial export economy and the political forces which it bred.

The rural political base of the CPP state

The development of the CPP state was determined in important respects by the cocoa price boom of the first post-war decade. The large surplus appropriated by the state, still largely under direct colonial control, laid the foundations for the massive growth of state institutions in the post-independence period. The way in which this growth was generated, however, simultaneously contributed to the underdevelopment of the political mass base of the regime. In doing so, it also reinforced the dependence of the new state on the traditional colonial export economy. The state grew at an extremely high rate, but without developing an organised political base capable of supporting, financially and politically, its maintenance and reproduction. This early expansion was achieved primarily as a result of windfall earnings in the international market, not from the mobilisation of local resources and political support.

The political costs of financing this rapid growth were kept low also by the service-oriented strategy of the government. New roads, schools, health centres and other public facilities were disseminated. No serious strains were imposed on the overall development of the economy. Imports of consumer goods were allowed to rise rapidly and prices could be kept stable, in spite of the sharp increase in urban and wage employment. The producer price of cocoa greatly improved in comparison with the preceding decades. The problem of how to ensure a continued flow of public services and imported consumer goods was not pressing. The international cocoa market supplied funds faster than the government was able to spend them. The accumulated reserves seemed to offer adequate security, even at the point when expenditure had caught up with revenue. The surplus-generating capacity of colonial exports was high enough to allow the rapid growth not only of the state but of the economy as a whole. A fundamental redirection of productive effort as envisaged in the programme of the CPP had no immediate urgency. Nor did such reorientation seem necessary in order to secure the future financial base of the new public service sector. A "consumptionist" approach to public funds was encouraged. Resources were committed without ensuring their long-term replacement. Free compulsory education, for example, involved heavy demands on future public resources. Expenditure was bound to rise as the number of age groups enrolled in the new schools multiplied.

The low political costs at which resources were acquired facilitated at the same time the advancement of the individual and class aspirations of the cadres of the new state. A "consumptionist" tendency was encouraged in this respect, as well. The political accountability of the cadres was low. They did not "owe" the resources at their disposal to political forces outside their own ranks. Intense competition between individuals and groups claiming a share in public funds may have imposed some restraints but, on the whole, state and party cadres were well placed to promote their own interests, collectively and individually. The resources under their control ensured a high degree of independence vis-à-vis the loosely organised movement which had brought them to power. As they entrenched themselves in the institutions of the state and the party, accountability declined even further and the ability of the broader movement to exercise influence deteriorated. The political mass forces released by the nationalist movement were actively underdeveloped by the state.

As long as the fortunes of the colonial export economy were favourable and the public cocoa surplus high, the contradictions between the cadres and the masses could be bridged or concealed by the welfare disseminating capacity of the state and the general "prosperity" of the economy. As the market deteriorated the contradictions were brought into the open. The tension between the overdeveloped superstructure and the underdeveloped mass base of the state was highlighted. Windfall appropriations had committed the state to the maintenance of a large bureaucracy, bred and nourished by the surplus of the good years. It could not be easily dismantled in the bad ones.

The colonial export economy was no longer able to sustain the flow of "low cost" resources to the public sector. Nor was it able to ensure the overall growth of the economy. The government responded by intensifying its hitherto half-hearted attempts to industrialise. It looked for fresh resources internally and externally to meet the cost of maintaining the established public sector and to finance the new investments in state production. Entirely new claims were made on the political organisation of the state. Cadres which so far had been engaged in receiving and distributing benefits were now expected to spearhead a national productive effort and to mobilise political support for a strategy which meant sacrifices and "austerity" for broad sections of the population. The decline in cocoa income and the shift in strategy reduced the capacity of the state to maintain its redistributive activities. The new state enterprises created fresh employment but this was done at a high cost, and the benefits were not as widely spread as those of the public service investments. The impressive structure of services erected was gradually undermined for lack of funds for its maintenance. Roads, hospitals, schools, and other facilities were all hit in due course by acute financial difficulties. Prices rose and real wages fell as imports were redirected to meet the needs of the public economy for production goods. The number of urban consumers continued at the same time to increase.

The CPP government overestimated its own popular support. The strong reactions to the 1961 "austerity" budget came as a shock. It was the first occasion when the consequences of the changed financial position of the state were impressed on the urban wage earners, previously an important source of CPP support. Cocoa farmers, on the other hand, had seen their fortunes fall already for several years. It was the third time the producer price was cut since independence. Certain steps had been taken to strengthen the political organisation of the party and to raise the ideological awareness of its cadres. An attempt was made, for example, to coordinate overall leadership by integrating the separate party wings more closely with the main body. Judging from the Farmers' Council, this was not particularly successful, at least not as far as this organisation was concerned. The weakness of the party as an organised political force capable of providing support for new strategies and demands was not overcome. Other aspects of political power were instead given greater attention. The repressive machinery of the state was developed. Greater political control over the courts was enforced. The state intervened with severity against the "illegal" strikers of 1961. Preventive-detention laws were used to imprison a growing number of real and suspected opponents and plotters.

The joint commercial and political monopoly of the Farmers' Council and its fictitious mobilisation of the cocoa farmers fall within the repressive pattern of political organisation developed during the last and "radical" phase of the CPP regime. The introduction of the commercial monopoly, which was designed to enhance the political control of the party in the cocoa areas, coincided in time with the first "austerity" budget. In the case of the Council, I have tried to show that its repressive functions did not work out in the interests of the political security of the state but rather served to provoke further political alienation and opposition. We have seen how political repression was manipulated and exploited by the new middlemen in pursuit of their own class aspirations at the expense of the producers. The underdevelopment of the political mass base, so glaring in the case of the Farmers' Council, was reinforced in the interest of the cadres and to the detriment of the national development effort.

The Farmers' Council records confirm the judgement of those who have emphasised the weak rural political organisation of the CPP. The Council, as the agricultural wing of the party, was entrusted the organisation of the peasantry, by far the most important mass stratum in Ghanaian society. The Council, however, concentrated its efforts on manipulating a narrow section of this stratum, the big and middle cocoa farmers, whose commercial aspirations provided the original platform for the organisation. The broad mass of agricultural producers were only marginally touched by its mobilising activities. These were fictitious in two ways: firstly, because support was pretended where opposition was expected, secondly, because the "captive" cadres recruited from this narrow stratum were presented as the representatives of the broad mass of the preasantry. The party manipulated those who were appropriated instead of basing its support on those who were supposed to benefit from the appropriations.

The cocoa appropriations of the state had important implications for the distribution of benefits and economic opportunities within the rural economy. The penetration of colonial export agriculture had led to a highly uneven development. The forest region was the centre of this economic activity, while the savannah was assigned the role as provider of cheap labour. By claiming and redistributing the surplus of the cocoa economy, the CPP state contributed to a new balance of regional and social forces. Prospects for improvements were held out for large sections of the population which had been disadvantaged under the traditional colonial economy. The Farmers' Council experience demonstrated the failure to organise politically the social forces intrested in redressing the unequal balance of the past.

The approach to the social organisation of the cocoa economy underlined this basic limitation of the party. The cleavage between a narrow stratum of big farmers and the mass of poor peasants and sharecrop labourers was not recognised in the political work of the Council. No attempts were made to seek allies among the mass forces of the cocoa economy who would have more to gain than to lose from a wider development of opportunities than those permitted by colonial export agriculture. Production was highly stratified. The bulk of the surplus, just as the bulk of total output, originated from the big and middle farmers. The mass of small farmers had more to gain from the new public economy but contributed less to its revenues. The big farmers, for example, had been able to buy education and privileges for their children. Now they were compelled to make the major contribution to a system of public education which put their children and those of the poor peasants on an equal footing. By treating the cocoa farmers as one group while in practice concentrating its attention on a minority of big and middle farmers, the CPP obstructed its own aims. The repressive and negative aspects of its organising efforts were enhanced. The positive possibilities were left unexplored.

The indifference of the party to the social organisation of production stands out clearly when we turn to the cleavage between labour and capital. The key role of hired labour, especially sharecroppers, in the cocoa economy was ignored. These were to an important extent the real cocoa producers. In spite of the socialist aspirations of the party, this large semi-peasant, semi-working class element was not identified as a potential ally.

The Farmers' Council not only failed to organise the farm labourers. It actively obstructed their attempts to organise themselves by claiming to be the only "accredited" representative for this section of the peasant community, as well. The Farmers' Council opposed proposals to improve farm-labourers' wages and to establish a "wage board" for this purpose.[2] The compulsory savings of 1961, as we have seen, would have transferred income from sharecroppers to farm-owners if they had actually been refunded.[3] The Farmers' Council appeared on these occasions as the representatives of the employers, not of the mass of agricultural producers which it claimed to organise.

The failure of the CPP to develop the Farmers' Council into an organisation capable of supporting positively the national redistribution of cocoa income was also demonstrated by the Council's limited reach outside the cocoa growing communities. Cocoa growers—farmers and labourers—constituted only one-third of the total agricultural population of Ghana. Even many who were registered in the census as cocoa growers had their main source of livelihood in other crops. Yet the organising efforts of the Council were concentrated on the cocoa farmers. Representatives of the non-cocoa regions or the large non-cocoa communities within the cocoa regions formed a small minority in the representative organs of the Council.

There was a distinct logic in the party's method of seeking out and organising those who had most to lose and least to gain from the redistribution of cocoa income. It was an exercise in *demobilisation*: the pacification of the potentially most dangerous source of organised resistance. By pretending that the "captive" farmer cadres of the Council represented not only the entire cocoa growing community but also the peasantry as a whole, the demobilising and repressive impact was extended to social strata which might have been able to provide the party with a genuine rural political base.

Two sources of underdevelopment in the political mass base of the CPP have been discussed above. One was located in the internal dynamics of the party and in the interaction and contradictions among the class forces which generated and were generated by the nationalist state. Sectional forces used the party and the state as platforms for their own class aspirations. They acquired vested interests in the continued underdevelopment of such broader political forces as might challenge or obstruct their appropriations. The process is illustrated by the performance of the Farmers' Council, where the commercial aspirations of its leaders contributed to the underdevelopment of the party's peasant base. The other source of political underdevelopment has been traced to the dependence of the CPP state on the dynamics of the colonial export economy. At one point this sector was able to provide resources for the rapid growth of state institutions, but it served at the same time to separate the regime from the mass forces of the nationalist movement which had brought it to power. The internal dynamics of the party interacted with the dynamics of the colonial economy in shaping the class character of the CPP state. The cocoa surplus was a natural target for sectional rivalries and aspirations. The independence of the new state vis-à-vis its mass base provided protection for the sectional pursuits of its cadres. The entrenchment of sectional interests in the state and party apparatus reinforced the political alienation of a broader following.

The national development effort was at the same time undermined. Attempts to develop new productive forces capable of overcoming the restraints imposed by the colonial economy were obstructed. The organised political forces required to support such attempts were left underdeveloped. The dependence of the regime on the surplus-generating capacity of the colonial economy increased. As cocoa prices slumped, the government sought protection in the repressive machinery of the state.

There were undoubtedly sources of the underdevelopment of the CPP rural base other than those discussed here. It would be necessary to examine the party itself more closely in order to reach a fuller understanding. Particular attention would have to be paid to the nature of the non-farmer cadres of the party and their position in relation to the social contradictions of rural society. The relations to community leadership, including chiefs and elders, would also have to be discussed.[4] The picture would have to include more subtle and complex patterns of political penetration and support than those represented by the fictitious mobilisation of the Farmers' Council. I believe, however, that the Farmers' Council occupied a sufficiently central position in the political organisation of the CPP state to allow certain general conclusions.

The Farmers' Council experience highlights the interaction and contradictions between national and sectional forces with relevance for the understanding of the class character of nationalist movements and state institutions, in Ghana and elsewhere. On the one hand, we find social forces concerned with establishing themselves in strategic positions within the framework of existing colonial relations in order to cash in on the surplus which these generate. These forces constitute the "new middlemen" of a neo-colonial economy. While often strongly nationalistic in their rhetoric, they become committed to the maintenance of the colonial structure of the economy on which they feed. On the other hand, we find those who seek to break the bounds imposed on national development by the colonial economy. Both of these forces are inherent in the colonial situation. Political solutions and institutions usually contain elements of both. Sectional and national aspirations at times reinforce each other; on other occasions, contradictions are clearly exposed. The Farmers' Council gives an example of this complex pattern of interaction—reinforcing as well as contradictory. The organisation was heavily marked by sectional aspirations, but its character cannot be uderstood in separation from the wider national objectives attached to the cocoa appropriations of the state.

The class character of the CPP state: a concluding note

The Farmers' Council experience highlights important issues raised in the wider discussion of the class character of the CPP regime. It allows in particular a close examination of the economic, social and political basis of the regime in the agricultural export economy. The specific relevance of the Farmers' Council experience in this respect should be clear from the above summary and discussion. I wish to conclude, however, by returning to some of the more general questions posed in the wider discussion of the CPP period and reported in Chapter 1. (See my summary of that discussion, pp. 19–29, above.) Does the Farmers' Council confirm the picture of a dependent, neo-colonial, petty-bourgeois regime painted by Fitch and Oppenheimer and others?[5] Or does my analysis lend support to Roger Genoud's interpretation of CPP strategy as representing a natural, national solution, reflecting the actual balance of social forces in Ghanaian society?[6] To what extent does the Farmers' Council illustrate the contradictions between neo-colonial and anti-imperialist forces for which Murray and Saul suggest that we should look?[7]

While remembering the obvious limitations in coverage of the present study, it seems nevertheless as if certain comments of more general relevance suggest themselves.

The financial dependence of the CPP regime on the cocoa economy was a central aspect of an international economic dependence inherited from the colonial period. Financial dependence meant political dependence, as well. We have seen how the underdevelopment of the political basis of the regime can be linked directly to developments in the international cocoa market. The methods used in organising the farmers demonstrate the failure of the CPP to establish an organised political base capable of reducing its political dependence on the cocoa market. It may be argued that the appropriations of cocoa income by the state (whatever the political methods used) served to build an alternative economic structure and thereby an alternative political base, as well. This would be in line with a common argument among those concerned with the prospects of a non-capitalist development.[8] The potential of the CPP strategy to achieve more basic structural changes may be open to question. However this potential is assessed, the period of transition was nevertheless bound to be long and difficult. The question of political organisation could therefore not be postponed. There was a need for solutions, adapted to the requirements of the transition period and rooted in the social organisation of the society. The fictitious mobilising efforts of the Farmers' Council highlight the vulnerable position of the government.

Yet, the CPP strategy—however misconceived—was an attempt to overcome the country's dependence on the colonial export economy. The failure to achieve more substantial results in the way of national economic liberation should not lead us to ignore the difference between this unsuccessful break-away attempt and the systematic consolidation of dependency and neo-colonial structures in most other newly independent African countries during the same period.[9] If Nkrumah's government was a dependent, neo-colonial regime it was at least one which attempted to liberate itself.

What about its petty-bourgeois character?[10] Looking at the Farmers' Council the strong petty-bourgeois element was obvious. The Council's origins were in the tradition of farmer-trader organisations, representing in particular the commercial aspirations of the large farmers: the most important entrepreneurial element in the rural economy. We have also observed the private entrepreneurship of many of the Council officers and the connections between private and official business activities. Cadres were recruited from the minority of large commercial farmers, rather than from the mass of small peasant producers of farm labourers. The Council even supported the interests of the capitalist farmers vis-à-vis the latter. I have tried to show how a section of this rural-based petty-bourgeoisie sought bureaucratic rather than private capitalist solutions to its social advancement.

The trading monopoly of the Farmers' Council, however, cannot, as we have seen, be reduced to an instrument for the self-advancement of such a bureaucratised petty-bourgeoisie. The class character of the monopoly was marked by its wider national role in the reallocation of cocoa income. This process had a broad social base in popular aspirations for social and economic development. The objectives of industrialisation, the modernisation of agriculture, wage employment, and social welfare were widely shared. The principal contribution of the study on this point is therefore to demonstrate the interaction of the sectional interests of this bureaucratised petty-bourgeoisie and the broader aspirations of the nationalist movement, reinforcing each other to a point, while simultaneously creating blockages in the way of further national advance.

It may be argued, with Genoud, that the state-bureaucratic solutions of the CPP were only natural (or inevitable) in view of the weakness of other class forces. Neither the local capitalist class, nor the working class was strong enough to make a major impact on national development. It is however, important not to ignore either the contradictions or the real choices facing the CPP leadership. These were clearly demonstrated in the struggle which preceded the decision to establish the Farmers' Council monopoly. The government's choice was not merely a reflection of the weakness of the private capitalist class but a decision which blocked the further advance of this class in a strategic sector of the economy.

The description of CPP strategy as a middleway between neo-colonial laissez-faire and a radical break-away attempt does not take account of the existance of several middleways. The regime faced a range of choices with widely different long-term implications for the development of the Ghanaian political economy. The failure to distinguish alternatives within such middle range is as unsatisfactory as the indiscriminate application of the neo-colonial/petty-bourgeois model of interpretation to regimes as different as, for example, Tanzania and Kenya or the Ivory Coast and Ghana (at the time of Nkrumah). It seems therefore justified to stress, with Murray and Saul, the contradictions within the leadership and the unsettled class character of the regime. The anti-imperialist forces may not have been particularly strong in the Ghanaian case but this fact does not exhaust the range of contradictions. National forces seeking to liberate themselves from colonial and neo-colonial restraints were clearly at work, pressing ahead in different directions. The leadership faced choices between solutions pointing in the direction of private capitalist development, state capitalism or non-capitalist development. The long-term implications of a particular solution were not always easy to forsee. The outcome was affected by the continued struggle between social forces, internal as well as external, seeking to turn a particular arrangement to their own advantage.

The Farmers' Council monopoly illustrates a case where a national solution was undermined by class forces using the monopoly for their own sectional aspirations, at the expense of national political objectives. It demonstrates the weakness of the organised political forces and their capacity to protect national goals against such distortions. It may be argued with some students that an important reason for the political failure was that an unsupportable burden was placed on the young party-state apparatus. It is easy to see how this argument fits the Farmers' Council experience. The Council was placed in charge of the country's entire cocoa crop in a period of rapidly expanding production. The relative success in maintaining the flow of the trade was achieved at the expense of the Council's wider political functions. It is similarly easy to see the relevance of arguments which emphasise the bureaucratisation of the party. The role of the Council as a party organisation was clearly submerged by its administrative and commercial functions. I have sought to demonstrate how this "overextension" and bureaucratisation were directly linked to the underdevelopment of the organised political base of the party and the self-advancement of state and party cadres.

Did this imply the rise of bureaucratic-bourgeoisie, as suggested by some students?[11] It depends, of course, on what one means by this term.[12] If used to indicate a new ruling class with substantial control over the means of production, it seems certainly premature: the degree of public ownership was limited. It seems similarly out of place to suggest that the bureaucracy played a significant role for the development of a local private capitalist class. The Farmers' Council experience rather confirms the picture given by those who have stressed the way in which local capitalism was frustrated by the state, at least in the later years of the CPP. It may be slightly more justified to speak of a bureaucratic-bourgeoisie, if we focus on the continued cooperation of the government with private foreign capital. The nationalisation of the cocoa trade, for example, did not prevent the major foreign interests from regrouping and engaging themselves even more deeply in the Ghanaian economy. It would be a great oversimplification, however, to reduce this new bureaucracy to an agent of more advanced forms of foreign penetration. The relative significance of the continued foreign participation must be judged in relation to the rapidly expanding public economy and the attempt to restrict the functions of foreign capital within the framework of state planning and control.

The experience of the Farmers' Council, however, points at an important aspect of this new public economy: its preoccupation with establishing control over the export-import trade. The CPP state acquired primarily an intermediary position in relation to the dominant forces of production operating in the Ghanaian economy. The new bureaucracy held a vulnerable position as middlemen in the relations between local primary producers and an erratic world market. This description applies, of course, most directly to the Farmers' Council itself, but it may also be used to characterise the CPP state in general. In important respects, the new state was rooted in international relations of exchange rather than in the social organisation of production in the country itself. The regime was placed largely at the mercy of these international relations. Fundamental limitations were imposed on its capacity to build a national economy.

Notes

Notes are used in this book for documentary purposes. I have avoided the inclusion of additional information or discussion. The method used in referring to particular documents requires explanation. Most of the references are to the *UGFC headquarters files.* When these are lettered and numbered this is stated in the note, for example, A 6, F 18 etc. Some files are in several volumes (B 18 vol. 2, A 65 vol. 5, etc.). Other files have headings, such as "Speeches and Messages", "National President", "Float", but no numbers. A few lack headings as well as numbers. Certain minutes of meetings were kept in separate minute books.

At the time I was given access to the UGFC files (1969–71) these had been "closed". They were kept in the offices of the successor organisation, the Purchasing Department (later the Produce Buying Agency) of the Cocoa Marketing Board. Judging from the file register the files were far from complete. Some had apparently been removed at the time of the 1966 military coup. The future of the files was most uncertain. (I urged that they should be moved to the National Archives). To facilitate future reference and research I have therefore also included references to my own excerpts (hand written research notes) from the files. A microfilm copy of these excerpts will be deposited at the Institute of African Studies, University of Ghana, Legon. It covers ten volumes of excerpts (I–X) from the UGFC files proper, an eleventh volume (XI) with notes from CMB minutes and files, and a separate volume on the deGraft-Johnson Committee files (see below).

Reference numbers in the notes below should therefore be read as follows: A 6/III:51 = UGFC file no. A 6, my note book no. III, p. 51.

The letters, petitions and memoranda submitted to the *deGraft-Johnson Committee* investigating the purchasing of cocoa in 1966 constitute the second most important source of documentation. Documents from the files of the Committee are marked GJ in the notes below without further specifying the date (May–June 1966) or the addressee (the Committee). The figure added (GJ/48, GJ/23 etc.) refers to the relevant page in my note book. The files were kept at the time of my investigation in the office of the Secretary of the Committee, Mr. P. K. Agboh.

Notes to Chapter 1. Ghana and the CPP: an Introduction

1. Richard Rathbone claims that "the CPP is undeniably the most written about of all African political parties". *Journal of Development Studies* 9:3, 1973 p. 391.

2. For Nkrumah's more important Pan-African writings see *I Speak of Freedom,* London 1961, and *Africa Must Unite,* London 1963. See also *The Challenge of the Congo,* London 1967.

3. Midnight pronouncement of Independence at the Polo Ground, Accra, 5–6 March 1957, in Nkrumah, *I Speak of Freedom,* London 1961, p. 107.

4. See, for example, A. L. Adu's assessment as quoted in W. Scott Thompson, *Ghana's Foreign Policy,* Princeton 1969, p. 414.

5. For an account of the economics of Pan-Africanism, see Reginald H. Green and Ann Seidman, *Unity or Poverty?,* Harmondsworth 1968. See also Samir Amin, *Neo-colonialism in West Africa,* Harmondsworth 1973: "Yet it can never be sufficiently stressed that the overriding obstacle to development is the division of the area into nine states". (p. 273).

6. A number of important studies of political development in the CPP period have been published, including David E. Apter's *The Gold Coast in Transition,* Princeton 1955 (and later

editions: *Ghana in Transition*) which covers the "transfer of power". Dennis Austin's *Politics in Ghana 1946-1960*, London 1964, remains the most important source of information on the party struggle. Maxwell Owusu's *Uses and Abuses of Political Power*, Chicago 1970, examines party politics in Swedru, a rural small town. Others, like Aristide R. Zolberg, *Creating Political Order: The Party-States of West Africa*, Chicago 1966, discuss the Ghanaian experience in a comparative context. These and other studies are introduced in a useful review article by Jon Kraus, 'On the politics of nationalism and social change in Ghana', *Journal of Modern African Studies* 7:1, 1969. See also Kraus' 'Political change, conflict and development in Ghana', in Philip Foster and Aristide R. Zolberg (eds.) *Ghana and the Ivory Coast: Perspectives on Modernization*, Chicago 1971. Highly critical studies of Nkrumah's "personal rule" have been written by Henry L. Bretton, *The Rise and Fall of Kwame Nkrumah*, London 1966, and T. Peter Omari, *Kwame Nkrumah: The Anatomy of an African Dictatorship*, Accra-Tema 1970. For sympathetic but equally "personal" interpretations, see Basil Davidson's *Black Star: A View of the Life and Times of Kwame Nkrumah*, London 1973, and Geoffrey Bing, *Reap the Whirlwind: An Account of Kwame Nkrumah's Ghana from 1950 to 1966*, London 1968.

7. The Ghanaian economy in the CPP period is unusually well documented, both in terms of available primary sources and of analysis. A basic source of information is the *Economic Survey* published annually by the Central Bureau of Statistics. A multitude of statistical information is also published by the Bureau. The 1957-63 period is well covered by the important *A Study of Contemporary Ghana*, vol. I: *The Economy of Ghana*, London 1966, edited by Walter Birmingham and others but written mainly by Tony Killick and Robert Szereszewski. *The Economic Bulletin of Ghana* has published numerous detailed studies, including a comprehensive survey of 'macro-economic data of Ghana' by T. Merrit Brown (Second Series, 2: 1 & 2, 1972). For two rather different interpretations of the CPP economic performance see Elliot J. Berg, 'Structural transformation versus gradualism: recent economic development in Ghana and the Ivory Coast' and Reginald H. Green, 'Reflections on economic strategy, structure, implementation and necessity: Ghana and the Ivory Coast', both in Foster and Zolberg 1971, as in note 6, above.

8. Nkrumah, *Africa Must Unite*, 1963, p. 108. See also his *Neocolonialism The Last Stage of Imperialism*, London 1965.

9. *Second Development Plan*, Accra 1959, p. 16.

10. Elliot J. Berg, (as in note 7, above) p. 204.

11. Recommendations in the preliminary report of the International Monetary Fund Mission as reported in the *1965 Supplementary Budget Statement*, Accra 1965; Also in *National Assembly* 10.9.65, vol. 40, cols. 717-718.

12. Reginald H. Green, 'Political independence and the national economy: an essay in the political economy of decolonisation', in Christoffer Allen and R. W. Johnson (eds.), *African Perspectives: Papers in the History, Politics and Economics of Africa Presented to Thomas Hodgkin*, Cambridge 1970, p. 273.

13. *Same*, p. 284. For a good summary of arguments why major structural changes are necessary, see same, pp. 277-288.

14. Bob Fitch and Mary Oppenheimer, 'Ghana: End of an illusion', Special Issue, *Monthly Review* 18:3, 1966 (also published separately in book form).

15. Roger Genoud, *Nationalism and Economic Development in Ghana*, New York 1969.

16. Roger Murray, 'Second thoughts on Ghana', *New Left Review* 42, 1967.

17. Jitendra Mohan, 'Nkrumah and Nkrumaism', *The Socialist Register 1967*, London 1967. Judith Marshall, *The Political Economy of Dependence: Ghana 1945-1966*, Institute of Social Studies, The Hague 1972 (mimeo).

18. See note 42, below.

19. Fitch and Oppenheimer, p. 82. They quote Charles Bettelheim, *Planification et croissance accelérée*, Paris 1965. See also André Gunder Frank, *Capitalism and Underdevelopment in Latin America*, New York 1967, and Samir Amin, *Le développement inégal*, Paris 1973 (the last chapter).

20. Fitch and Oppenheimer, pp. 82-84.

21. *Same,* pp. 113–120.
22. Mohan, p. 206 and 219.
23. Marshall, pp. 240 ff. For Green's indicators see his 1970 essay, pp. 289–290.
24. Marshall, pp. 249–51. This observation is in line with a common criticism of import-substitution strategies from widely different quarters. See for example, Ian Little, Tibor Scitovsky and Maurice Scott, *Industry and Trade in Some Developing Countries,* London 1970. See also H. Myint, *The Economics of the Developing Countries:* London 1964, pp. 156–57, and Samir Amin 1973 (as in note 19, above) pp. 181–83.
25. Marshall, pp. 283–84. She refers on this point to Teotonio Dos Santos and Tamás Szentes. For a useful summary of the argument on technological dependence see Jan Annerstedt and Rolf Gustavsson, *Towards A New International Economic Division of Labour?,* Roskilde 1975.
26. Marshall, pp. 259–60. See also Fitch and Oppenheimer, pp. 123–26.
27. Green 1971, p. 264.
28. Marshall, pp. 279–82.
29. For a rebuttal of Fitch and Oppenheimer in these terms, see Pat Sloan, 'Ghana without illusions', *Political Affairs* (New York), October 1966, pp. 16 ff.
30. Green 1971. See also Green 1970.
31. Genoud, pp. 148–50.
32. *Same,* pp. 215–16, 130.
33. *Same,* pp. 100 ff.
34. The model, outlined primarily against a Latin American background, has also been directly applied by Seers to the Ghanaian case. See 'The stages of economic development of a primary producer in the middle of the twentieth century', *Economic Bulletin of Ghana* VII:4, 1963.
35. Genoud, pp. 94 ff., 227–28.
36. *Same,* pp. 185 ff.
37. Murray, as in note 16, above. Compare Fitch and Oppenheimer, p. 128.
38. Murray, p. 27.
39. John Saul, 'The state in post-colonial society: Tanzania', *Socialist Register 1974,* London 1974. For a discussion of various theoretical issues involved see also my 'The post-colonial state: autonomy and social determination', Paper to the Nordic Political Science Conference, Århus 1975, Mimeo/Uppsala 1975.
40. Saul, p. 358.
41. Murray, p. 31.
42. For this school of analysis, see for example the journal of the Moscow Institute for World Economy, *Mirovja Ekonomika i Mezhdunarodnyye Ontnostreniya* (MEiMO). For summaries of this and other relevant material see the publications of the Central Asian Research Centre, London. Most of the references quoted here are drawn from their journal *Mizan* (replaced in 1971 by *USSR and Third World*). Much material is also published in English and other foreign languages in their country of origin. For a brief bibliography, see *Den Ny Verden* (Journal of the Institute of Development Research, Copenhagen) 9:4, 1975, pp. 120–22. For a collection of representative contributions, see *Developing Countries on the Non-Capitalist Road,* Proceedings of the Third International Conference of Africanist Marxists of the Socialist Countries, Varna 1971, Sofia 1974.
43. See David Morison, 'Africa and Asia: some trends in Soviet thinking', *Mizan* 10:5, 1968, and by the same author, 'USSR and the Third World: Ideal and reality', *Mizan* 12:1, 1970.
44. See Vasilii G. Solodovnikov, 'Problems of non-capitalist development', in *Developing Countries on the Non-Capitalist Road.*
45. See N. A. Simoniya in *Aziya i Afrika Segodnya* (1967), as quoted in *Mizan* 1968, 10:5 p. 168, and V. Tyagunenko in *MEiMO* 1962 (see note 42, above), as quoted in the same source.
46. V. Kondrat'yev in *MEiMO* 1965 (see note 42, above), as quoted in *Mizan* 7:11, 1965, pp. 12–14.
47. *MEiMO* 1968 (see note 42, above), as quoted in *Mizan* 10:5, 1968, p. 175. See also *Mizan* 12:1, 1970, p. 11–13.

48. A. Sterbalova, *MEiMO* 1968, and Ostrovityanov, *MEiMO* 1968, as quoted i *Mizan* 10:5, 1968, pp. 172, 180-81.

49. Tamás Szentes, *The Political Economy of Underdevelopment,* Budapest 1971, p. 317.

50. Claude Meillassoux, 'A class analysis of the bureaucratic process in Mali', *Journal of Development Studies,* January 1970, quoted and discussed in Saul, pp. 355-56.

51. Mohan, pp. 208, 221.

52. Genoud, pp. 201-22.

53. Saul, pp. 357 and 365.

54. See as different observers as Fitch and Oppenheimer (p. 53), Genoud (pp. 162 ff.) and David Apter, *Ghana in Transition* (second rev. ed.), Princeton 1972, pp. 376 ff.

55. Frantz Fanon, *The Wretched of the Earth,* Harmondsworth 1967 (First French ed., Paris 1961); Chapter 3: 'The pitfalls of national consciousness'.

56. See Thomas Hodgkin, *African Political Parties,* Harmondsworth 1961, pp. 68 ff., for the distinction between mass and elite parties.

57. See for example Mohan, p. 191 and Fitch and Oppenheimer, chapt. 1. See also A. A. Iskenderov on the "silence and indifference of the popular masses" as quoted in *Mizan* 2:1, 1970, p. 10.

58. Genoud, pp. 160 ff. and 194. Compare Apter 1972, and Zolberg 1966.

59. Fitch and Oppenheimer, p. 107. Mohan, p. 206. For Genoud's criticism of this position, see his pp. 175-77.

60. Apter 1972, p. 385.

61. Maxwell Owusu (as in note 6, above), pp. 249 and 268.

62. See Fitch and Oppenheimer, p. 107 and Genoud, pp. 160 ff.

63. Genoud, pp. 184 and 201-02.

64. Mohan, pp. 212, 219.

65. Ruth First, *The Barrel of a Gun: Political Power in Africa and the Coup d'Etat,* Harmondsworth 1972, p. 185. See also Fitch and Oppenheimer and Mohan.

66. Marshall, pp. 327 and 329.

67. Saul, p. 366.

68. Apter 1972, p. 377.

69. Fitch and Oppenheimer, p. 107.

70. Marshall, pp. 333-34, 336 and 325.

71. First, pp. 200 and 189.

72. Owusu, p. 272.

73. Mohan, p. 219.

74. Genoud, pp. 226-27.

75. Green 1971, pp. 247-48.

76. See for example contributions by S. M. Lipset and others under the heading "social stratification" in the *International Encyclopaedia of the Social Sciences,* vol. 15, 1968.

77. For a definition of class on these lines, see V. I. Lenin, 'A great beginning', *Collected Works,* vol. 29, Moscow 1965, p. 421.

78. For a distinction within the Marxist tradition between class as an abstract concept applicable to all post-tribal societies and as a concept referring to self-conscious groups of people specific of modern capitalist society, see E. J. Hobsbawm in *Daedalus,* Winter 1971, p. 36.

79. Mohan, p. 192.

80. The social composition of the CPP activists is documented by above all Dennis Austin, *Politics in Ghana,* 1963, which is a common source for those who discuss the petty-bourgeois base of the CPP.

81. Compare Lenin's inclusion of independent agricultural producers who produce for a market in which contradictions between capital and labour have developed although they do themselves not hire labour. V. I. Lenin, *The Development of Capitalism in Russia,* 1899, in *Collected Works* vol. 3, Moscow 1964, p. 177.

82. Tamás Szentes, *The Political Economy of Underdevelopment,* Budapest 1971, p. 312. He draws heavily on this point from Ignacy Sachs, *Patterns of Public Sector Economy in Under-*

developed Economies, London 1964.

83. Sachs pp. 68–69, as quoted in Szentes p. 312.

84. Sachs, pp. 79–80.

85. Annerstedt and Gustavsson, as in note 25, above.

86. For a fuller discussion of the structure of a dependent colonial economy, see Green 1970.

87. For figures, see *Economic Surveys.* There is at times an attempt to reduce the West African public sector economies to the selfenrichment of the political class or bribes to its supporters. See, f. ex. Douglas Rimmer, 'The public sector in West Africa: a characterisation', *African Public Sector Economies:* Proceedings of a Seminar at the Centre of African Studies, Edinburgh 1970, pp. 23 ff.

Notes to Chapter 2. The Cocoa Economy

1. For the historical development see Robert Szereszewski, *Structural Changes in the Economy of Ghana 1891–1911,* London 1965. See also A. Baron Holmes, 'The Gold Coast and Nigeria', in W. A. Lewis (ed.), *Tropical Development 1880–1913,* London 1970, and Kwamina B. Dickson, *A Historical Geography of Ghana,* Cambridge 1969. The 1900–60 developments are well covered by the collection of documents and statistics edited by G. B. Kay, *The Political Economy of Colonialism in Ghana,* Cambridge 1972. For the structure of the economy in the late 1950's and the early 1960's, see Birmingham and others (eds.) *The Economy of Ghana,* 1966.

2. Robert Szereszewski in Birmingham, p. 104.

3. Birmingham, pp. 18–19.

4. For the high level of economic organisation and commerce in West Africa before colonial occupation, see A. G. Hopkins, *An Economic History of West Africa,* London 1973.

5. Birmingham, p. 126. E. N. Omaboe summarises the special position of Ghana in an introductory survey, same, pp. 17 ff.

6. *Same,* pp. 19–20, 45, 106 ff., and 192 ff.

7. For a regional analysis of the Ghanaian economy, see Robert Szereszewski in Birmingham, pp. 62 ff.

8. For the constitutional history of colonial occupation, see Martin Wight, *The Gold Coast Legislative Council,* London 1948.

9. For a fuller discussion of the Ghanaian cocoa economy, see my *Social Differentiation and Class Formation in the Cocoa Economy,* (Forthcoming, 1976).

10. For the historical development of cocoa production in Ghana, see Polly Hill, *The Migrant Cocoa Farmers of Southern Ghana: A Study in Rural Capitalism,* Cambridge 1963, pp. 170 ff.

11. Birmingham, p. 372. For the structure of world production, see for example, FAO, *Review of the World Cocoa Economy,* FAO, CCP/Cocoa/63/6, 1963, and L. A. Are and D. R. G. Gwynne-Jones, *Cacao in West Africa,* Ibadan 1974. For Ghanaian figures see Kay as in note 1, above, and Ghana, *Statistical Yearbook.*

12. Birmingham, p. 333 and *Economic Surveys* for 1963–66.

13. Birmingham, p. 60. For a closer discussion of the dependence of the economy on cocoa, see Reginald H. Green, 'The Ghanaian cocoa industry: an examination of some current problems', *Economic Bulletin of Ghana* 5:1, 1961.

14. Birmingham, pp. 126 and 238–39.

15. See Polly Hill 1963, as in note 10.

16. F. R. Bray, *Cocoa Development in Ahafo, West Ashanti,* Achimota 1959 (mimeo).

17. Two important pioneering studies by W. H. Beckett cover well these different patterns: *Akokoaso: A Survey of a Gold Coast Village,* London 1944, and *Koransang: A Gold Coast Farm,* Accra 1945.

18. For the organisation of production, see Are and Gwynne-Jones, as quoted in note 11. For a general survey, especially of labour arrangements, see Polly Hill, *The Gold Coast Cocoa Farmer*, London 1956. See also Hill's and others contributions on cocoa in J. Brian Wills (ed.) *Agriculture and Land Use in Ghana*, London 1962.

19. For a discussion of the distinction between capitalist farming and peasant farming, see my forthcoming study, as in note 9, above.

20. Polly Hill, *The Occupation of Migrants in Ghana*, Ann Arbor 1970, pp. 73-74, and Birmingham, p. 239.

21. For a full discussion see my study, as in note 9, above. For a preliminary report on the purchasing records see my *The Distribution of Cocoa Income*, Staff Seminar Paper No. 13/1969/70, Department of Economics, Univeristy of Ghana, Legon 1970.

22. See Hill 1956, 1962 and 1963, and Beckett 1944 and 1945.

23. As in note 9, above.

24. These issues can be studied in, for example, the household surveys undertaken in the cocoa areas in the mid-1950's by the Ghana Government Statistician, *Survey of Population and Budgets of Cocoa Producing Families*, (1) *Oda-Swedru-Asamankese Area* and (2) *Ashanti*, Accra 1958 and 1960.

25. P. T. Bauer, *West African Trade*, Cambridge 1954, p. 224. See also the 1938 *Report* (as in note 27, below), pp. 48-49.

26. See, for example, S. LaAnyane, *Ghana Agriculture*, London 1963, pp. 98 and 102, and Sam Rohdie, 'The Gold Coast Cocoa Hold-up of 1930-31', *Transactions of the Historical Society of Ghana*, vol. IX, Legon 1968.

27. *Report of the Commission on the Marketing of West African Cocoa*, Cmd 5845, 1938 ("The Nowell Commission").

28. *Same*, p. 71.

29. *Same*, p. 147.

30. *Same*, pp. 157-61. See also extract in Kay, pp. 252-67.

31. Colonial Office, *Report on Cocoa Control in West Africa and Statements on Future Policy*, Cmd 6554, 1944. Also in Kay, pp. 267-72.

32. *Same*. For a summary and discussion of the war-time marketing scheme see V. D. Wickizer, *Coffee, Tea and Cocoa: An Economic and Political Analysis*, Stanford 1951, pp. 328-40, and Bauer 1954, pp. 246-59.

33. As in note 31, above. For the final proposals, see Colonial Office, *Statement on Future Marketing of West African Cocoa*, Cmd 6950, 1946, a slightly revised version of the 1944 proposals.

34. *The Gold Coast Cocoa Marketing Ordinance*, No. 16 of 1947. See also extract in Bauer 1954, pp. 280-81.

35. The 1946 *Statement*, pp. 6-8.

36. *Same*, p. 8.

37. See Appendix A, Table 2.

38. For Cocoa Marketing Board investments, see *Annual Reports*, Accra 1948 ff.

39. Fitch and Oppenheimer, pp. 42-47.

40. For expenditure against the reserves of the Board, see *Annual Reports*.

41. The 1946 *Statement*, as in note 33, p. 4.

42. For an account of the American opposition, see Wickizer, as in note 32, pp. 357-66.

43. Norman Edwards, *Cadbury on the Gold Coast*, Birmingham (Bournville) 1955 (mimeo), pp. 41-42.

44. The 1938 *Report* (as in note 27, above), p. 191.

45. For a discussion of pre-war trading in West Africa, see Bauer 1954, as in note 138; On the integration of merchandise trade and produce buying, see for example, p. 206.

46. Compare Bauer 1954, pp. 56-57, and Charles Wilson, *Unilever*, London 1968, Chapt. 9.

47. See for example, *The Colonies: The Labour Party's Post-war Policy for the African and Pacific Colonies*, London 1943.

48. Cocoa Marketing Board, *Annual Reports* 1950 and 1951.

49. The 1938 *Report,* p. 153.
50. *Same,* p. 154.
51. Compare Bauer 1954, pp. 251–52.
52. See Wickizer, p. 352.
53. *Same,* p. 352. See Wickizer's reference to the *Economist* 1945, p. 866.
54. Compare Austin, *Politics in Ghana,* pp. 66–72.
55. The 1938 *Report,* p. 29.
56. *Same,* p. 28.
57. Bauer 1954, p. 265.
58. See Roger Southall, *The "New" Political Economy Cadbury, Cocoa and the CPP,* unpublished manuscript, Centre of West African Studies, Birmingham 1975.
59. The 1938 *Report,* pp. 158–59.
60. The 1944 *Statement,* §§ 36 and 38 (Also in Kay, p. 269); The 1946 *Statement,* pp. 4–7.

Notes to Chapter 3. The Origin of the Farmers' Council

1. Nkrumah, *Dawn Broadcast,* 8.4.61, p. 1.
2. On the place of the *Dawn Broadcast* in the "1961 crisis", see Legum, 'Socialism in Ghana', in Friedland and Rosberg (eds.), *African Socialism,* 1964, pp. 143–49.
3. *Dawn Broadcast,* p. 6.
4. *Same,* p. 8.
5. I rely primarily on Austin, *Politics in Ghana 1946–1960,* 1964, for the account of post-war party politics and elections.
6. Richard Rathbone, *The Transfer of Power in Ghana 1945–57,* Doctoral Thesis, London 1968.
7. UGFCC, *A Decade of Existence,* 1963, p. 9.
8. *Report of the Commission of Enquiry into the Affairs of the Cocoa Purchasing Company Limited,* 1956 (*Jibowu Report*), p. 29.
9. Austin, p. 65.
10. *Same,* pp. 66 and 116–17.
11. National Farmers Union of (Gold Coast) Ghana, Eastern Division, *Petition to the National Liberation Council,* 1966, signed by Kwabena Koranteng Ameyaw (Gen. Secr.) (GJ/83).
12. *Report of the Commission on the Marketing of West African Cocoa,* 1938 (Nowell Report), p. 69.
13. UGFCC, *A Decade of Existence,* p. 9.
14. H. A. Dodoo, *Memorandum on Cocoa Buying,* 13.5.66 (GJ/45).
15. Bafour Peasah I and A. E. Ayisi, "for and on behalf of entire Farmers Groups in Suhum", 1966 (GJ/82).
16. UGFCC, *A Decade of Existence,* p. 9.
17. Ashanti Confederacy Council, *Minutes,* June 1948, as quoted in Austin, p. 80.
18. *Same.*
19. Austin, pp. 167–68. For Dzenkle Dzewu's and Mate Kole's relation to the Ghana Farmers' Congress, see *Jibowu Report,* p. 29.
20. Ghana Farmers' Congress, *Minutes,* October 1950, as quoted in Austin, p. 116.
21. CPP, *Towards the Goal. The Manifesto of the CPP for the Forthcoming General Election,* 1951, p. 3.
22. National Farmers Union, Eastern Division, *Petition to the National Liberation Council,* as in note 11, above.

23. J. C. deGraft-Johnson, *African Experiment,* 1958, p. 114.
24. *Same,* p. 115.
25. National Farmers Union, Eastern Division, *Petition to the NLC,* as in note 11.
26. *Same.*
27. *Same.*
28. Appiah Danquah, *Revitalization of the UGFC,* Undated memo (1955) (Speeches and Messages/IX:35).
29. *Same.*
30. Legislative Assembly 3.3.54, as quoted in *Jibowu Report,* p. 26.
31. For the history of the Cocoa Purchasing Company, see *Jibowu Report,* p. 17 ff.
32. See table p. 78.
33. Colonial Office, *Statement on Future Marketing of West African Cocoa,* 1946, p. 6.
34. *Jibowu Report,* p. 21.
35. Appiah Danquah, *Revitalization of the UGFC,* as in note 28.
36. First General Conference of the UGFC, *Minutes,* Kumasi, September 1953 (Minute Book of the General Delegates Conference/X:88).
37. Finance and Staff Committee, *Minutes* 20.3.54 (Minute Book of the Working Committee III:33).
38. Working Committee, *Minutes,* 2nd emergency meeting, 1955 (III:39).
39. *Same.*
40. On the Loans Scheme, see *Jibowu Report,* 1956, p. 20 ff.
41. Much of the *Jibowu Report* relates to irregularities in the administration of the Loans Scheme. See espec. p. 32. Some of the loans dockets were examined by Hill in "The Pledging of Ghana Cocoa Farms", *Cocoa Research Series* No 15, 1958.
42. *Jibowu Report,* p. 59.
43. *Same,* p. 60.
44. Finance and Staff Committee and Working Committee, Joint Meeting *Minutes* 17.2.54 (Minute Book of the Working Committee III:33).
45. Appiah Danquah, *Points raised against the UGFC in the Report of the Jibowu Commission,* Undated Memo (1956) (Speeches and Messages/IX:44). Compare *Jibowu Report,* p. 30.
46. See section "The political struggle 1954-56", below.
47. Supreme Council, *Message during the Colony Regional Trek,* March 1955 (Speeches and Messages/IX:33B).
48. Working Committee, *Minutes* 18.1.55 and 23.2.55 (III:37).
49. *Same,* 23.2.55.
50. Appiah Danquah, *Points raised against the UGFC,* as in note 45, above; compare *Jibowu Report,* pp. 26-30.
51. For the elections and the general political development, 1954-56, see Austin, *Politics in Ghana,* Chapters V-VII.
52. Austin, pp. 253-57.
53. Austin, p. 257.
54. *Jibowu Report,* p. 30.
55. Appiah Danquah, *The Necessity of Approaching the Government Constitutionally for a Review of the Cocoa Price,* Undated speech (August 1955) (Speeches and Messages/IX:37).
56. *Same.* Compare Austin, p. 255.
57. See note 55; Finance and Staff Committee and Working Committee, Joint Meeting, *Minutes* 3.9.54 (Minute Book of the Working Committee/III:36).
58. See Chapter 7, below, section "Producer prices and public development: the pre-independence experience".
59. For a brief discussion of the National Liberation Movement and the cocoa issue see Richard Crook, The decolonisation period in Ghana: political and economic factors in Ashanti underlying NLM opposition to the CPP, 1950-1957. Seminar Paper, Centre of West African Studies, Birmingham, 1972 (mimeo).

60. National Delegates Conference, *Minutes,* Okadjakrom, November 1954 (X:91).
61. UGFCC, *A Decade of Existence,* p. 11.
62. Kwame Boateng, Abuakwa, to UGFC, No date (I:2).
63. Kwame Borgor, Mankranso, to CMB Loans Department 25.1.60 (F 26/I:94).
64. Working Committee, *Minutes* 25.5.55 and 4.1.56 (III:38, 40).
65. *Same,* 20.4.55 (III:37).
66. UGFCC, *A Decade of Existence,* p. 11.
67. *Resolution by the Colony Region of the UGFC,* Agona Swedru 17.9.54. (CMB/XI:23).
68. Appiah Danquah, *Revitalization of the UGFC,* as in note 28.
69. Working Committee, *Minutes* 25.5.55 (III:38).
70. *Same* and 4.1.56, 24.2.56, and 20.3.56 (III:40, 41).
71. As in note 36, above.
72. Working Committee, *Minutes* 22.6.56 (III:42). See also UGFCC, *A Decade of Existence,* p. 14.
73. For the plebiscite, see Austin, p. 310.
74. UGFC, *Petition to Osagyefo* 25.3.61 (Memorandum File/IX:180).
75. Working Committee, *Minutes* 22.6.56 (III:42).
76. Appiah Danquah, *Points raised against the UGFC,* as in note 45, above.
77. *Government Proposals in regard to the Future Constitution and Control of Statutory Boards and Corporations in the Gold Coast,* 1956, p. 7.
78. *Jibowu Report,* pp. 23–25.
79. CMB, *Minutes* 2.5.57 (XI:34).
80. *Jibowu Report,* pp. 18 and 21.
81. *Same,* p. 21.
82. Working Committee, *Minutes* 30.8.56 (III:43).
83. *Same.*
84. Appiah Danquah, *Memorandum on the New Marketing Policy of the Cocoa Purchasing Company, Ltd.,* no date (1956) (Memorandum File/IX:172).
85. UGFCC, *A Decade of Existence,* p. 14. See also Appiah Danquah's statement in 1973, in Appendix B.
86. Working Committee, *Minutes* 4.1.56 (III:40); UGFC Agency, *Minutes of the Board of Directors,* various dates from 13.2.56 to 24.7.58 (Minute Book/X:81).
87. Working Committee, *Minutes* 28.5.57 (III:47).
88. UGFC, Special Conference, *Minutes,* Accra, May 1957 (Minute Book of the National Delegates Conference/X:92).
89. CMB, *Minutes* 2.5.57 (XI:34).
90. *Jibowu Report,* p. 19.
91. Ghana Farmers' Marketing Association, *Minutes of the Board of Directors* 24.5.57 (X:83).
92. Working Committee, *Minutes* 2.9.57 (III:48); CMB, *Minutes* 15.10.57 (XI:41).
93. CMB, *Minutes* 4.3.58 (XI:44).
94. Ghana Farmers' Marketing Association, *Minutes of the Board of Directors* 18.9.58 (X:87).
95. CMB, *Annual Reports.*
96. See Austin, *Politics in Ghana,* Chapter VIII.
97. UGFCC, *A Decade of Existence,* p. 14.
98. *Same.*
99. Working Committee, *Minutes* 26.6.58 (III:51).
100. See table p. 78, below.
101. Working Committee, *Minutes* 26.6.58 (III:51).
102. UGFCC, *A Decade of Existence,* p. 11.
103. *The General Secretary's Annual Report 1959,* p. 1.
104. National Delegates Conference, *Minutes,* Nsawam, January 1959 (X:96).
105. UGFC, *The Farmers' Year-book 1960–61,* pp. 10 ff., 16.
106. As in note 104, above.
107. National Delegates Conference, *Minutes,* Kpandu, January 1960 (X:108).

108. UGFCC, *A Decade of Existence*, p. 11. See Chapter 7, below.
109. National Delegates Conference, *Minutes*, Bibiani, September 1960 (X:112); *The General Secretary's Annual Report 1960*, pp. 33–34.
110. National Delegates Conference, *Minutes*, Nsawam, January 1959 (X:93).
111. Working Committee, *Minutes* 15.1.57 (III:45).
112. *Same*, 26.1.57 (III:46).
113. *The General Secretary's Annual Report 1959*, p. 14.
114. *Same*, p. 3.
115. Working Committee, *Minutes* 26.6.58 (III:51).
116. UGFCC, *A Decade of Existence*, p. 11.

Notes to Chapter 4. The Struggle for Monopoly

1. CMB, *Minutes* 21.4.59 (XI:55).
2. Supreme Council, *Petition of the CMB* 18.4.59 (J 29/IX:155).
3. *Policy Statement by the General Secretary*, District Chief Farmers' Conference, Kumasi, 6.11.55 (Speeches and Messages/IX:40).
4. *Jibowu Report*, 1956, p. 27.
5. John Dunn and A. F. Robertson, *Dependence and Opportunity: Political Change in Ahafo*, Cambridge 1973, p. 325.
6. Interview with W. R. Feaver, Bournville, Birmingham, June 1974. For a full discussion see Roger Southall, *The "New" Political Economy: Cadbury, Cocoa and the CPP*. Unpublished manuscript, Centre of West African Studies, Birmingham 1975.
7. As in note 3, above.
8. Appiah Danquah, *Points raised against the UGFC in the Report of the Jibowu Commission*, Undated memo (1956) (Speeches and Messages/IX:42).
9. Appiah Danquah, *Memo on Price Differentials* 14.6.58 (CMB/XI:11).
10. *The General Secretary's Annual Report 1959*, pp. 18–20.
11. National Delegates Conference, *Minutes*, Nsawam, January 1959 (X:93).
12. CMB, *Minutes* 25.2.60 (XI:68).
13. *Same*, 16.7.59 (XI:60).
14. Southall, as in note 6, above. It is not clear whether these figures refer to the UAC group as a whole, including, for example, G. B. Ollivant.
15. As in note 13, above. See also table, p. 78, above.
16. *The General Secretary's Annual Report*, Kpandu, 1960, p. 22.
17. UGFC, Sefwi Wiawso District, to the District Commissioner 14.7.60 (A 91 vol. 3/IX:69).
18. CMB, *Minutes* 25.2.60 (XI:68).
19. *National Assembly* 20.7.60, vol. 20, col. 123.
20. *Same*, 25.7.60, col. 246.
21. CMB, *Minutes* 30.8.60 (XI:73).
22. *Same*, 19.8.60 (XI:70).
23. National Delegates Conference, *Minutes*, Bibiani, September 1960 (X:111).
24. *Financial Times* (London) 14.9.60. On the increasingly nationalistic mood of public opinion see *West Africa* 20.8.1960, p. 945: 'Economic nationalism urged in Ghana'.
25. See Bauer, *West African Trade*, 1954, pp. 22, 67–68, 107–108 and Wilson, *Unilever*, 1968, Chapter 9.
26. Interview with H. D. Popplewell of Cadbury & Fry, Accra 1970.
27. Wilson, *Unilever*, pp. 215–19.
28. For a discussion of the profitability of the cocoa trade, see Southall, as in note 6, above.
29. CMB, *Minutes* 21.3.57 and 15.10.57 (XI:33, 41).
30. Popplewell, as in note 26. For details see Southall.

31. Southall, as in note 6.
32. W. R. Feaver, a Cadbury manager, was appointed Personal Adviser to the President on Cocoa. Correspondence in relation to his assignment is contained in UGFC File G 7.
33. *Report of the Committee of Enquiry on the Local Purchasing of Cocoa,* Accra 1967 (*deGraft-Johnson Report*), p. 151. See chapter 5, below, section 'The achievement'.
34. UGFC, *The Farmers' Year-book 1960–61,* pp. 106, 21.
35. *Report of the Commission on the Marketing of West African Cocoa,* 1938, Chapter IV.
36. See *The General Secretary's Annual Report 1959,* p. 19, and *The General Secretary's Radio Talk on the New Local Cocoa Purchasing Policy,* no date (1961) (Speeches and Messages/IX:46).
37. Supreme Council, *Petition,* 1959, as in note 2.
38. *Same.*
39. Appiah Danquah, *Memo,* 1958, as in note 9.
40. See Chapter 5, below.
41. *The General Secretary's Annual Report 1959,* p. 20.
42. National Delegates Conference, *Minutes,* Bibiani, September 1960 (X:112).
43. Supreme Council, *Petition,* 1959, as in note 2.
44. *Policy Statement by the General Secretary,* 1955, as in note 3.
45. *The General Secretary's Radio Talk,* 1961, as in note 36.
46. Supreme Council, *Petition,* 1959, as in note 2.
47. *CMB,* Minutes 15.10.57 (XI:40).
48. For the conditions see CMB, *Minutes* 21.5.57 (XI:35).
49. CMB, *Annual Report,* 1952, p. 4.
50. CMB, *Minutes* 21.5.57 (XI:35).
51. *Same,* 30.7.57 (XI:39).
52. *Same,* 15.10.57 (XI:46).
53. *Same,* 23.9.58 (XI:48).
54. *Same.*
55. *Same,* various dates from 21.5.57 to 15.10.57 (XI:35–40).
56. *Same,* various dates from 21.5.57 to 3.9.57 (XI:35–39).
57. *Same,* various dates from 16.4.58 to 21.4.59 (XI:44–55).
58. *DeGraft-Johnson Report* p. 159.
59. *CMB,* Minutes, 25.8.59 (XI:60).
60. DeGraft-Johnson, *African Experiment,* 1958, chapters 4 and 6. Also the 1938 *Report,* Chapter V. On the anti-trader bias of the colonial government, see also Dunn & Robertson, p. 57.
61. Registrar of Cooperative Societies, *Memorandum to the Committee on Cocoa Purchases* 27.5.66 (GJ/50).
62. DeGraft-Johnson, *African Experiment,* p. 78.
63. *Same,* p. 115.
64. *Policy Statement by the General Secretary,* 1955, as in note 3.
65. Ghana Farmers' Marketing Association, *Minutes of the Board of Directors* 3.7.57 (X:83).
66. *The General Secretary's Annual Report 1959,* p. 3.
67. *Ghana Farmers' Marketing Cooperatives Joins Ghana Cooperative Movement,* undated memo (1959) (Memorandum File/IX:176).
68. *Report of the Registrar of Cooperative Societies for 1959–60.* p. 4.
69. *Same,* p. 10. See also the statement by the Minister for Economic Affairs, *National Assembly* 2.3.60, vol. 18, cols. 181–82.
70. Registrar of Cooperative Societies, *Memorandum.* 1966, as in note 61.
71. *The General Secretary's Annual Report,* Bibiani, 1960, pp. 32–33.
72. National Delegates Conference, *Minutes,* Bibiani, September 1960.
73. Interview with N. A. Dennis, who was a senior official in the Cooperative Marketing Association at the time of the merger.
74. *National Assembly* 23.2.61, vol. 22, cols. 224–25.
75. Advisory Committee on the Marketing of Cocoa, *Minutes* 30.11.60 (VI:45).
76. *Same,* 28.12.60 (VI:52).

77. *Same,* 25.1.61 (VI:53).
78. *Suggestions on the Organisation of Cocoa Marketing,* Memo by D. T. Ackah, Ag. Registrar of Cooperative Societies 3.1.61. (File: Advisory Committee on the Marketing of Cocoa/VI:61).
79. Advisory Committee on the Marketing of Cocoa, *Minutes* 7.2.61 (VI:55-60).
80. General Purpose Committee, *Minutes* 4.1.61 (Minute Book of the Working Committee/III:59).
81. As in note 79.
82. Meeting on New Organisation for Purchasing Cocoa, *Minutes* 3.2.61 (File: Advisory Committee on the Marketing of Cocoa/VI:54).
83. *National Assembly* 23.2.61, vol. 22, cols. 210-11.
84. *Same,* cols. 192-228.
85. *Same,* cols. 226 and 220.
86. *Same,* cols. 195 and 216.
87. *Same,* cols. 206 and 208.
88. *Same,* col. 211. See for example the 1958 theft case involving a senior officer and £30,000 and the indiction of the Chief Marketing Officer in 1959; Working Committee, *Minutes* 27.11.58 and *Circular Letter,* 1959 (A 33/IX:19). See also *Audit Report* on misappropriated funds outstanding, March 1961 (J 56/IX:91).
89. *National Assembly* 23.2.61, vol. 22, cols. 195, 204, 207, and 212.
90. *Same,* col. 206.
91. *Same,* cols. 196, 208, 211, 212, 220, and 204.
92. *Ghanaian Times* 27.2.61; the editorial was quoted in full in the *National Assembly* 27.2.61, vol. 22, cols. 268-70.
93. *Same,* cols. 270-75.
94. *Same,* 7.3.61, vol. 22, cols. 297-98.
95. Interview with G. G. Mensah, Chief Marketing Officer of the UGFC, formerly with Cadbury & Fry.
96. Nkrumah, *Dawn Broadcast,* 1961, p. 6.
97. *Same,* p. 8.
98. UGFC, *Petition to Osagyefo* 25.3.61 (Memorandum File/IX:180).
99. ABASCO, *Petition to Osagyefo* 17.4.61 and *Points of Discussion with Osagyefo* 21.4.61 (Reports from the President's Office/VII:40, 48).
100. A. Mate-Johnson, *Letter to the President,* no date (place in file: April 1961) (Reports from the President's Office/VII:42).
101. *West Africa* 11.3.61, p. 271.
102. Commissioner of Police (Special Branch), *Letter to the President* 12.5.61 (Reports from the President's Office/VII:49).
103. See *National Assembly,* 17.5.61, vol. 23, cols. 677-87; 6.7.61, vol. 24, cols. 911-15.
104. UGFC, *The Farmers' Year-book 1960-61,* pp. 80-81.
105. *Same,* pp. 88-89. Also in *deGraft-Johnson Report,* pp. 4-5.
106. See Southall on Cadbury, as in note 6 above, section "The firm and its African employees".
107. Registrar of Cooperative Societies, *Memorandum,* 1966, as in note 61. See also *deGraft-Johnson Report,* p. 5. The fate of the cooperative assets was investigated after the coup by a committee, chaired by E. N. Moore.

Notes to Chapter 5. The New Middlemen

1. See the section on internal marketing in the Proceedings of the Cocoa Economics Research Conference, Legon, April 1973: R. A. Kotey, C. Okali and B. Rourke (eds.), *The Economics of Cocoa Production and Marketing,* Legon 1974.

2. Petition to the National Liberation Council submitted by Ghana Cooperative Marketing Association (signed by J. K. Aboagye and others), 20.4.66 (GJ/48).

3. *Government Statement on the Report of the Committee of Enquiry on the Local Purchasing of Cocoa,* 1967, p. 2.

4. *DeGraft-Johnson Report,* pp. 6, 7, 9, 33.

5. Harry A. Dodoo, *Memorandum on Cocoa Buying,* 13.5.66 (GJ/45).

6. *DeGraft-Johnson Report,* p. 151.

7. W. R. Feaver, "Review of two cocoa seasons", in UGFC, *A Decade of Existence,* 1963, pp. 50-51.

8. W. R. Feaver, *Aid Memoire,* 21.6.66 (GJ/62).

9. *DeGraft-Johnson Report,* p. 9. See H. A. Dodoo, *Memorandum,* as in note 5, for time series on grade I purchases, 1951-1966.

10. See for example letter from the Administrative Secretary, UGFC, to the Regional Officer, Koforidua, 13.3.63 (F 39/I:39).

11. *The General Secretary's Annual Report 1965,* p. 32. The National Delegates Conference, Kumasi, 1962, decided that the percentage defective beans tolerated for grade I should be lowered from 5 to 3 per cent. (A 65 v.2/X:27).

12. A spirited defense of the Farmers' Council achievement was presented at the Cocoa Economics Conference at Legon, April 1973, by the former General Secretary, Mr. Appiah Danquah. His speech is reprinted in my Appendix B.

13. See memorandum from the Secretary Receivers Employees Union, Ashanti (GJ/24).

14. *DeGraft-Johnson Report,* p. 20.

15. See particularly the debate on the 1963 *Budget* (agriculture, estimates), *National Assembly* 30-31.10.63, vol. 34, cols. 445-511. For the growing public criticism of the Farmers' Council, see also the section on "Cooperative democracy and government control", Chapter 6, below.

16. Bauer and Paish, *Economic Journal,* vol. 64 (1954), p. 709.

17. CPP Executive and others, Chiraa, *Resolution of Non-Confidence in the UGFC Committee in Chiraa,* 21.6.63 (File: National President/IX:165).

18. E. A. Prempeh, Sunyani (GJ/11).

19. *DeGraft-Johnson Report,* p. 15. See also memorandum from the Ashanti Farmers Union, signed by Opanin Kojo Owusuansah and others (GJ/25).

20. Nana Komera Ababioh, Manso-Mim, via Bekwai (GJ/21). See also the *deGraft-Johnson Report,* p. 15. See, for example, a petition against the Secretary Receiver, Nargode Society, Nyinahin, 30.5.64 (A 65 v.5/X:52).

21. Opanin Kofi Forfieh, Ahafo Maaban (GJ/13).

22. *DeGraft-Johnson Report,* pp. 16-17.

23. See for example memorandum from Peter K. Kunah, Techiman (GJ/6).

24. See the Chiraa "resolution", 1963, as in note 17.

25. CMB, *Minutes* 1.8.61 and 16.8.61 (XI:83-84).

26. Note of meeting on cocoa buying in the office of Kojo Botsio, the Minister for Agriculture, 27.7.61 (Proposals on Local Cocoa Marketing/VIII:13).

27. See, for example, *Notes and Instructions for the Guidance of Assistant Chief Marketing Officers,* April 1963 (Memo — Presidential Adviser on Cocoa/VI:22).

28. Petition to the District Officer, Juaso, from the Odikro, Elders and others, Atta-Mampong, 9.11.65 (F 68/II:222).

29. Ghana Commercial Bank, *Memorandum on Cocoa Financing,* 1966 (GJ/73).

30. J. S. Sarkodie and others, Pokukrom (GJ/14).

31. *DeGraft-Johnson Report,* p. 17.

32. *Same.*

33. Opanyin Yaw Hima, Sunyani (GJ/4).

34. Daniel Mensah and others, Tepa (GJ/40).

35. *The General Secretary's Radio Talk on the New Local Cocoa Purchasing Policy,* No date (1961) (Speeches and Messages/IX:46).

36. See for example memorandum from Daniel Mensah and others, as in note 34.

37. *The General Secretary's Annual Report 1963,* p. 17.
38. The Executive Committee of the CPP, Eastern Region, to the Minister of Agriculture, 14.6.63 (A 65 v.4/X:45).
39. Statement by Appiah Danquah reported in *West Africa* 28.4.62, p. 469.
40. *DeGraft-Johnson Report,* p. 33. See various memoranda in the files of the committee; J. K. Agyei, Hwidiem; "Cocoa Farmers", Wenchi; "Chiefs and People", Bechem; Kwaku Dormaah and others, Tepa; J. B. K. Owusu, Kumasi; Daniel Mensah and others, Tepa; A. Agbanyo and others, Kadjebi (GJ/3, 4, 9, 23, 29, 41, 91). See also the Chiraa "resolution" of 1963, as in note 17.
41. See for example *Minutes of the General Meeting of UGFC, Bogoso District,* 14.6.62 (F 38/I:173).
42. See memoranda from Opanin Kwabena Buansi, Sabronum, Ahafo-Ano; Jonas Abebrese and others, Woraso-Kumawu; Ashanti Farmers Union; Okyeame Kwame Owusu, Akim Oda (GJ/19, 20, 26, 77). See also letters to the *Daily Graphic,* Accra, forwarded to the Farmers' Council: "Worried Farmer", Damang-Wassaw, Oct. 1961. and "Save the Farmers from Them", Sept. 1962 (H 13/I:4).
43. See memorandum from Y. Aduamoyaw-Asiedu, Ex-Produce Examiner, Ministry of Agriculture (GJ/43).
44. The Administrative Secretary, UGFC, to the Regional Officer, Koforidua, 13.3.63 (F 39/I:39).
45. Letter to the *Daily Graphic,* "Save the Farmers from Them", 1962, as in note 42.
46. *DeGraft-Johnson Report,* p. 18.
47. *The New Farmer,* Journal published by the UGFC, Oct. 1965.
48. Motion by M. O. Kwatia, National Assembly, 21.5.64 (B 18 v.2/IX:54).
49. The committee was established in January 1966. See letter from the Principal Secretary, Ministry of Agriculture, to J. P. Acquah, UGFC, 18.1.66 (F 36/I:87).
50. G. Amo-Manu, Tanoso, Sunyani (GJ/1).
51. *The General Secretary's Annual Report 1965,* p. 31.
52. Ministry of Agriculture, Swollen Shoot Campaign, *Monthly Progress Reports by the Propaganda Secretaries,* 1964-65 (B 18 v.2/IX:56-61). Quote from report from Dunkwa Zone, Feb. 1965.
53. Correspondence between I.C.I., Tema and UGFC, Ministry of Agriculture, and Ministry of Trade, Aug. 1963, Feb. 1964, and Feb. 1966 (A 55/III:72; Correspondence — I.C.I./VII:106; F 72/VI:64).
54. *The General Secretary's Annual Report,* Kpandu, 1960, pp. 7, 30.
55. *Same,* Bibiani, 1960, pp. 12-16. See also minutes of meeting on the Capsid Control Scheme between UGFC and the Ministry of Agriculture, 23.2.59 (B 17/IX:30).
56. *Same.* See also *deGraft-Johnson Report,* pp. 49-50, on the Council's role in the distribution of spraying machines.
57. Agriculture Production Drive Committee, *Minutes,* 30.1.63 (The Committee/VII:145).
58. *The General Secretary's Annual Report 1963,* p. 16.
59. *Same,* 1965, p. 31.
60. *Same,* pp. 103-05.
61. Meeting at the Ghana National Trading Corporation (GNTC) with UGFC on the distribution of cutlasses, *Minutes,* 7.9.62 (F 41/I:6). See also GNTC, *Circular,* 7.9.62, to all District Agents/Managers (Summons and Minutes — Corporations and State Enterprises/V:1).
62. Address by Krobo Edusei to an Emergency Meeting of UGFC, Central Region, Cape Coast, 14.9.62, *Minutes* (F 38/I:167).
63. *DeGraft-Johnson Report,* pp. 50-51.
64. *Extract from Minutes of the Cabinet,* 27.4.65 (F 36/I:78).
65. *DeGraft-Johnson Report,* pp. 18, 50-51. See for example memoranda from Sabronum/Ahafo-Ano, Manso-Mim/Bekwai, and Akim-Awenade in 1966 (GJ19, 21, 79). See also petitions from Nargode/Nyinahin, 1964 (A 65 v.5/X:52), and from Kukurantumi and Juaso in 1965 (F 68/II:219, 223); See also reports from the Swollen Shoot Campaign, 1964-65, as in note 59.

66. *Report of the Commission on the Marketing of West African Cocoa,* 1938, pp. 34–35. Also quoted in *deGraft-Johnson Report,* p. 47.
67. See Chapter 4. Compare H. A. Dodoo, *Memorandum on Cocoa Buying,* 13.5.66 (GJ/45).
68. *DeGraft-Johnson Report,* pp. 14–15, 47.
69. *Same,* p. 19.
70. Kwasi Ampofo, Kumasi (GJ/35).
71. Abdulai Tuba, Nyinahin (GJ/36).
72. District Internal Auditor, UGFC, Sunyani, Report on the Nsuatre Society, Oct. 1964 (J 25 v.2/99).
73. *DeGraft-Johnson Report,* pp. 14–15.
74. Opanin Asare Boaten and others, on behalf of "700 Ashanti Farmers" meeting at Kumasi, 30.4.66 (GJ/27).
75. Kofi Dade and others, Kyremasi, via Akim Akroso, to the General Secretary, 5.6.65 (F 68/II:222).
76. As in note 20.
77. See for example petitions from Dodi Papase, 1961, Chichiwere via Kumasi, 1962, Kumawu, 1963, Nyinahin, 1964, and Kukurantumi, Akim-Manso, Akim Akroso, and Sefwi Anhwiaso, 1965 (A 65 v.2/X:21, 31; B 41/III:82; A 65 v.5/X:52; F 68/II:220, 221, 222; Corresp. — Misc./VII:73).
78. Opanin Kwadjo Okyere, Gyasehene, Kukurantumi, 28.6.65 (F68/II:220).
79. As in note 75.
80. *DeGraft-Johnson Report,* p. 17.
81. *The General Secretary's Annual Report 1963,* p. 18.
82. *Same,* p. 20.
83. General Purpose Committee, *Minutes* 30.6.64 (Minute Book/VI:78). See also *Circular,* 23.7.64, signed by the Administrative Secretary (A 33 v.2/IX:148).
84. As in note 38.
85. James G. Edusei, Kumasi, to Osagyefo, 25.4.64 (Reports from the President's Office-/VII:56).
86. *The General Secretary's Annual Report 1963,* p. 18.
87. See note 62.
88. Nana Kwasi Nyarko I and others, Dodi Papase (GJ/92).
89. See for example Meeting of Senior Officers, Koforidua, 28.4.64, *Minutes* (F 38/II:185).
90. J. F. Boabang, Berekum (GJ/13).
91. Regional Executive Meeting, Ashanti, 3.10.63, *Minutes* (F 38/I:178).
92. Treasury Committee, *Minutes* 10.3.64 (Summons and Minutes/VI:73).
93. As in note 41.
94. Ntotroso 19.12.61 (A 65 v.2/X:23).
95. *Minutes.* 2.2.62 (J 56 v.1/IX:90).
96. 9.2.65 (J 56 v.2/IX:93).
97. W. R. Feaver, *Report,* 21.3.62 (G 7/III:69).
98. Kwaku Dormaah and others, for Tepa Area Farmers Union (GJ/23). See also Petition against Committee and Secretary Receiver, Nkum, to the District Commissioner, Nyakrom, from Chiefs and Elders, No date (A 65 v.4/X:61).
99. B. A. Acheampong, CPP Branch Chairman, Kumawu, and others (signed or thumbprinted by 40 farmers), to Regional Officer, UGFC, Kumasi, 12.8.63 (B 41/III:81).
100. Ben A. Y. Ocloo, Wudome-Peki (GJ/89).
101. E. Y. Dzebu, Chief Farmer, Avenui-Awudome (GJ/90).
102. Kwasi Akyeampong, Kumasi and E. B. Ohemeng Baah, Akim Awenade (GJ/22, 79).
103. *DeGraft-Johnson Report,* p. 15.
104. "Complaint against Members of the Committee and the Secretary Receiver", Badu Society, Brong-Ahafo, June 1963 (A 65 v.4/X:48).
105. D. K. Pie, Kyiremfaso, Mampong/Ashanti (GJ/17).
106. Petition against the Secretary Receiver, Nargode Society, Nyinahin, 30.5.64 (A 65

v.5/X:52).

107. GJ/26.

108. Audit Report on Kpedze District, 30.3.62 (J 25 v.1/IX:98).

109. Daniel Mensah and others, Tepa (GJ/40).

110. Jonas Abebrese and others, Worasa-Kumawu (GJ/20).

111. *DeGraft-Johnson Report,* p. 15.

112. E. A. Prempeh, Sunyani (GJ/11).

113. Petition from farmers, Akropong, Ashanti, 4.12.61 (A 65 v.2/X:21).

114. Petition from Atta-Mampong, Juaso, 1965, as in note 28.

115. Memoranda from Dormaa Cooperative Union and from "Chiefs and People", Bechem Traditional Area (GJ/7,9).

116. Kwasi Acheampong, Ex-Chief Farmer, Kubin, Nkawie (GJ/43).

117. Opanin Osei Assibey Nsiah, New Tafo (GJ/32).

118. See note 11.

119. *DeGraft-Johnson Report,* p. 16.

120. "Worried Farmer", Damang-Wassaw, 1961, as in note 42.

121. "Save the Farmers from Them", 1962, as in note 42.

122. As in note 104.

123. Okyeame Kwame Owusu, plus one, Akim Oda (GJ/77).

124. Nsiah, New Tafo, as in note 117.

125. Kwadwo Mensah, Badu/Wenchi (GJ/5).

126. *DeGraft-Johnson Report,* p. 16.

127. Meeting of Senior Officers, Koforidua, 19.6 and 29.9.64, *Minutes* (F 38/II:187, 198).

128. Central Regional Auditor to Chief Audit and Inspection Officer, 3.10.61 (J 25 v.1/IX:96).

129. E. A. Prempeh, Sunyani (GJ/11).

130. *DeGraft-Johnson Report,* p. 17. See for example memoranda from Ashanti Farmers Union and B. K. Safo, Kumasi (GJ/26, 16).

131. E. K. Asubonteng, Assistant Chief Marketing Officer (Supervisory), at Meeting of Senior Officers, Koforidua, 16.1.64, *Minutes* (F 38 v.2/II:192).

132. See petition from Nkum, as in note 98, and memorandum from Mampong, as in note 105.

133. See petition against the Committee and Secretary-Receiver of the Santrokofi Society, Benua, Volta Region 24.6.64 (A 65 v.5/X:54).

134. Opanin Kwabena Buani, Sabronum, Ahafo-Ano (GJ/19).

135. See for example petition from Nkum, as in note 98, and memoranda from B. K. Safo, Kumasi, and Kwasi Bonsam, Ahafo-Ano (GJ/16, 18).

136. As in note 17.

137. J. O. Ansah, Kwaboanta, Kroboa-Coaltar, to the General Secretary, 23.9.65 (Corresp. Misc./VII:76).

138. Atta-Mampong/Juaso, 1965, as in note 28.

139. Kukurantumi, 1965, as in note 78.

140. E. B. Ahemeng Baah, Akim Awenade (GJ/78).

141. Swollen Shoot Campaign, *Monthly Progress Report,* Dunkwa, Feb. 1965, as in note 52.

142. *Same,* Western Region, March 1965. See also petition from Nargode Society, Nyinahin, 1964, as in note 20.

143. As in note 20.

144. 22.2.62 (A 69/IX:139).

145. For a discussion of problems of "bureaucratic self-control", see Chapter 6.

146. J 25 v.1/IX:94-96.

147. "On Secretary Receivers' Transfer", 12.2.64 (A 32/IX:126).

148. Meeting of Senior Staff, Koforidua, 2.3.64, *Minutes* (F 38 v.2/II:195).

149. See for example a report from the Regional Auditor, Volta Region, 12.5.62: "There is a great gap between the principles of Bookkeeping and the Secretary-Receivers. The average standard of Accounting Books is very low." (J 25 v.2/IX:98).

150. *Same,* Kpedze District, 30.3.62.

151. The General Secretary Meeting with Senior Staff, Sunyani 31.7.64, *Minutes* (F 38/II:188).
152. E. A. Prempeh, Sunyani .(GJ/12).
153. J. K. Agyei, Hwidiem (GJ/3).
154. Kwame Asase-Asa, Bontomuroso, Ntoboroso (GJ/17).
155. *DeGraft-Johnson Report,* p. 20.
156. Depot Keepers, Kumasi, to the President, 30.4.64 (A 38/IX:31). Also in *DeGraft-Johnson Report,* Appendix XIV.
157. 31.7.64 (J 31/IX:71).
158. Kofi Kwarteng and others, Agogo (GJ/18).
159. "Resolution adopted at a Mass Meeting", Mangoase, Techiman, signed Kofi Amankwaah, the Odikro (GJ/6).
160. GJ/7.
161. See for example memoranda from "Cocoa Farmers", Aboabo No. I, Dormaa-Ahenkro, I. E. Adade and Kwadwo Tieku, Goaso, Asare Boaten and others, Kumasi, and Anthony Kwasi, Jasikan (GJ/7, 10, 28, 92). Compare also *deGraft-Johnson Report,* p. 20.
162. Kwasi Bonsam, Ahafo-Ano (GJ/18).
163. Daniel Mensah and others, Tepa (GJ/41).
164. Peter K. Kunah, Techiman (GJ/6).
165. "Adomasah Farmers" (GJ/16).
166. Farmers' Chambers, Dormaa/Ahenkro (GJ/8).
167. Theophulus R. Bamfo, Accra (GJ/44).
168. Anthony Kobena Essel, Besease (GJ/68B).
169. D 2/I:146.
170. Circular, 7.12.62: "Transport Agreement" (Float/II:234).
171. *Progress Report on the Ghana Cooperative Movement,* Sept. 1962, and comments by R. M. Taylor, Financial Adviser, UGFC, 3.4.63 (F20/II:209–11). Compare J. Obuobi, Registrar of Cooperative Societies, *Memorandum to the Committee on Cocoa Purchases,* 27.5.66 (GJ/54).
172 Kojo Nimo, Berekum, to the General Secretary, 24.11.62 (H 13/I:5).
173. Ashanti Farmers Union (GJ/27).
174. Daniel Mensah and others, Tepa (GJ/41).
175. *The General Secretary's Annual Report,* 1965, p. 106.
176. Daniel Adomako Sarpong, Abenase, near Onwi (?) (GJ/42).
177. 20.3.63 (Float/II:231).
178. 23.4.63 (Float/II:232).
179. *Report of the Jiagge Commission Appointed to Enquire into the Assets of Specified Persons,* vol. I, Accra 1969, pp. 213–32.
180. See Chapter 2.
181. Hill, *Migrant Cocoa Farmers of Southern Ghana,* 1963, pp. 190–92.
182. *Report of the Jiagge Commission,* I, pp. 222–23.
183. *Same,* p. 231.
184. *DeGraft-Johnson Report,* Appendix XIII.
185. *Statistical Yearbook 1965–66,* p. 127. Establishments with less than 10 employees are only partially covered.
186. About 840 out of a total of 13,140 were employed in non-cocoa regions, or in the Agricultural Production Unit and the Other Crops and Livestocks Departments. As in note 184.
187. *Population Census 1960,* vol. IV, table 3.
188. *UGFCC Staff Pay,* 1966 (Unmarked/I:107–09).
189. *Report of the Jiagge Commission,* I, p. 213.
190. Finance Board, *Minutes,* 25.8.60 (F 50/I:154).
191. *Incremental Decisions,* Feb. 1963 (J 31/IX:69).
192. J 31/IX:72.
193. J 31 v.2 and P/J 13.
194. National Headquarters, *Establishment List,* 1963.
195. Ashanti Regional and District Establishment, *Salaries and Increments,* 1963.

196. Average for 1964: 153, and for 1965: 196. *Economic Survey,* 1961, p. 83, and 1965, p. 147.
197. *Incremental Policy,* no date (1964) (J 31/IX:73).
198. Finance Board, *Minutes,* 1960, as in note 190, and UGFC, *Conditions of Service,* June 1965 (A 32/IX:127).
199. *DeGraft-Johnson Report,* p. 11.
200. *UGFCC Staff Pay,* 1966, as in note 188.
201. See for example Senior Staff Meeting, Koforidua, 28.4.64, *Minutes* (F 38/II:185).
202. Working Committee, *Minutes,* 26.6.58 (III:52).
203. Finance Board, *Minutes,* 23.6.59 (F 50/I:153).
204. *Same,* 21.12.61 (F 50/I:156).
205. As in note 195.
206. As in notes 194 and 204.
207. Auditor General to UGFC, 19.5.64 (3.56 v.2/I:20).
208. CMB to UGFC, 19.6.61 (F 15/I:150).
209. See for example protests against such levies from the District Commissioner, Mampong, to the Regional Chief Farmer, Kumasi, 3.1.62 (F 39:I:36), and from Nargode Society, as in note 20. See also memoranda from S. W. Gyasie, Nkrankrom/Techiman, and J. F. Boabang, Berekum (GJ/6, 13).
210. Assistant Regional Officer, Volta Region, to all Secretary Receivers and District Officers, 7.1.64 (A 33 v.2/IX:147).
211. *DeGraft-Johnson Report,* p. 20. Compare memorandum from J. Kumi and others, Adumasaa/Suhum (GJ/78).
212. The figure suggested was £19,000. Its basis is unknown. Memorandum from Kenyasi No. I, Brong-Ahafo (GJ/10).
213. Circular from the Administrative Secretary, 24.10.63 (Float/II:226).
214. GJ/20.
215. Asare Boaten and others, as in note 74.
216. Working Committee, *Minutes,* 20.3.56 (III:41).
217. Kwame Annan, Agona Bobikuma (GJ/70).
218. Henry Akuamoah, Dodo Amanfrom (GJ/89).
219. *The General Secretary's Annual Report,* Bibiani, 1960, p. 24.
220. General Purpose Committee, *Minutes,* 4.1.61 (Minute Book of the Working Committee/III:60).
221. UGFC, *The Farmers' Yearbook 1960–61,* p. 84, and *The General Secretary's Radio Talk,* 1961, as in note 35.
222. *Cocoa Buying Instructions,* 1961–62 (A 33/IX:23).
223. UGFC, *The Farmers' Year-book 1960–61,* p. 106.
224. UGFC, *Petition to Osagyefo,* 25.3.61 (Memorandum File/IX:180).
225. Circular to all Regional Officers, 18.7.62 (A 33/IX:24).
226. *Same,* 1965, Appendix I; The 1961–62 crop was 409,000 tons.
227. For crop statistics see Appendix A, Table 1.
228. General Purpose Committee, *Minutes,* 24.1.62 (Minute Book of the Working Committee/III:64); *The General Secretary's Report 1963,* p. 17; Budget Meeting, 14.7.64, *Minutes* (F 33/I:55).
229. *The General Secretary's Radio Talk.* 1961, as in note 35.
230. Cadbury Brothers, Birmingham, to the Committee on Cocoa Purchases, 25.5.66, in *deGraft-Johnson Report,* p. 151. See also same report, p. 31.
231. The financial records of the Council were partly removed to the Auditor General's Department. An audit report produced by the latter has not been released.
232. *Budget Statement,* 22.2.66, pp. 19–20.
233. *deGraft-Johnson Report,* pp. 10–11.
234. W. R. Feaver, *Report,* 21.3.62 (G 7/III:69).
235. UGFC, *A Decade of Existence,* 1963, Appendix IV. See also *deGraft-Johnson Report,* p. 8, for the 1961–62 and 1965–66 figures.

236. See Chapter 2 for references to the purchasing records.
237. *DeGraft-Johnson Report,* p. 8. See circular, 13.3.63 (Float/II:233), and *Petition to the President on the Cooperative Movement,* 30.6.64 (Memoranda File/X:6). The latter mentioned that 576 societies handled less than 200 tons.
238. *The General Secretary's Annual Report 1965,* p. 8.
239. 28.7.65 (J 25 v.2/IX:101).
240. *DeGraft-Johnson Report,* pp. 27, ff.
241. Southall, as in Chapter 4, note 6, above.
242. *DeGraft-Johnson Report,* p. 29.
243. See note 39, above.
244. *The 1965 Budget Statement.* Also quoted in *deGraft-Johnson Report,* p. 28.
245. UGFC, *Accounts.* I am grateful to the Chief Accountant, Mr. K. A. Morrison for his assistance on this point.
246. *UGFC Budget 1964/65 (A 89).*
247. *DeGraft-Johnson Report,* p. 28.
248. *Report of the Commission on Kwame Nkrumah's Properties (Apaloo Report), 1967, pp. 50–51.*

Notes to Chapter 6. Participation and Political Control

1. See Chapter 5, p. 139, above.
2. *Budget statement* 22.2.66, p. 19.
3. *Same,* p. 20.
4. *Same.*
5. For the 1959 constitution see *The Farmers' Year-book 1960–61,* pp. 110 ff. For the 1965 constitution, see *The General Secretary's Annual Report 1965,* pp. 63 ff.
6. *The General Secretary's Annual Report 1965,* p. 65.
7. See also section "Social cleavages and monopoly representation", below.
8. *The General Secretary's Annual Report 1959,* p. 11.
9. *Same.*
10. Working Committee, *Minutes,* 27.11.58 (III:57).
11. National Delegates' Conference, *Minutes,* Nsawam, 1959 (X/93–94).
12. Compare Chapter 3, p. 73.
13. *The Farmers' Year-book 1960–61,* p. 117.
14. *The General Secretary's Annual Report,* Kpandu, 1960, p. 34.
15. See also section "Cooperative democracy and government control", below.
16. National Delegates' Conference, *Minutes,* Bibiani, 1960 (X:113).
17. See Chapter 4, p. 102.
18. *The Farmers' Year-book 1960–61,* pp. 88–89.
19. General Purpose Committee, *Minutes,* 4.1.61 (III:59). Compare Chapter 4, pp. 94–95.
20. General Secretary, Circular to all Regions, Districts, and Societies, 9.6.62 (A 33/IX:24).
21. Emergency General Meeting, Bogoso District, *Minutes,* 9.4.63 (F 38/I:117).
22. Administrative Secretary, Staff Circular, "Destoolment of Chiefs", 19.6.63 (A 33 v.2/IX:145–46).
23. *Same.*
24. As in note 21.
25. Emergency Meeting of the UGFC, Central Region, Cape Coast, *Minutes,* 14.9.62 (F 38/I:169).
26. *The General Secretary's Annual Report 1965,* p. 70.

27. Elections to district and regional offices were to be held "soon" according to the Regional Executive Meeting, Brong Ahafo, *Minutes,* 23.3.65 (F 38 v.2/II:201).
28. Petition to the Supreme Council from Agogo Society, 7.5.63 (Float/II:237).
29. K. A., Atwima Mfensi, 17.10.62 (A 65 v.3/X:37).
30. Ben A. Y. Ocloo, Wudome-Peki, 1966 (GJ/89).
31. *The Farmers' Year-book 1960–61,* p. 115; *The General Secretary's Annual Report 1965,* p. 74.
32. *Same,* second reference, pp. 72–74.
33. Regional Organising Secretary, Ashanti, to District Officer, Ejisu and Konongo, 19.6. and 26.6.62 (F 39/I:37).
34. See for exampel "Letter of Removal" from Regional Officer, Eastern Region, to District Chief Farmer, Asikuma, 14.1.63 (F 39/I:39).
35. For exemple, Regional Executive Committee, Sunyani, *Minutes,* 9.4.64 (F 38/II:186).
36. For example, Regional Executive Meeting, Kumasi, *Minutes,* 28.1.65 (F 38 v.2/II:200).
37. Petition from Asekyerewa/Bodwease, 19.6.63 (A 65 v.4/X:47).
38. As in note 25.
39. Portrait of E. K. Damoah, Regional Officer, Brong Ahafo, in *New Farmer,* July, 1964.
40. E. B. Ohemeng Baah, Akim Awenade, 1966 (GJ/78).
41. See p. 149, above. See also the 1965 constitution as in note 31, second reference, p. 73.
42. Meeting of the Executive of the National Cooperative Rubber Growers Association, UGFC, Takoradi, 23.3.63 (F 38/I:175).
43. For a fascinating account of the interaction between stool politics and party politics, see Dunn and Robertson, *Dependence and Opportunity: Political Change in Ahafo,* 1973.
44. See for example, Austin, *Politics in Ghana,* 1964, chapter VI: "The struggle for power, 1954–56".
45. Regional Officer, Ashanti, to Deputy General Secretary, 14.2.63 (A 65 v.3/X:41).
46. See for example address by the General Secretary, *The Place of the Country's Farmers in Self-governing Gold Coast:* "The farmer has found . . . how unfairly the chiefs and the governing bodies have treated him." No date (1956) (Speeches and Messages/IX:42).
47. See for example roll-call at the National Delegates' Conference, Kpandu, 1960, *Minutes* (X:97).
48. See speech by B. K. Senkyire, Minister of Cooperatives, at the Kwame Nkrumah Institute, Winneba 2.8.65 (F 36/I:86).
49. CPP Secretary, Akim Manso, to UGFC, Accra, 28.6.65 (F 68/II:221).
50. Assistant Regional Officer, Western Region, to Deputy General Secretary, 15.8.63 (A 18/IX:163).
51. See for example the Kpandu conference, 1960, *Minutes.*
52. Standing Committee, Brong Ahafo House of Chiefs, to UGFC, National Delegates' Conference, Kpandu, 1960 (F 7/I:158).
53. Kwabena Buansi and others, Sabronum, Ahafo-Ano, 1966 (GJ/19).
54. *Rent (Cocoa Farms) Stabilization Act,* 1962, and subsequent regulations and amendment. The files of the UGFC, especially D 1 vols. 1–3 ("Farms and Lands (Act)"), contain considerable material on this issue.
55. See the appeal against the expulsion; Nana Kwamina Ampina-Boateng, Odobeng, to General Secretary, 13.9.62 (F 39/I:38).
56. General Purpose Committee, *Minutes,* 12.4.61 (III:61).
57. W. R. Feaver to Kojo Botsio, Memo for meeting 10.8.61 (Proposals on Local Cocoa Marketing/VIII:11).
58. "He directed that the election of officers which had been confirmed in our constitution to be done yearly should be repealed". See note 25. See also circular: "I order that the election of officers . . . should be suspended". As in note 20.
59. Working Committee, *Minutes,* 4.1.56 (III:40).
60. National Delegates' Conference, *Minutes,* Kpandu, 1960 (X:109). See also the 1959 and 1965 constitutions as in note 31, pp. 124 and 81, respectively.

61. See also below p. 179.
62. See Chapter 3, p. 73, above.
63. Nkrumah to Appiah Danquah, 18.6.60 (CPP etc./IV:25).
64. CPP Headquarters to Appiah Danquah, September 1962 (CPP etc./IV:26).
65. Nkrumah, *Dawn Broadcast,* 1961, p. 8.
66. UGFCC *Accounts.*
67. *DeGraft-Johnson Report,* Appendix VIII. See also Chapter 5, p. 134, above.
68. See Chapter 5, p. 126.
69. *Same.*
70. Emergency Regional Executive Meeting, Kumasi, *Minutes,* 7.5.62 (F 38/I:164–65).
71. *Same.*
72. Auditor General to Secretary, Cabinet, 21.2.63 (J 56/IX:92).
73. See for example memo from K. B. Oppong, Tepa, 1966 (GJ/39).
74. Report from the Ministry of Agriculture, Produce Inspection and Grading Division, 17.6.66 in the files of the deGraft-Johnson Committee (GJ/60).
75. Proposals by the Senior Staff of the Internal Audit Division to General Secretary, 11.2.65 (J 25 v.2/IX:100).
76. Abdulai Tuba, Nyinahin, 1966 (GJ/36).
77. CMB, Purchasing Department, Oral information.
78. See Chapter 5, pp. 126–27.
79. Osei Assibey Nsiah, New Tafo, Kumasi, 1966 (GJ/33).
80. Treasury Committee, *Minutes,* 7.6.64 (Summons & Minutes/VI:74).
81. Petition from Society Chief Farmers, Twifu District, 9.12.64 (F 68/II:217).
82. Kwame Boateng, Asosomaso/Abuakwa, No date (I:2).
83. Letter from General Secretary terminating appointment, 27.10.65 (Recommendations-/VII:103).
84. J. Y. Ghann to Appiah Danquah, 5.11.65 (Recommendations/VII:103).
85. See especially unlettered file "Recommendations".
86. Appiah Danquah to J. Y. Ghann, 12.1.66 (Recommendations/VII:104).
87. Kwasi Ampofo, Kumasi, 1966 (GJ/36).
88. Petitions against the Secretary-Receiver, Nargode Society, Nyinahin District, 30.5.64 and 22.6.64 (A 65 v.5/X:52).
89. Report on a meeting with the Regional Council, Koforidua, 27.8.65 (A 65 v.5/X:60).
90. See also Emergency Meeting of the Regional Executive Council, Koforidua, *Minutes,* 24.10.63 (F 38/II:180).
91. Petition from Yaw Kusi, Akorabuokrom, Nyinahin District, to General Secretary, 18.6.65 (F 68/II:218).
92. See Austin, *Politics in Ghana,* 1964, pp. 381–82 and Kesse-Adu, *The Politics of Political Detention,* 1971.
93. Report from a Farmers' Conference held at Tepa, 24.9.64 (F 66/II:214).
94. E. B. Ohemeng Baah, Akim Awenade, 1966 (GJ/78).
95. Daniel Anyomi, Hodzo Aviefe, 1966 (GJ/90).
96. Samuel M. Y. Oteng, Kadjebi, 1966 (GJ/90). See also E. Y. Dzebu, Awenui-Awudome (GJ/90).
97. Daniel Mensah and others, Tepa, 1966 (GJ/41).
98. James G. Edusei, Kumasi, to Osagyefo, 25.4.64 (Reports from the President's Office/VII:59).
99. Petition from Santrokofi, Benua, Volta Region, 24.6.64 (A 65 v.5/X:54).
100. Peter K. Narh to Nkrumah, 10.6.61 (F 23/IX:50). See also petition from Chiraa, Brong Ahafo, 21.6.63 (National President/IX:165).
101. E. A. Prempeh, Sunyani, 1966 (GJ/12).
102. Petition from Atta-Mampong Society to District Officer, Juaso, 9.11.65 (F 68/II:222).
103. See for example memoranda submitted to the deGraft-Johnson Committee in 1966 by Kofie Forfieh, Ahafo-Maaban; Kwasi Bonsam, Ahafo-Ano; Kofi Kwarteng, Agogo; Kwabena

Buansi, Sabronum/Ahafo-Ano; S. O. Bonsu, Jamasi; Asare Boaten, Kumasi; J. B. K. Owusu, Kumasi; D. B. Dua, Agona-Akrofonso; A. Y. Ocloo, Wudome-Peki (GJ/13, 18, 19, 21, 28, 32, 39, 89).

104. Resolution from Kyekyewere, Offinso District, 27.7.63 (A 65 v.4/X:47).

105. "Unrest in Akim Swedru", Report by the Security Officer, CMB, to General Secretary, UGFC, 17.6.62 (A 69/IX:143).

106. As in note 20, above.

107. Emergency Meeting of UGFC, Cape Coast, 14.9.62, as in note 25.

108. District Commissioner, Mampong, to Regional Chief Farmer, Kumasi, 3.1.62 (F 39/I:36).

109. Awisahene Nana Kwabena Sarpong, Mampong/Ashanti, to General Secretary, 3.9.65 (Recommendations/VII:100).

110. C. K. Amanor, Assistant Secretary, Manya Krobo Constitutency Council, to General Secretary, UGFC, 29.12.60 (CPP—Regional and District Branches/IV:65).

111. A. Mate Johnson, M. P., to Tette Donkoh, Regional Secretary, UGFC, Koforidua, 22.12.60 (Same file/IV:65).

112. A. Mate Johnson, M. P., to General Secretary, UGFC, 22.12.60 (Same file/IV:67).

113. Executive Committee, CPP, Agogo, to Regional Officer, UGFC, Kumasi, 3.4.63 (National President/IX:167).

114. Eastern Region Steering Committee, CPP, to Minister of Agriculture, 14.6.63 (A 65 v.4/X:45).

115. See notes 49-50, above.

116. Petition from Nsokote Society, Adansi Banka, 5.6.65 (F 68/II:218).

117. Akim Manso, 28.6.65, as in note 49.

118. Petition from Kofi Dade and others, Kyremasi, via Akim Akroso, to General Secretary, 5.6.65 (F 68/II:221).

119. Nana Yao Duodu VIII, Omanhene, Nsaba Traditional Area, and Elders, to General Secretary, No date (E 8/IX:52).

120. See for example petition from Chiraa, as in note 124, below.

121. Atta Mampong/Juaso, 9.11.65, as in note 102.

122. See for example petition against the committee and Secretary-Receiver, Nkum, to District Commissioner, Nyakrom, from Chiefs and Elders, Nkum, no date (A 65 v.4/X:61).

123. Akim Swedru, 1962, as in note 105.

124. Resolution of Non-Confidence in the Ghana Farmers' Marketing Cooperative Committee in Chiraa, 21.6.63 (National President/IX:165).

125. See p. 145, above.

126. Working Committee, *Minutes,* 30.8.56 (III:44).

127. Ghana Farmers' Marketing Association, Board of Directors, *Minutes, 30.8.57 (Minute Book/X:85*).

128. *Working Committee, Minutes,* 26.6.58 (III:53).

129. "Ghana Farmers' Marketing Cooperatives Join Ghana Cooperative Movement", Memorandum, no date (1959) (Memorandum File/IX:176).

130. *Same.*

131. *Report of the Registrar of Cooperative Societies for 1959-60,* p. 4.

132. Commercial Committee, *Minutes,* 11.9.59 (F 59/II:207).

133. *The General Secretary's Annual Report 1959,* p. 14.

134. *Annual Report of the Registrar of Cooperative Societies for 1966-67,* p. 1. See also Nkrumah, *Dawn Broadcast,* 1961.

135. The General Secretary's Radio Talk on the New Local Cocoa Purchasing Policy, no date (1961) (Speeches and Messages/IX:47).

136. For the fate of the farmers' "profits", the commissions, see Chapter 5, pp. 137-39.

137. The amount collected this way was not great. Paid-up shares in the 1963 Balance Sheet were worth £63,501. *The General Secretary's Annual Report 1965,* p. 118.

138. *Report,* p. 10, as in note 131, above.

139. See for example the Committee to Advise on the Marketing of Cocoa, *Minutes,* 30.11.60

(Advisory Committee — Marketing of Cocoa/VI:51).

140. Registrar of Cooperative Societies, *Memorandum to the Committee on Cocoa Purchases,* 27.5.66 (GJ/58).

141. *Same.* See also *National Assembly* 27.9.63, vol. 33, col. 274.

142. Minister of Labour and Social Welfare, *Progress Report on the Ghana Cooperative Movement,* September 1962 (F 20/II:209).

143. *Same.* See also note 146.

144. *Same.*

145. Osei Owusu-Afriyie, *Cabinet Memorandum: Progress Report on the Ghana Cooperative Movement,* 19.3.63 (F 36/I:73).

146. R. M. Taylor, Financial Adviser, UGFCC, *Comments on Progress Report on the Ghana Cooperative Movement,* 3.4.63 (F 20/II:210).

147. R. M. Taylor, *Memorandum on the Progress Report on the Ghana Cooperative Movement,* 27.3.63 (F 20/II:211).

148. Nkrumah, Inaugural Address by Osagyefo the President at the Ninth Annual National Delegates' Conference of the UGFC, Kumasi, 26.3.62, in *Osagyefo in Kumasi,* 1962, p. 12..

149. *National Assembly* 11.5.62, vol. 27, cols. 128–47.

150. *Same,* col. 135.

151. *National Assembly* 30 & 31.10.63, vol. 34, cols. 445–511.

152. For the Council's role in the agricultural development programme, see *The General Secretary's Annual Reports.*

153. Letter from the Committee on Cooperatives to Appiah Danquah, 23.4.64, including an extract from a special meeting of the Cabinet, 2.4.64 (F 36/I:75).

154. Cabinet Committee on the Reorganisation of the Cooperative Movement in Ghana, *Report,* no date (1964) (A 29/IX:124).

155. Committee on the Reorganisation of the Ministry of Agriculture (Krobo Edusei), *Report,* 29.6.64 (A 29/IX:123).

156. General Purpose Committee, Emergency Meeting, *Minutes,* 14.8.64 (F 33/I:58).

157. UGFCC, Petition to the President on the Cooperative Movement in the Country and the UGFCC, No date (1964) (Memorandum File/X:6).

158. *Same.*

159. *Same.*

160. See Chapter 4, note 98.

161. As in note 156.

162. *The General Secretary's Annual Report 1965,* pp. 6–9.

163. *Same.*

164. CPP, Farmers and Cooperative Organisations Committee, *Minutes,* 19.3.65 (File with this title/IV:10).

165. *Annual Report of the Registrar of Cooperative Societies for 1966–67,* pp. 1 and 11.

166. *National Assembly* 30.8.65, vol. 40, cols. 203–05. See also B. K. Senkyire, Minister of Cooperatives, *Speech at the Opening of a Training Centre for Staff of the Ministry of Cooperatives and the Central Cooperative Council,* Winneba 2.8.65 (F 36/I:84).

167. *Same,* second reference.

168. Ministry of Cooperatives, Summons to Meetings of Committee to Consider Proposals for Reorganisation of the Cooperative Movement, 15.9., 22.9., 13.10.65 (Summons and Minutes — Ministries/V:76).

169. See Chapter 7, section "The Seven-year Plan and the 1965 cocoa crisis".

170. Victoria Nyame (Kintampo), *National Assembly* 3.9.65, vol. 40, col. 427.

171. A. B. Quartey (Nkwanta), *Same,* col. 428.

172. J. K. Twum (Asankrangwa), *Same,* cols. 433–34.

173. As in note 171, col. 431.

174. As in note 172, col. 437.

175. Kofi Baako (Minister of Defense), *Same,* col. 457.

176. K. Amoako-Atta, *Budget Statement 1966,* pp. 19–20.

177. See for example statement by Kofi Baako, Minister of State for Parliamentary Affairs, *National Assembly* 23.2.61, vol. 22, cols. 219–20.
178. See Farmers' Council file F 26 and unnumbered files: CPP Farmers' and Cooperative Organisations Committee and Correspondence with CPP Headquarters, regional and district branches (I:94 ff., III:97 f., IV:1 ff.).
179. See Nkrumah, *Dawn Broadcast* 1961 and related correspondence in file F 26 (I:96–97).
180. A. B. Quartey, Deputy Chief Administrative Officer, CPP Headquarters, to General Secretary, UGFC, May 1961 (F 26/I:95).
181. T. Adamafio, CPP Headquarters, to General Secretary, UGFC, 1.5.61 (F 26/I:96).
182. See for example file on CPP Farmers' and Cooperative Organisations Committee, as in note 178. Compare reference to "vast networks of committees which did not meet" in Apter, *Ghana in Transition,* 1972, p. 377.

Notes to Chapter 7: Cocoa and the Public Economy

1. For the marketing board system, see Chapter 2.
2. For an account of the development of production and prices, see Killick, "The economics of cocoa", in Birmingham, a.o. (eds.), *A Study of Contemporary Ghana,* vol. 1, 1966.
3. See Kay, *The Political Economy of Colonialism in Ghana,* 1972, Table 21b, p. 336: Volume of major exports, 1900–60.
4. For values of cocoa exports, see Kay, Table 21a, p. 334. For import prices, see same, Table 20c, p. 332.
5. See Killick in Birmingham, p. 370. For producer prices from 1947/48 onwards, see my Appendix A, Table 4.
6. On the price stabilising function of the marketing board system, see Chapter 2.
7. See Kay, Tables 21a and 24c, pp. 334 and 359.
8. Cocoa Marketing Board, *Annual Reports.*
9. *Same.*
10. *Same.*
11. For government expenditure, see Kay, Table 23, p. 345. For consumer price index, see Birmingham, p. 141. The Accra retail price index commences in May 1939, which is here taken to represent roughly the general price level of the late 1930's.
12. *Same.*
13. Appendix A, Table 3.
14. *Same.*
15. Department of Agriculture, *Intensive Survey of the Cocoa Producing Areas of the Gold Coast,* Mimeo, Dec. 1955, p. 7 and Appendix I, Report prepared by E. Tanburn.
16. See discussion of determinant of supply by Killick in Birmingham, pp. 377–79, and by Bateman in Kotey, Okali and Rourke (eds.), *Economics of Cocoa Production and Marketing,* Legon 1974. See also Ady, Supply functions in tropical agriculture, *Bull. Oxf. Inst. Econ.* 1968, 30 (2).
17. See "Determinants of demand", Killick in Birmingham, p. 373. For world cocoa statistics, see FAO, *Review of the World Cocoa Economy and the Work of the FAO Study Group,* 1963.
18. Kay, Table 23, p. 344.
19. To 47% (official figures—App. A Tab. 3) or 42%, if Kay's revised figures are used.
20. Official figures (App. A Tab. 3), show only a very slight deficit at this stage.
21. See Ahmad, *Deficit Financing, Inflation and Capital Formation: The Ghanaian Experience 1960–1965,* 1970.
22. For composition of CMB reserves and investments, see CMB, *Annual Reports.*

23. For import prices, see Birmingham, p. 345 (up to 1962) and *Economic Survey 1963* and *1964* (not included in later issues). For domestic prices see *Economic Surveys* and Birmingham, p. 141 (370).
24. Figures for 1964/65 in Appendix A, Table 3, are estimates due to changes in accounting periods. See *Explanatory note to tables,* Appendix A.
25. As in note 22.
26. *Same.* See also Kay, Table 42, "CMB receipts and payments, 1947/48–1959/60", p. 411.
27. As in note 22.
28. Estimate, see *Explanatory note to tables,* Appendix A.
29. CMB, *Annual Reports* and Birmingham, pp. 247–48.
30. H. Millar Craig, Principal Secretary, Ministry of Finance, to Principal Secretary, Ministry of Trade, 7.8.60 (CMB XI:6).
31. Joe S. Annan, Princ. Secr., Ministry of Trade, to Millar Craig, 24.8.60 (Same file).
32. Millar Craig to Annan 30.8.60 (Same file).
33. Annan to Cocoa Marketing Board, 3.9.60 (Same file).
34. H. A. Dodoo, Managing Director, CMB, to E. C. Tansley, Managing Director, Cocoa Marketing Company, 20.9.60 (Same file).
35. Ag. Permanent Secretary, Ministry of Trade to CMB, 16.3.60 (Same file).
36. K. A. Gbedemah, Minister of Finance, to Akin Deko, Minister of Agriculture, Western Region, Nigeria, and Dr. Tosta Filho, Brazil, 25.2.60 (Same file/XI:8).
37. Cocoa Marketing Company (= the sales subsidiary of the CMB), *Minutes of Board of Directors,* various dates, 3.7.62 onwards (UGFC: no number/VIII:109 ff.).
38. See for ex., FAO, *Review of the World Cocoa Economy and the Work of the FAO Study Group,* 1963.
39. CMB, *Annual Report 1963,* p. 3. See also J. N. Seward, *The Cocoa Hold-up of 1964: Lessons for Policy,* Seminar paper, Department of Economics, University of Ghana, Legon, (1970).
40. Cocoa Marketing Board, *Minutes,* 15.10.64 (XI:133).
41. *Same,* 18.2.65 (XI:134).
42. See Seward, as in note 39, above.
43. Ghana did not have sufficient foreign exchange reserves to support an extended hold-up of cocoa sales. Imports were interrupted, inflation aggravated. Imports were also increasingly financed by expensive short-term credits. See *Economic Survey 1965* and *1966.*
44. See Chapter 9.
45. See Bauer, *West African Trade,* 1954, pp. 251–52.
46. Wickizer, *Coffee, Tea and Cocoa: an Economic and Political Analysis,* 1951, p. 352.
47. Same. Wickizer refers to a report in *The Economist,* 1945, p. 866.
48. See Killick in Birmingham, p. 370.
49. See Chapter 2.
50. *The Economist,* 1945, p. 866.
51. See "CMB net surplus" and "Public cocoa income" in Appendix A, Table 2.
52. See Fitch and Oppenheimer, *Ghana: End of an Illusion* 1966, pp. 42–46. See also Chapter 2.
53. See Chapter 3, pp. 51 ff.
54. CPP, *Towards the Goal. The Manifesto of the CPP for the Forthcoming General Election,* 1951.
55. *Same.*
56. *Same,* p. 3.
57. Killick in Birmingham, p. 370.
58. Chapter 3, pp. 53–54.
59. Chapter 3, pp. 65 ff.
60. See expenditure against surplus and reserves in CMB, *Annual Reports.*
61. CMB, *Annual Reports.*
62. Customs Ordinance, 30 of 1947 as substituted by No. 33 of 1951, first schedule, part II. For an amount below £110 per ton: one-ninth or £10 per ton, whichever is less. Note inaccuracy in the text of the Ordinance: amount per ton equal to excess over £100. Judging from actual payments it

should read *half* of the excess. Compare 1954 Ordinance, as in note 65 below.

63. For an explanation of the sharp rise, see for ex. J. L. Cadbury, Address to Annual General Meeting of Cadbury Brothers: Press Release 29.6.54 (CMB/XI:29).

64. Ordinance 25 of 1954.

65. Ministry of Finance, *Explanatory Statement on the Cocoa Duty and Development Funds (Amendment) Ordinance*, 1954 (XI:26).

66. *Same.*

67. See retail price index for Accra in Birmingham, p. 370.

68. Ghana, *Statistical Year Book* and Kay, as in note 3.

69. As in note 65.

70. See D. K. Greenstreet, Public Administration: A Comparative Analysis of the 1950 Colonial Ten-Year Plan and the 1951 'Nkrumah' Development Plan for the Gold Coast (Ghana). *Economic Bulletin of Ghana* (Sec. ser.) 1973, vol. 3:4, for ex. p. 37.

71. *Same.*

72. As in note 65.

73. CMB to Permanent Secretary, Ministry of Commerce and Industry, 7.5.54 (CMB/XI:29).

74. CMB, Managing Director to General Manager, 8.5.54 (Same file/XI:30).

75. Extract from CMB, *Minutes* 10–11.8.54 (Same file/XI:24).

76. *Same.*

77. Chapter 3, pp. 65 ff.

78. *Same.*

79. *Resolution by the Colony Region of UGFC at Delegates' Conference, Agona Swedru,* 17.9.54 (CMB/XI:23).

80. Same. See quotation, Chapter 3, p. 67.

81. *The General Secretary's Annual Report 1963,* p. 1.

82. *Extract from CMB Minutes* 2–3.9.54 (CMB/XI:24).

83. For opposition politics, see Austin, *Politics in Ghana,* 1964, Chapter 6.

84. For CMB expenditure on rehabilitation and compensation, see CMB, *Annual Reports.*

85. See an address with this title. Undated (place in file: mid-1955) (Speeches and Messages/IX:37).

86. Working Committee, *Minutes* 20.3.56 (III:41).

87. *Memorandum on Producer Price Mid-crop 1957 and Main Crop 1957/58* by H. A. Dodoo, Gen. Manager, CMB (CMB/XI:19).

88. Nkrumah, "Statement on Government Policy on Development", *National Assembly* 28.5.57 vol. 6 cols. 278 ff., and "Statement on National Resources and Development", Same, 20.2.58 vol. 9 cols. 69 ff.

89. As in note 88, first reference, cols. 280–83; second reference cols. 80–81.

90. CMB, *Minutes* 2.5.57 (XI:34). See also *Memorandum* by General Manager, as in note 87, where he even suggested a lower price than 72/—"if the Board wishes to maintain all its present development projects".

91. CMB, *Minutes* 21.5.57 (XI:35).

92. M. K. Apaloo (F.Y.O.-Anlo South) in debate on the "Budget Statement", *National Assembly* 3.6.57 vol. 6 col. 419. See also Victor Owusu (N.L.M.-Agona Kwabre), Same, cols. 447 ff.

93. Togoland National Farmers' Union, Telegram to CMB, July 1957, signed Chief Farmer Salomon Adzei (CMB/XI:16).

94. The Ashanti based National Farmers' Union was reported to demand a sharp increase in the producer price in early 1958, *West Africa* 31.5.58 p. 523.

95. *Gold Coast Cocoa Marketing Ordinance,* No. 16 of 1947.

96. The Ghana Cocoa Marketing Board (Amendment) (No. 2) Bill, Second Reading, *National Assembly* 10.7.57 vol. 6 cols. 1877 ff.

97. A. Casely Hayford, Minister for Commerce and Industry, as in note 96, cols. 1877–78.

98. Main speakers for the opposition were Kodzo Ayeke (Togoland Congress) B. F. Kusi (National Liberation Movement), and M. K. Apaloo (Federation of Youth Organisations).

99. Kodzo Ayeke (T.C.-Ho West), as in note 96, col. 1880.
100. B. F. Kusi (N.L.M.-Atwima Nwabiagya), as in note 96, col. 1884.
101. *Same.*
102. A. Casely Hayford, as in note 96 (97), col. 1898.
103. CMB, *Minutes* 9.5.58 (XI:45).
104. *Same,* 17.6.58 (XI:45).
105. K. A. Gbedemah, Minister of Finance, "Government's Financial Policy" (Budget Statement), *National Assembly,* 7.7.58 vol. 11 cols. 214-15.
106. Victor Owusu (U.P.-Agona Kwabre), *National Assembly* 18.7.58 vol. 11 col. 650. See also R. R. Amponsah (U.P.-Sekyere West), *same,* col. 670.
107. *National Assembly* 29.7.58 vol. 11 cols. 939 ff.
108. The Cocoa Duty and Development Fund Bill, Second Reading, *National Assembly* 25.8.58 vol. 11 cols. 1795 ff.
109. J. D. Wireko (U.P.-Amansie East), as in note 107, col. 1797.
110. See B. K. Adama (U.P.-Wala South), as in note 107, col. 1801. See also M. K. Apaloo (U.P.-Anlo South), *same,* cols. 1899-80.
111. *Second Development Plan 1959-64,* Accra 1959. See also Motion in *National Assembly* 4.3.59 vol. 14 cols. 189 ff.
112. *Same,* first reference, pp. 1-3.
113. *Same,* p. 16.
114. *Same,* pp. 4 ff.
115. As in note 111, second reference, col. 192.
116. *Same,* cols, 199-200.
117. Statement on Cocoa Price, *National Assembly* 19.3.59 vol. 14 cols. 404-06 (including the document submitted by UGFC).
118. *Same.* For the government's promise to maintain the producer price see also Minister of Finance in *National Assembly* 8.10.62 vol. 29 col. 163.
119. *National Assembly* 20.3.59 vol. 14 cols. 464-65.
120. *Same,* cols, 467-68.
121. P. K. K. Quaidoo, Minister of Trade and Industries, *same,* col. 468.
122. W. K. Aduhene (CPP-Sefwi Wiawso), A. Mate Johnson (Parliamentary Secretary to the Ministry of Justice and Local Government), and P. K. K. Quaidoo, *same,* cols. 465-67.
123. The loan was granted on 1 October 1959 according to a report by the Auditor General, quoted in *National Assembly* 17.5.61 vol. 23 cols. 679-80.
124. UGFC, *A Decade of Existence,* 1963, p. 11. Compare chapter 3, pp. 71 ff.: "Official recognition".
125. D. D. Duncan, "The Ghana Farmer and the Second Five Year Development Plan", Draft press release, no date (H4/II:67).
126. National Agricultural and Plantation Workers Union, to Minister of Trade and Industries 6.5.59: Resolution on the Pegging of Cocoa Price (CMB:C/6XI:14).
127. As reported in *West Africa* 23.5.59 p. 499.
128. As in note 94, above.
129. *National Assembly* 9-13.7.59 vol. 16 cols. 333 ff.
130. B. F. Kusi (U.P.-Atwima-Nwabiagya), as in note 129, col. 405.
131. See Victor Owusu (U.P.-Agona Kwabre), as in note 129, cols. 336 ff.
132. The Development Fund (Members' Contributions) Bill, Second Reading, *National Assembly* 31.7.59 vol. 16 cols. 1233 ff.
133. Victor Owusu, as in note 131 (129), col. 1234.
134. J. E. Appiah (U.P.-Atwima-Amansie), as in note 132, cols. 1249-50, 1262.
135. J. E. Appiah, Victor Owusu, and Abayifaa Karbo (U.P.-Lawra-Nandom), *same,* cols. 1265, 1240 and 1245.
136. Appiah, Karbo, and J. D. Wireko (U.P.-Amansie East), *same,* cols. 1262, 1241 and 1259.
137. See *National Assembly* 17.6.60 vol. 19 cols. 107-08.
138. CMB, *Annual Report 1959,* p. 5.

139. See, for example, CMB, *Annual Report 1961,* pp. 28–29, on the changing structure of CMB investment in this period.

140. *National Assembly* 2.7.59 vol. 16 col. 128.

141. See CMB to Ministry of Trade and Industries 21.4.59, with reference to Board meeting the same day and to directions from the Ministry 4.5.59 (CMB/XI:15).

142. See Birmingham, Chapter 16: The Volta River Project.

143. See *Economic Survey 1961,* pp. 115 ff. See also Birmingham, Chapter 14: External Trade.

144. *Same.*

145. See Nkrumah, Sessional Address, *National Assembly* 4.7.61 vol. 24 cols. 11 ff. and Minister of Finance, "Government's Financial Policy" (The Budget Statement), same, cols, 117 ff. For a summary of main points, see *Economic Survey 1961,* pp. 105–06.

146. See *Economic Survey 1961,* pp. 112–13.

147. See Sessional Address as in note 145, col. 16, and *Economic Survey 1961,* p. 113.

148. Compulsory Savings Bill, Second Reading, *National Assembly 24.7.61 vol. 24 cols. 633 ff.*

149. See St. Clair Drake, J. G. and L. A. Lacy, Government versus the unions: The Sekondi-Takoradi Strike, 1961, in Carter, G. M. (ed.) *Politics in Africa,* New York 1966. See also Jeffries, R. *Trade Unions and Politics in Ghana,* Doctoral thesis, London 1974.

150. Nkrumah, Sessional Address, *National Assembly* 4.7.61 vol. 24 col. 24.

151. A. E. A. Ofori Atta, Minister of Local Government, *National Assembly* 6.7.61 vol. 24 col. 56.

152. Krobo Edusei, Minister of Transport and Communications, *same,* col. 57.

153. Kofi Baako, Minister for Parliamentary Affairs, same, col. 49. See also Krobo Edusei, *same,* cols. 94–95.

154. Krobo Edusei, *same,* col. 95.

155. B. F. Kusi (U.P.-Atwima Nwabiagya), *same,* col. 57. See also A. W. Osei (U.P.-Ahafo), cols. 76–77. See also debate on Compulsory Savings Bill, same, 24.7.61 cols. 678–79.

156. Victor Owusu (U.P.-Agona Kwabre), *National Assembly 10.7.61 vol. 24 col. 161.*

157. P. K. K. Ouaidoo (CPP-Amenfi-Aowin), *same,* 24.7.61 col. 645. See also A. E. A. Ofori Atta (Minister of Local Government), *same,* col. 656.

158. Nkrumah, Speech at the Opening of Cocoa House, 19.11.60, in UGFC, *The Farmers' Year-book 1960–61,* p. 21.

159. See Chapter 4, section "The State Agency scheme and the triumph of monopoly".

160. See R. O. Amuako-Atta (CPP-Obuasi), *National Assembly* 17.5.61 vol. 23 cols. 677–87.

161. General Purpose Committee, UGFC, *Minutes* 5.7.61 (IX:21).

162. Draft (?) press release, "It's your own budget" (H4/III:65).

163. *Same.*

164. See for example monthly reports by the Regional Organising Secretaries (F38/I:161–62).

165. For the details on the Compulsory Savings Scheme, see Statement by Minister of Finance on the National Development Bonds, *National Assembly* 25.10.61 vol. 25 cols. 207–09.

166. Resolution by Emergency Regional Delegates' Conference, UGFC, Eastern Region 21.2.62 (A75v.2/X:26).

167. See for example "unanimous decision" by the Regional Delegates Conference at Okadjakrom, Volta Region, 10.3.62. Reference in letter from Regional Executive and Marketing Officer, Volta Region, to the Senior Auditor, Ho (National President/IX:167).

168. Resolution no. 11 passed by the National Delegates Conference, Kumasi, March 1962 (A65v.2/X:27).

169. Nkrumah, Inaugural address at the Ninth Annual Conference of the UGFC, Kumasi 26.3.62, in *Osagyefo in Kumasi,* 1962, p. 13.

170. *National Assembly* 11.5.62 vol. 27 cols. 128 ff. For quotation, see col. 131.

171. Only £44,000 were given up, including £32,600 by Members of Parliament and 4,200 by the President out of a total non-cocoa contribution of some £12 million (by May 1964). *National Assembly* 21.5.64 vol. 36 col. 141.

172. Principal Secretary, Ministry of Finance, to UGFC 24.2.65, acknowledging receipt of letter of 26.1.65 (A54/VI:4).

173. See *Annual Reports of the Central Revenue Department* for 1962-65 (including arrears collected during 1963/64 and 64/65).

174. For the 1963 budget, see the "Budget Statement", *National Assembly* 22.10.63 vol. 34 cols. 171 ff. and *The Financial Statement 1963-64.*

175. *Same,* first reference, col. 184.

176. *Same,* col. 184.

177. J. A. Braimah (CPP-Gonja East), *National Assembly* 28.10.63 vol. 34 col. 314.

178. C. E. Donkoh (CPP-Wenchi West), *same,* cols. 329-30.

179. I. J. Adomako-Mensah (CPP-Atwima-Amansie), *same* col. 386, and in the debate on the Compulsory Savings (Abolition) Bill 30.10.63, col. 465.

180. A. W. Osei (U.P.-Ahafo), *same,* col. 353. See also B. F. Kusi (U.P.-Atwima-Nwabiagya), cols. 381 and 448.

181. Kofi Baako, Minister of Defence and Leader of the House, *same,* col. 404.

182. Rev. C. K. Dovlo (CPP-Keta), *same,* col. 370.

183. See report in *National Assembly* by Minister of Agriculture, vol. 34 col. 511. See also UGFC, *A Decade of Existence* and *The General Secretary's Annual Report 1963.*

184. Sub-committee of the Regional Executive Council, Koforidua 1.11.63 (F38/I:180).

185. Report on the Settlement of Misunderstanding ... at the Office of the Regional Commissioner, Eastern Region, Koforidua 14.1.63 (F38/I:174).

186. *National Assembly* vol. 34 cols. 420 ff.

187. See in particular M. O. Kwatia (CPP-New Juaben), *same,* cols. 458 ff., J. A. Owusu-Ansah (CPP-Offinso-Kwabre), cols. 470 ff., C. E. Nimo (CPP-Akim Abuakwa), cols. 479 ff., and for the official opposition, B. F. Kusi (UP-Atwima Nwabiagya), cols, 447 ff. See also the debate on the budget, cols, 331 (C. E. Donkoh), 355 (A. W. Osei), 382 (B. F. Kusi). For the issues involved, see Chapter 5.

188. J. A. Owusu-Ansah, *same,* col. 474.

189. B. F. Kusi, *same,* col. 449.

190. J. D. Wireko (CPP-Amansie East), *same,* cols. 497-98. See also J. K. Essien (CPP-Bibiani), col. 495.

191. See *Economic Survey 1962* and *1963.*

192. Notes on meeting with Minister of Finance and Trade 21.8.63 (Summons & Minutes—Ministries/V:35). Among those present were also the Minister of Agriculture, the Managing Director, CMB, and the Gen. Secr. UGFC. See also "Budget Statement" 1962, *National Assembly* vol. 29 cols. 161-63. The £18.6 millions arrears on this account were paid in 1963/64 only by cancelling part of the Board's loans to the government. See Minister of Finance, January 1965, *National Assembly* vol. 38 col. 182.

193. CMB, *Annual Reports.*

194. As in note 192, first reference. A committee was appointed to examine the matter. No change of this kind was taken.

195. Managing Director, CMB, to Executive Secretary, State Enterprises Secretariat, 17.5.65 (UGFC:Correspondence—CMB/VII:118).

196. As in note 192, first reference.

197. *Seven-Year Development Plan,* 1964, pp. v-xxii.

198. *Programme of the CPP for Work and Happiness,* 1962, p. 4.

199. *National Assembly* 11.3.64 vol. 35 col. 21. Also in the Plan itself, p. ix.

200. *Seven-Year Development Plan,* p. 249.

201. *Same,* pp. 255. See also pp. 256, 274.

202. *Economic Survey 1964* and *1965.*

203. Minister of Finance (Kwasi Amoako-Atta), "Government's Financial Policy" (The Budget Statement), *National Assembly* 21.1.65 vol. 38 cols. 179 ff. Also separate in *The Budget 1965,* pp. 4-5.

204. Nkrumah, Sessional Address, *National Assembly* 12.1.65 vol. 38 col. 15. See also Agricultural Credit and Cooperative Bank Bill, Second Reading, *National Assembly* 9.3.65 vol. 38 cols. 1220 ff.

205. Agricultural Credit and Cooperative Bank Bill as printed in UGFC, *General Secretary's Annual Report 1965,* pp. 58-59. See also *National Assembly* vol. 38 col. 192.

206. General Secretary's Annual Report 1965, pp. 3-4.

207. *Economic Survey 1964,* p. 151, *1965,* p. 148.

208. See Minister of Finance, "Government's Financial Policy", *National Assembly 10.9.65 vol. 40 col. 716.*

209. See Chapter 1, p. 18.

210. As in note 208, cols. 717-18.

211. *The New Farmer,* July 1965.

212. *Same.* See also Martin Appiah Danquah in *National Assembly* 3.9.65 vol. 40 cols. 443-44.

213. Speech by Sir Tsibu Darku, Chairman of the CMB, no date (2nd quarter 1965) (CMB/XI:4).

214. H. A. Dodoo, Managing Director, CMB, to Executive Secretary, State Enterprises Secretariat 17.5.65 (UGFC: Correspondence—CMB/VII:118). See also CMB, *Minutes* 25.5.65 (XI:136).

215. *Same,* second reference.

216. State Enterprises Secretariat, *Draft Cabinet Memo: Subject: State Cocoa Marketing Board,* signed J. V. L. Phillips 3.6.65 (UGFC:Correspondence—Corporations and State Enterprises/VIII:63) and *Notes on Meeting with the Special Finance and Foreign Exchange Committee* 9.6.65 (F36/I:80). Present at this meeting were K. Amoako Atta (Chairman), W. M. Q. Halm, Ayeh Kumi, Tsibu Darku, J. V. L. Phillips, J. H. Frimpong-Ansah, H. A. Dodoo, U. K. Hackman, G. Gyasi-Twum, and by invitation P. E. Amuah and Appiah Danquah.

217. J. V. L. Phillips, *Financial Position of the Cocoa Marketing Board* (Note on meeting of the Special Finance Committee 2.6.65) (UGFC:Correspondence—Corporations and State Enterprises/VIII:61).

218. Meeting 9.6.65, as in note 216.

219. Report of the 11th National Delegates Conference, UGFC, Tamale, April 1965 (F7/I:159).

220. Meeting 9.6.65, as in note 216. See also report from meeting 2.6.65, as in note 217.

221. Editorial, *The New Farmer,* July 1965. See also *National Assembly* 3.9.65 vol. 40 col. 424.

222. Nkrumah, Sessional Address, *National Assembly* 24.8.65 vol. 40 col. 24.

223. D. E. Asafo-Agyei (Manhyia), moving the motion, *National Assembly* 3.9.65 vol. 40 cols. 423 ff.

224. See Chapter 6, pp. 175-76.

225. *National Assembly* vol. 40 col. 453.

226. See Chapter 6, pp. 176-77.

227. Average index numbers for October-March 1964/65 were 128.7 (March 1963 = 100), and for same period 1965/66, 165.3. *Economic Survey 1964,* p. 15, *1965,* p. 148, and *1966,* p. 140.

228. *Same.* This tendency is clearly seen in the Primary price statistics available at the Central Bureau of Statistics, Accra.

229. I am grateful to Mr. G. G. Mensah, of the Purchasing Department, CMB, for this information.

230. See Are and Gwynne-Jones, *Cacao in West Africa,* 1974, p. 2.

231. Kankam Boadu, *Memorandum on Cocoa Price,* 1966 (CMB/XI:2). See also report on meeting between CMB officials and farmers in Sunyani 27.9.66 (CMB/XI:1).

232. Techiman Traditional Council 31.5.66 (GJ/5).

233. Kwaku Mensah, Ahafo-Kenyasi No. 1 (GJ/7).

234. Osei Tutu, Kenyasi No. 1, Brong-Ahafo (GJ/10).

235. Kweku Appiah, Abodom, Bekwai (GJ/14).

236. A. K. Donkor, Accra ("former produce buyer and farmer"), (GJ/43).

237. See, for example, also Nana Kwaku Gyasi, Bedomasi/Ashanti, Mercy E. Ankrah, Anyinam/Eastern Region, and Ben A. Y. Ocloo, Wudome-Peki/Volta Region (GJ/18, 78, 89).

238. See Chapter 2.

239. See unpublished survey material in the Division of Economics and Marketing, Ministry of Agriculture, Accra. I am grateful to Mr. B. P. Acquah for his assistance on this point. See also

Rourke, B. E. *Wages and Incomes of Agricultural Workers in Ghana,* 1971, and Rourke, B. E. and S. K. Sakyi-Gyinae, Agricultural and Urban Wage Rates in Ghana, *Economic Bulletin of Ghana,* 1972 (2:1).

240. Regional Commissioner's Office, Sunyani, to UGFCC, Accra 8.11.61, including petition from Farm Caretakers' Committee, Kenyase No. 1 and 2, to District Commissioner, Goaso 4.11.61 (A65v.2/X:16). Regional Commissioner, Sunyani, 28.12.61 and Petition from Caretakers' Union, Mpasatia, Nkawie, Ashanti (A65v.2/X:24).

241. Economic Surveys, as in note 227.

242. UGFCC, *Conditions of Service,* June 1965 (A32/IX:127).

243. Cocoa Marketing Company, *Minutes* 1.10.65 (Summons and Minutes—CMC/VIII:148).

244. Compare the discussion of blockages in national development caused by labour migration in both West Africa and in the "labour reserve economies" of Southern Africa in Samir Amin, *Le Développement Inégal,* 1973, pp. 287–90.

245. See report by the Managing Director of the Cocoa Marketing Company to C.M.B. 31.3.59 (CMB/XI:12).

246. The negative impact on economic growth of most import-substituting industrialisation is a common view among Western economists. See for example Ian Little and others, *Industry and Trade in Some Developing Countries,* 1970.

Notes to Chapter 8. The Farmers' Council and the CPP: Summary and Conclusions

1. See petition from the "Farm Labour Movement Affiliated Part of the T.U.C. (Ghana)", Eastern Region, to Secretary General, T.U.C., 28.6.60 (A 91v.3/X:67).

2. See letter from the Commissioner of Labour, Accra, to Regional Labour Officer, Kumasi, 4.9.62 and letter from Regional Commissioner, Kumasi, to Principal Secretary, Ministry of Agriculture, 14.2.63 (A 30/X:26–27).

3. See Chapter 7, p. 220.

4. For the complex pattern of local politics, see for example, Dunn and Robertson, *Dependence and Opportunity: Political Change in Ahafo,* 1973. Also Austin, *Politics in Ghana,* 1964.

5. Fitch and Oppenheimer, *Ghana: End of an Illusion,* 1966. See also Mohan and Marshall, as in Chapter 1, section "The political economy of the CPP: The wider discussion".

6. Roger Genoud, *Nationalism and Economic Development in Ghana,* 1969.

7. Roger Murray, 'Second thoughts on Ghana', *New Left Review* 42, 1967; John Saul, 'The state in post-colonial society: Tanzania', *Socialist Register 1974.*

8. For Non-capitalist Development writings, see Chapter 1, p. 23.

9. For the comparison between Ghana and the Ivory Coast, see Foster and Zolberg 1971.

10. For the concept "petty-bourgeoisie", see Chapter 1, p. 30.

11. See for example Fitch and Oppenheimer, Mohan and Saul.

12. See my discussion of terms like "state capitalism" and "bureaucracy", Chapter 1, pp. 31–32.

Appendix A: Tables

Table 1. *Production, Sales, Prices and Marketing Costs*

	Production '000 ton	Sales £ million	Average selling £ per ton[1]	Marketing costs £ million	Same as per cent of sales
1947/48	208	41.5	201.2	1.9	5
1948/49	278	37.5	136.7	3.3	9
1949/50	248	45.1	178.4	3.0	7
1950/51	262	70.3	268.5	3.7	5
1951/52	211	51.6	245.1	3.4	7
1952/53	247	57.1	231.4	4.3	8
1953/54	211	74.7	358.7	3.6	5
1954/55	220	77.5	355.1	3.8	5
1955/56	229	52.3	221.8	4.5	9
1956/57	264	50.7	189.5	5.2	10
1957/58	206	62.9	304.3	4.1	7
1958/59	255	70.9	280.5	5.5	8
1959/60	317	69.9	225.5	6.7	10
1960/61	432	71.6	174.9	10.9	15
1961/62	409	69.0	158.4	10.0	14
1962/63	422	69.1	163.8	10.3	15
1963/64	421	76.9	178.6	11.2	15
1964/65	572[3]	71.2[2]	124.6[2]	18.8[2]	26
1965/66	406[3]	n.a.[4]	n.a.[4]	n.a.[4]	—

Source: Cocoa Marketing Board, *Annual Reports.*
Notes: 1. Calculated on tonnage sold/shipped, not tons produced.
2. Estimate. See *Explanatory note,* below.
3. From UGFCC purchasing records. Compare *Economic Survey 1966,* pp. 46 and 49.
4. Not available due to change of accounting periods.

Table 2. *Public Cocoa Income* (£ million)

	Current CMB-payments to government[1]	Same as per cent of sales value[2]	CMB net surplus/deficit	Public cocoa income[3]	Same as per cent of sales value[2]
1947/48	0.4	1	24.1	24.5	59
1948/49	1.7	5	-0.1	1.6	4
1949/50	3.5	8	18.0	21.5	48
1950/51	13.4	19	20.1	33.5	48
1951/52	14.7	29	3.5	18.2	35
1952/53	16.0	28	6.4	22.4	39
1953/54	34.0	46	11.1	45.1	60
1954/55	38.4	50	8.1	46.5	60
1955/56	14.6	28	0.8	15.4	29
1956/57	12.0	24	-5.2	6.8	13
1957/58	26.3	42	6.5	32.8	52
1958/59	26.2	37	8.5	34.7	49
1959/60	26.3	38	4.2	30.5	44
1960/61	24.7	35	-5.9	18.8	26
1961/62	27.4	40	-10.2	17.4	25
1962/63	27.6	45	-9.5	18.1	26
1963/64	21.8	28	3.1	24.9	32
1964/65	17.8[4]	25	-22.9[4]	-5.1	(7)

Source: Cocoa Marketing Board, *Annual Reports.*

Notes: 1. Export and local duty, voluntary contributions, compulsory savings/farmers' income tax. Loans and grants against reserves not included. (See Table 3.)

2. From Table 1.

3. Current CMB payments plus/minus CMB surplus/deficit.

4. Estimated. See *Explanatory note,* below.

Table 3. *Government Revenue, Expenditure, and Cocoa Income* (£ million)

Budget year[1]	Revenue	Expenditure	Surplus/ deficit	Cocoa income Current & capital[2]	Same as per cent of expenditure[3]
1947/48	10.2	10.2	0.1	0.1	7
1948/49	11.6	10.4	1.3	2.0	20
1949/50	18.1	13.4	4.7	3.8	28
1950/51	20.9	17.2	3.7	15.5	90
1951/52	38.9	23.0	16.0	16.4	72
1952/53	43.0	37.6	5.3	22.0	59
1953/54	50.4	47.9	2.6	40.3	84
1954/55	80.5	45.5	35.0	43.0	94
1955/56[4]	64.3	63.6	0.1	23.0	36
1956/57	49.7	55.3	-5.6	15.0	27
1957/58	60.1	52.8	7.3	29.2	55
1958/59	66.9	62.0	4.9	30.0	48
1959/60	70.1	78.2	-8.1	48.6	62
1960/61	83.2	107.9	-24.7	26.2	24
1961/62[4]	97.1	150.4	-53.3	27.4	18
1962/63	82.5	132.3	-49.7	51.1	39
1963/64[4]	146.7	188.6	-41.9	26.8	14
1965	142.0	180.8	-38.8	19.4[5]	11

Sources: Economic Surveys, Financial Statements, and *CMB Annual Reports.* For cocoa income, see note 2 below. Compare revised figures in Kay, *The Political Economy of Colonialism in Ghana,* 1972, pp. 344 ff.

Notes: 1. As distinct from crop years used in other tables.

2. Estimated, including current CMB payments as in Table 2, plus CMB grants and loans according to *CMB Annual Reports.* Note discrepancies between CMB and government accounting periods.
3. Indicates relative size, not how actual expenditure was financed.
4. 15 months period.
5. Estimated. See *Explanatory note,* below.

Table 4. *Producer Prices and Payments to Producers*

	Producer Prices[1] £ per ton	Producer Prices[2] Shilling per load	Payments to producers £ million	Same as per cent of sales value[3]
1947/48	74.7	40	15.5	37
1948/49	121.3	65	33.8	90
1949/50	84.0	45	20.8	46
1950/51	130.7	70	34.3	49
1951/52	149.3	80	31.5	61
1952/53	130.5	70	32.2	56
1953/54	134.4	72	28.3	38
1954/55	135.0	72	29.7	38
1955/56	148.5	80	34.0	65
1956/57	149.2	80	39.4	78
1957/58	134.2	72	27.7	44
1958/59	131.9	72	33.7	48
1959/60	112.2	60	35.6	51
1960/61	112.0	60	48.4	68
1961/62	100.8	54	41.3	60
1962/63	100.8	54	42.5	62
1963/64	100.8	54	42.4	55
1964/65	100.6	54	57.5	81
1965/66	74.7	40	30.3	n.a.[4]

Source: Cocoa Marketing Board, *Annual Reports.*
Notes: All figures are net of voluntary contributions, compulsory savings/income tax, and contribution to Agricultural Credit and Cooperative Bank.

1. Weighted average when different price for main and mid crop.
2. Main crop only.
3. As in Table 1.
4. Not available due to change in accounting periods.

Explanatory note to tables

Most figures in the tables above are based on the *Annual Reports* of the Cocoa Marketing Board. Until 1964 the reports were based on crop years, ending 30 September. In 1964/65, however, the accounting period was changed to calendar years. It meant that the 1964/65 crop was split between two reports, one covering the period 1.10.64 to 28.2.65, the other covering the remaining months of 1965, thus including also much of the 1965/66 crop. It is therefore difficult to obtain figures for the last two seasons (1964/65, 65/66), which can be compared with those of earlier seasons. The 1964/65 figures are estimates. The 1965/66 figures have had to be partly omitted. This note is concerned with explaining the method used in estimating the 1964/65 figures.

Table 1. Sales 1964/65. Total tonnage sold by 28.2.65 was 149,603 tons. This has been reduced by the remainder of the 1963/64 crop (9,972 tons) and multiplied by the weighted average selling price for the period 1.10.64-28.2.65 (£171.8 per ton) = £24.0 million. I add the remainder of the 1964/65 crop (571,722 tons less 139,631), multiplied by the weighted average selling price for the period 1.3.65-31.12.65 (£114.6 per ton) and obtain a total of £71.2 million. It should be noted that the estimate is based on weighted average prices and takes no account of changes in prices during the accounting periods.

Table 1. Marketing costs 1964/65. Marketing costs in the CMB accounts are calculated on the basis of tonnage shipped (including local deliveries—See the "Financial and Statistical Records" in the CMB *Reports*). This figure amounted for the two periods to 589,694 tons, or only slightly higher than the figure for the tonnage bought from the farmers in the purchasing records (571,722). I have therefore used the cost figures for these two periods as roughly equal to the costs of marketing the 1964/65 crop, the sale of which was much delayed. The following items are included: Other shipping Expenses, Buying Agents Allowance, Administrative Expenses, and Other Expenses (excl. Loans Department).

Table 2. Current CMB payments to government 1964/65. I have added the export and local duty for the first period (£4.9 million) and 90 per cent of the same item for the second (£6.3 million) – £11.3 million. To this is added the farmers' "income tax" of £11.2 per ton for the 571,722 tons bought (£6.4 million), and, finally, the £7.5 per ton contribution to the Agricultural Credit and Cooperative Bank for the mid-crop 1965 (£113,100). Grand total: £17.8 million.

Table 2. CMB net surplus/deficit 1964/65. Value of sales (Table 1) less the sum of the payments to the producers (Table 4), current CMB payments to the government (Table 2) and marketing costs (Table 1).

Table 3. Government cocoa income current and capital 1965. This figure is the sum of (1) Export and local duties, farmers' income tax, contributions to ACCB as estimated in Table 2, above, and recalculated on a calendar year basis (£14.3 million); (2) Other CMB payments to the central government (cocoa rehabilitation scheme, produce inspection) (£1.0 million); (3) Government compensatory stock acquired by the CMB (£4.1 million). Total: £19.4 million.

Appendix B

Statement by Martin Appiah Danquah, former General Secretary of the United Ghana Farmers' Council, to the Cocoa Economics Research Conference, Legon, April 1973.

Note: This written statement was read to the session on internal marketing and included in the printed proceedings of the conference. The purpose of the staatement was to defend the Council's achievement and, more specifically, to show that "the opinions expressed by the writer (Beckman) did not give the true picture of the affairs of the United Ghana Farmers' Co-operative Council". I had submitted a paper, "The internal marketing monopoly of the Farmers' Council", with essentially the same approach as the present study, although less elaborated and documented. It seems therefore appropriate to give the readers access to Mr. Appiah Danquah's statement for the purpose of comparison. I have not entered into a discussion of detailed points where our views differ. The purpose of our two interpretations are rather different. My own concern is to explain the wider functions of the Council in the political economy of Ghana.

Source: Kotey, Okali and Rourke (eds.), *The Economics of Cocoa Production and Marketing,* Legon 1974, pp. 403–11.

The statement

Before 1953 each village in the country had a Chief Farmer appointed by the Chief of the village to represent the farmers of that village. The farmers had no say in the appointment of a Chief Farmer representing their interest. We felt then that this was wrong, and that the farmers of this country should be so organized that they should have the chance of electing their leaders. At that time the exploitation of farmers had reached such an alarming height that we felt they could only be rescued from this form of exploitation, when they had their own organization to talk on their behalf.

The representatives of farmers throughout the country met on two occasions in 1950 and 1951, and elected a Committee to draft a Constitution for a democratic farmers organization in the country. On 26th September, 1953, delegates of farmers throughout the country met again, and gave approval to the Constitution.

Some of the aims and objects of the Council were as follows:

(i) To foster and promote unity of purpose and action amongst farmers and the improvement of agriculture and allied industries in Ghana.

(ii) To safeguard the interest of the farmers of the country, so that they may work and enjoy the fruits of their labour and to aspire towards complete emancipation from all forms of exploitation and economic servitude.

the farmers, and kept the stock in their homes as they did not possess special storage accommodation.

(ii) The activities of the Brokers continually contributed towards the lowering of the quality of cocoa, as they always rushed and bought cocoa, notwithstanding its dryness or quality.

(iii) They also mercilessly exploited the farmers by weighing the cocoa above the authorized weight, and also by giving them loans at 50 to 100 per cent interest rates. Often the farmers were not able to repay such loans, and they suffered from the sale of their cocoa farms.

In 1956 delegates of the Farmers' Council travelled to the consumer countries to listen to complaints levelled against our cocoa, and also to see to what use the manufacturers put the cocoa. It was there that we found that the quality of our cocoa was so bad that a sharp turn for improvement was necessary. We therefore mounted a vigorous campaign during the years 1956 to 1960 for the improvement of the quality of the country's cocoa. We could not achieve our aim because of the presence of Brokers in the field.

We therefore decided at a conference at Kpandu during the early months of 1960 that an approach should be made to the Government to make the United Ghana Farmers' Co-operative Council the sole buying agent of the Cocoa Marketing Board, because of the following reasons:

(i) To improve upon the quality of the country's cocoa.

(ii) To get the farmers into one organization so that they could discuss together their common problems affecting the cocoa industry.

(iii) To protect farmers from exploitation by Brokers.

(iv) To provide a platform for officials of the Government and the Cocoa Marketing Board to talk to farmers and vice versa.

(v) To facilitate the compilation of statistics of production of cocoa concerning individuals, villages, districts, etc. so that if at any given time, we wanted to know the number of farmers who produced 1 to 50 loads of cocoa throughout the country, it would easily be known.

The Council had to struggle more against the machinery of the Convention Peoples Party and the Government to achieve this high objective, than opposition from the other firms. All that the writer said about the political movement having had something to do with the United Ghana Farmers' Co-operative Council becoming the sole Buying Agent of the Cocoa Marketing Board cannot therefore be justified.

When the Government approved the ideal plan of the farmers to monopolize the sale of their own cocoa, a consultation meeting was held between representatives of the United Ghana Farmers' Co-operative Council and the Co-operatives, and we agreed on terms of merger. It was wrong for the writer to have written that the terms of our merger were entirely those of the Farmers' Council.

All the officials of the outgoing Licensed Buying Agents, including the officers and the leadership of the two co-operative movements were absorbed. The leadership of the co-operatives was made up as follows:

(i) The late Mr. Asubonteng led the Ashanti, Brong-Ahafo, Sefwi Co-operative Association, (ABASCO) and on absorbtion, he was made the Regional Officer of our operations in Ashanti.

(ii) Mr. N. A. Dennis, currently the Deputy Managing Director of the Produce Buying Agency Ltd., was the leader of the Southern Co-operative Movement, and he was absorbed into the United Ghana Farmers' Co-operative Council as the Chief Accountant.

It was wrong, therefore, for the writer to have said that though all the staff of the co-operatives was absorbed, the leadership of the co-operatives was not absorbed into the organization.

Mr. Chairman, at this stage I would like to speak about the results of the operations of the United Ghana Farmers Co-operative Council as the sole Buying Agent of the Cocoa Marketing Board. The results were praiseworthy, and even Mr. Feaver of Cadbury and Fry made this point aboundantly clear to the late President Dr. Kwame Nkrumah.

The following were the achievements of the Council.

(i) Quality of the country's cocoa: The Council was able to raise the quality of the country's cocoa from under 50 per cent Grade I to over 98 per cent Grade I.

(ii) Loans to farmers: The Council successfully operated short term loans to farmers which helped them at the off season to have money to maintain their farms.

(iii) Arbitration awards to foreign buyers: Arbitration awards to foreign buyers which ranged between £250,000 and £500,000 per year were completely eliminated as the requirements of weight and quality were strictly adhered to by the Council.

(iv) Building of sheds: When the Council took over the entire operation of cocoa marketing, storage accommodation in the country was inadequate. The farmers of the country, knowing the importance of storage accommodation, decided to use their bonuses and profits to make provision for this important item of the cocoa marketing system in the country, and now hundreds of such sheds are built throughout the country with the farmers' own money.

(v) Smuggling of cocoa: Smuggling of cocoa was reduced to the barest minimum, as the Council always had special operations against smuggling at our border socieities.

(vi) Combat against capsids: When in 1969 the Government decided to introduce free spraying of cocoa farms under the advice of the renowned economist, Professor Arthur Lewis, the Council objected to the scheme and suggested that the farmers should be educated to spray their own farms, and that the spraying equipment should be sold to the farmers at subsidized prices. We gave many tangible reasons which compelled the Government to agree to our proposal, and in three years the Council led the farmers to spray their farms, and this helped to raise the production of the country's cocoa from

200,000 to 500,000 tons. The Council used its machinery to distribute the spraying equipment to the farmers of the country, and the entire audio-visual aids machinery of the Council, i.e. cinema vans, drama troupes and bands were also used to send these important messages forcibly to the farmers.

(vii) Reduction of capital outlay: As we were the sole Buying Agent, we helped to reduce the number of scales, tarpaulins, safes, etc. needed for the operation of cocoa marketing. These items were purchased with foreign exchange and thus we made substantial savings in the utilization of the country's foreign exchange.

(viii) Clean business: During the five years that we operated as the sole Buying Agent of the Cocoa Marketing Board, we never closed any season with a deficit, rather we ended always with a considerable profit margin, which as I have said earlier was used to provide storage accommodation for the cocoa industry.

(ix) Feeder roads: As feeder roads were essential to a successful cocoa buying operation, we succeeded in getting the farmers to become aware of this, and to get them to create a fund by themselves to provide this important item of our development. They decided at a National Conference to set aside 1 lb. weight of each bag of cocoa that each farmer sold to create this fund.

(x) Transportation of cocoa stocks from up country: As transportation of cocoa from up country was vital, and we were not readily getting private transport owners to cart such stocks from difficult places, the Council helped to organize co-operative transport organization by giving loans to provide trucks for this important service. The transport co-operatives which we helped to establish, are the only ones in the country today.

(xi) Expansion of Ghana Commercial Bank: The Council decided to send monies throughout the country through the Ghana Commercial Bank. Since the Bank at that time had not opened branches at our various district headquarters, we advised the Commercial Bank to establish banks throughout the country and our operations earned them about 90 per cent of their operational income. We thus helped considerably in the expansion of the Ghana Commercial Bank.

(xii) Tailoring co-operatives: We also organized tailors of this country into co-operatives by giving them loans to operate and to sew school uniforms of the farmer's children. These were distributed through the societies to the farmers for their children as loans in kind. They paid back these loans during the season. This organization unfortunately died with the Council.

(xiii) Establishment of solid ties between farmers and the Government: The Council provided a solid platform for the officers of the Government and the Cocoa Marketing Board, to explain vital agricultural policies to the farmers. The farmers used the same platform to bring to the notice of the Government problems that assailed them in the discharge of their agricultural activities, which were always remedied by the Government. This helped to harmonize the relations between the Government and the farmers, a situation which is vital for the development of the country's agriculture.

(xiv) Distribution of matchets: The Council used its machinery to distribute matchets to the farmers at cost price. The farmers therefore had a regular flow of this essential tool for their farming operations.

Mr Chairman, I do not mean by giving this long list of achievements of the Council that there were no shortcomings in our operations. We were a human organization, and by all means the human element in the whole system created some shortcomings here and there, but we had machinery to check them. It was also true that the Council was an integral wing of the disbanded Convention Peoples Party, but this did not in any way create contradictions between our political affiliation and our efforts to fight for the welfare of the farmers. The two cases cited by the writer to prove his case, namely, the voluntary contributions to the Second Development Plan, and the compulsory savings, were used by the Council to forestall the plan of the Ministry of Finance to reduce the cocoa price to a level which would have been much lower than what we received after this action of the Council. The top officials of the Cocoa Marketing Board could bear that out. After the 1966 *Coup,* it was discovered that salaries of our staff were much lower than those of the Cocoa Marketing Board, and even now, after having made efforts to increase the salaries of the Produce Buying Agency Ltd. staff, the salaries level is still below that of the Cocoa Marketing Board.

Finally, Mr. Chairman, I would like to state that the opinions expressed by the writer (Beckman) did not give the true picture of the affairs of the United Ghana Farmers' Co-operative Council.

Bibliography

A. Books and articles

Ady, P., "Supply functions in Tropical Agriculture", *Bull. Oxf. Inst. Econ.* 30:2, 1968.

Ahmad, Naseem, *Deficit Financing, Inflation and Capital Formation: The Ghanaian Experience 1960–1965,* Munich 1970.

Allen, Christopher and R. W. Johnson (eds.), *African Perspectives: Papers in the History, Politics and Economics of Africa Presented to Thomas Hodgkin,* Cambridge 1970.

Amin, Samir, *Neo-colonialism in West Africa,* Harmondsworth 1973 (French original: L'Afrique de l'Ouest Bloquée, Paris 1971).

—, *Le développement inégal,* Paris 1973.

Annerstedt, Jan and Rolf Gustavsson, *Towards a New International Economic Division of Labour?,* Roskilde 1975.

Apter, David E., *The Gold Coast in Transition,* Princeton 1955. Second revised ed.: *Ghana in Transition,* Princeton 1972.

Are, L. A. and D. R. G. Gwynne-Jones, *Cacao in West Africa,* Ibadan 1974.

Austin, Dennis, *Politics in Ghana 1946–1960,* London 1964.

Baron Holmes, A., "The Gold Coast and Nigeria", in W. A. Lewis (ed.), *Tropical Development 1880–1913,* London 1970.

Bauer, P. T., *West African Trade,* Cambridge 1954.

Bauer, P. T. and F. W. Paish, "The reduction of fluctuations in the income of primary producers further considered", *Economic Journal* vol. 64, Dec. 1954.

Beckett, W. H., *Akokoaso: A Survey of a Gold Coast Village* (London School of Economics, Monographs on Social Anthropology No. 10), London 1944.

—, *Koransang. A Gold Coast Farm,* Accra 1945 (Also in Beckett, *Koransang Cocoa Farm, 1904–1970,* Legon (ISSER, Tech. Publ. No. 31), 1972.

Beckman, Björn, "The distribution of cocoa income", Staff Seminar Paper No. 13, Department of Economics, Univ. of Ghana, Legon 1970.

—, "The internal marketing monopoly of the Farmers' Council", in Kotey, R. A., C. Okali and B. E. Rourke (eds.), *Economics of Cocoa Production and Marketing,* Legon 1974.

—, "The post-colonial state: autonomy and social determination", Paper to the Nordic Political Science Conference, mimeo, Uppsala 1975.

—, *Social Differentiation and Class Formation in the Cocoa Economy,* forthcoming, Uppsala, 1976.

Berg, Elliot J., "Structural transformation versus gradualism. Recent economic development in Ghana and the Ivory Coast", in Foster and Zolberg 1971, see below.

Bing, Geoffrey, *Reap the Whirlwind: An Account of Kwame Nkrumah's Ghana from 1950 to 1966,* London 1968.

Birmingham, W., I. Neustadt and E. N. Omaboe (eds.), *A Study of Contemporary Ghana,* vol. I: *The Economy of Ghana,* London 1966.

Bray, F. R., *Cocoa Development in Ahafo, West Ashanti,* mimeo, Achimota 1959.

Bretton, Henry L., *The Rise and Fall of Kwame Nkrumah,* London 1966.

Crook, Richard, "The decolonization period in Ghana: political and economic factors in Ashanti underlying NLM opposition to the CPP 1950–57", Seminar Paper, Centre of West African Studies, Birmingham 1972.

Davidson, Basil, *Black Star. A View of the Life and Times of Kwame Nkrumah,* London 1973.

DeGraft-Johnson, J. C., *African Experiment: Co-operative Agriculture and Banking in British West Africa,* London 1958.

Den Ny Verden, (Journal of the Institute of Development Research, Copenhagen) 9:4, 1975. Special issue on "U-landsteori i Østeuropa" (Development theory in Eastern Europe).

Developing Countries on the Non-capitalist Road. Proceedings of the Third International Conference of Africanist Marxists of the Socialist Countries, Varna 1971. Sofia 1974.
Dickson, K. B., *A Historical Geography of Ghana,* Cambridge 1969.
Dunn, John and A. F. Robertson, *Dependence and Opportunity. Political Change in Ahafo,* Cambridge 1973.
Edwards, Norman, *Cadbury on the Gold Coast,* Publ. Dept. Bournville, mimeo Birmingham 1955.
Fanon, Frantz, *The Wretched of the Earth,* Harmondsworth 1967 (French original: Les damnés de la terre, Paris 1961).
FAO (Food and Agriculture Organization), *Review of the World Cocoa Economy,* CCP/Cocoa/63/6, 1963.
First, Ruth, *The Barrel of a Gun. Political Power in Africa and the Coup d'Etat,* Harmondsworth 1972 (first publ. 1970).
Fitch, Bob and Mary Oppenheimer, "Ghana: End of an illusion", *Monthly Review* 18:3, 1966. Also as a book, New York 1966.
Foster, Philip and Aristide R. Zolberg (eds.), *Ghana and the Ivory Coast. Perspectives on Modernization,* Chicago 1971.
Frank, Andre Gunder, *Capitalism and Underdevelopment in Latin America,* New York 1967.
Genoud, Roger, *Nationalism and Economic Development in Ghana,* New York 1969.
Green, R. H., "The Ghana cocoa industry. An examination of some current problems", *Economic Bulletin of Ghana* 5:1, 1961.
—, "Political independence and the national economy: an essay on the political economy of decolonisation", in Allen and Johnson (eds.), *African Perspectives,* 1970. See above.
—, "Reflections on economic strategy, structure, implementation and necessity: Ghana and the Ivory Coast", in Foster and Zolberg, 1971. See above.
Green, R. H. and Ann Seidman, *Unity or Poverty? The Economics of Pan-Africanism,* Harmondsworth 1968.
Greenstreet, D. K., "Public administration: A comparative analysis of the 1950 colonial ten-year development plan and the 1951 'Nkrumah' development plan of the Gold Coast (Ghana)", *Economic Bulletin of Ghana,* 3:4 (second series), 1973.
Hill, Polly, *The Gold Coast Cocoa Farmer,* London 1956.
(Hill, Polly), University of Ghana, Economics Research Division, *Cocoa Research Series,* 1-19, Legon 1957-60 (mimeo, 3 vols.).
Hill, Polly, "Social factors in cocoa farming", in Wills, *Agriculture and Land Use in Ghana,* 1962 (below).
—, *The Migrant Cocoa-farmers of Southern Ghana,* Cambridge 1963.
—, *The Occupations of Migrants in Ghana,* Ann Arbor 1970.
Hobsbawm, E. J. "From social history to the history of society", *Daedalus* vol. 100, Winter 1971.
Hodgkin, Thomas, *African Political Parties,* Harmondsworth 1961.
Hopkins, A. G., *An Economic History of West Africa,* London 1973.
Jeffries, R., *Trade Unions and Politics in Ghana,* Doctoral thesis, London 1974.
Kay, G. B. (ed.), *The Political Economy of Colonialism in Ghana. A Collection of Documents and Statistics 1900-1960,* Cambridge 1972.
Kesse-Adu, K., *The Politics of Political Detention,* Accra-Tema 1971.
Kotey, R. A., C. Okali and B. E. Rourke (eds.), *Economics of Cocoa Production and Marketing.* Proceedings of cocoa economics research conference Legon, April 1973. Legon 1974.
Kraus, Jon, "On the politics of nationalism and social change in Ghana", *Journal of Modern African Studies* 7:1, 1969.
La Anyane, S., *Ghana Agriculture. Its Economic Development from Early Times to the Middle of the Twentieth Century,* London 1963.
Legum, Colin, "Socialism in Ghana: A Political Interpretation", in W. H. Friedland and Carl G. Rosberg, Jr. (eds.), *African Socialism,* Stanford 1964.
Lenin, V. I., *The Development of Capitalism in Russia* (1899), in *Collected Works* vol. 3, Moscow 1964.
—, "A great beginning", *Collected Works* vol. 29, Moscow 1965.
Lipset, S. M., "Social class", *International Encyclopedia of the Social Sciences* vol. 15, 1968.

Little, Ian, Tibor Scitovsky and Maurice Scott, *Industry and Trade in Some Developing Countries,* London 1970.
Marshall, Judith Maureen, *The Political Economy of Dependence. Ghana 1945–1966.* The Hague (Inst. Soc. Stud., mimeo), 1972.
Meillassoux, Claude, "A class analysis of the bureaucratic process in Mali", *Journal of Development Studies,* Jan. 1970.
Mohan, Jitendra, "Nkrumah and Nkrumasim", *The Socialist Register 1967,* London 1967.
Murray, Roger, "Second thoughts on Ghana", *New Left Review* 42, 1967.
Myint, Hla, *The Economics of Developing Countries,* London 1964.
Nkrumah, Kwame, *I speak of Freedom,* London 1961.
—, *Africa Must Unite,* London 1963.
—, *Neo-colonialism. The Last Stage of Imperialism.* London 1965.
—, *Challenge of the Congo,* London 1967.
Omari, T. Peter, *Kwame Nkrumah: The Anatomy of an African Dictatorship,* Accra 1970.
Owusu, Maxwell, *Uses and Abuses of Power. A Case Study of Continuity and Change in the Politics of Ghana,* Chicago 1970.
Rathbone, Richard, *The Transfer of Power in Ghana 1945–57,* Doctoral thesis, London 1968.
—, "Businessmen in politics: party struggle in Ghana, 1949–57", *Journal of Development Studies* 9:3, 1973.
Rimmer, Douglas, "The public sector in West Africa: a characterisation", in *African Public Sector Economics,* Proceedings of a seminar, Centre of African Studies, Edinburgh 1970.
Rohdie, Sam, "The Gold Coast Hold-up of 1930–31", *Transactions of the Historical Society of Ghana,* vol. IX, 1968.
Rourke, B. E., *Wages and Incomes of Agricultural Workers in Ghana,* Legon (ISSER, Tech. Publ. No. 13), 1971.
Rourke, B. E. and S. K. Sakyi-Gyinae, "Agricultural and urban wage rates in Ghana", *Economic Bulletin of Ghana* 2:1 (Second series) 1972.
Sachs, Ignacy, *Patterns of Public Sector Economy in Underdeveloped Economies, London 1964.*
Saul, John S., "The state in post-colonial societies: Tanzania", *Socialist Register 1974,* London 1974.
Seers, Dudley, "The stages of economic development of a primary producer in the middle of the twentieth century", *Economic Bulletin of Ghana* 7:4, 1963. Also in Robert I. Rhodes (ed.), *Imperialism and Underdevelopment. A Reader,* New York 1970.
Sloan, Pat, "Ghana without illusions", *Political Affairs,* Oct. 1966.
Southall, Roger, "The 'New' political economy: Cadbury, cocoa and the CPP", unpublished manuscr., Centre of West African Studies, Birmingham 1975.
St Clair Drake, J. G. and L. A. Lacy, "Government versus the unions: The Sekondi-Takoradi strike, 1961", in G. M. Carter (ed.), *Politics in Africa,* New York 1966.
Szentes, Tamás, *The Political Economy of Underdevelopment,* Budapest 1971.
Szereszewski, Robert, *Structural Changes in the Economy of Ghana 1891–1911,* London 1965.
Thompson, W. Scott, *Ghana's Foreign Policy,* Princeton 1969.
Wickizer, V. D., *Coffee, Tea and Cocoa: An Economic and Political Analysis,* Stanford 1951.
Wight, Martin, *The Gold Coast Legislative Council,* London 1948.
Wills, J. Brian (ed.), *Agriculture and Land Use in Ghana,* London 1962.
Wilson, Charles, *Unilever, 1945–1965: Challenge and Response in the Post-war Industrial Revolution,* London 1968.
Zolberg, Aristide, *Creating Political Order: The Party-states of West Africa,* Chicago 1966.

Periodicals

The Economic Bulletin of Ghana. Quarterly published by the Economic Society of Ghana.
Mizan. Published by the Central Asian Research Centre, London.
West Africa. Weekly, London.

B. Published official documents (Government and Party)

1. *Annual or recurrent documents*

Central Bureau of Statistics, *Economic Survey* (yearly from 1957).
Central Bureau of Statistics, *Statistical Year Book* (1962, 1963, 1964, 1965–66).
Cocoa Marketing Board, *Annual Report and Accounts* (yearly from 1948).
Ministry of Finance, *Budget Statements* (also in National Assembly, see below).
Ministry of Finance, *Financial Statements* (published with the Budget Statements).
National Assembly, Official Report (Parliamentary Debates).
Registrar of Co-operative Societies, *Annual Reports*.

2. *Documents listed in chronological order*

1938 *Report of the Commission on the Marketing of West African Cocoa*, cmd 5845, 1938 (Nowell Commission).
1943 *The Colonies: The Labour Party's Post-war Policy for the African and Pacific Colonies*, London 1943.
1944 Colonial Office, *Report on Cocoa Control in West Africa and Statements on Future Policy*, cmd 6554, 1944.
1946 Colonial Office, *Statement on Future Marketing of West African Cocoa*, cmd 6950, 1946.
1947 *Gold Coast Marketing Board Ordinance*, No. 16 of 1947.
1951 Convention People's Party, *Towards the Goal. The Manifesto of the CPP for the forthcoming election*, Accra 1950/51.
1954 Ministry of Finance, *Explanatory Statement on the Cocoa Duty and Development Funds (Amendment) Ordinance*, Accra 1954.
1955 Department of Agriculture, *Intensive Survey of the Cocoa Producing Areas of the Gold Coast*, mimeo, Accra 1955.
1956 *Government Proposals in regard to the Future Constitution and Control of Statutory Boards and Corporations in the Gold Coast*, Accra 1956.
1956 *Report of the Commission of Enquiry into the Affairs of the Cocoa Purchasing Company Limited*, Accra 1956 (*Jibowu Report*).
1958 Office of the Government Statistician, *Survey of Population and Budgets of Cocoa Producing Families in the Oda-Swedru-Asamankese Area, 1955–56*, Accra 1958.
1959 *Second Development Plan 1959–64*, Accra 1959.
1960 Office of the Government Statistician, *Survey of Cocoa Producing Families in Ashanti, 1956–57*, Accra 1960.
1960 *Population Census* (various years of publication).
1961 Nkrumah, Kwame, *The Dawn Broadcast*, Accra 1961.
1962 Nkrumah, Kwame, *Osagyefo in Kumasi*, Accra 1962.
1962 Convention People's Party, *Programme of the CPP For Work and Happiness*, Accra 1962/63.
1964 *Seven-year Plan for National Reconstruction and Development*, Accra 1964.
1967 *Report of the Committee of Enquiry on the Local Purchasing of Cocoa*, Accra 1967 (*deGraft-Johnson Report*).
1967 *Report of the Commission on Kwame Nkrumah's Properties*, Accra 1967.
1969 *Report of the Jiagge Commission Appointed to Enquire into the Assets of Specified Persons*, vol. I, Accra 1969.

3. *UGFCC publications*

The General Secretary's Annual Reports, 1959–65 (mimeo or printed).
The Farmers' Year-book 1960–61, 1961.
A Decade of Existance, 1963.
The New Farmer (periodical).

C. Unpublished documents

(See also the comments introducing the *Notes* to the chapters, above)

1. Cocoa Marketing Board, *Minutes of the Board of Directors* and other documents.
2. Letters, memoranda etc. submitted to the deGraft-Johnson *Committee of Enquiry on the Local Purchasing of Cocoa* and kept in the files of the Committee.

3. *The Headquarters' records of the United Ghana Farmers' Council,* including:

(a) Minutes and other records from district, regional or national meetings, committees, and conferences of the UGFCC.
(b) Policy statements, resolutions, circulars, speeches, messages, memoranda and other documents prepared by Council officials.
(c) Accounts, budgets, audit reports, salary scales, establishment lists and other documents referring to the financial administration of the Council.
(d) Reports and petitions from local Council officers, committees or farmers and other correspondence.
(e) Documents (minutes, reports, speeches, statements etc.) affecting the Farmers' Council produced by other government, public or party bodies, including the Cocoa Marketing Board, the Registrar of Co-operatives, various Ministries, the Cabinet, the President's Office, CPP Headquarters, local CPP branches etc.

Abbreviations and Glossary of terms

Abusa. Twi word meaning tripartite division. A system under which an abusa labourer is paid one-third of the value of the cocoa harvested. See Chapter 2, pp. 37–38.

Accumulation. Capital formation. See Chapter 1, p. 32.

Ashanti. (1) Powerful kingdom in central Ghana. (2) British colonial dependency governed as part of the Gold Coast. (3) Region in independent Ghana. Divided in 1959: Brong-Ahafo set up as a separate region. See map p. 10.

ABASCO *Ashanti, Brong-Ahafo and Sefwi Cooperative Marketing Association.* Breakaway section of the Ghana Cooperative Marketing Association set up as a separate organisation in 1960 and "merged" with the UGFCC in 1961. See Chapter 4, pp. 91 ff.

Broker. Local trader acting as an intermediary between cocoa-buying firms and the farmers. See Chapter 2, pp. 46 ff.

Bureaucracy. As a category of class, see Chapter 1, p. 31.

Buying Allowance. Payment by the Cocoa Marketing Board to its Licensed Buying Agents covering predetermined costs per ton bought plus a variable allowance for profit related to the level of the producer price. See Chapter 2, pp. 40 ff.

Cadbury Brothers, Cadbury & Fry. Birmingham chocolate and confectionary manufacturers and their subsidiary cocoa purchasing company in Accra. See Chapter 2, p. 43 and Chapter 4, pp. 79 ff.

Capsids. A crop pest affecting cocoa. A massive anti-capsids spraying campaign was launched in the late 1950's.

Caretaker. "A semi-permanent arrangement under which an owner hands over a farm on an abusa crop-sharing basis, the caretaker undertaking all the farm work and receiving one-third of the proceeds." (W. H. Beckett, *Koransang Cocoa Farm 1904–1970,* 1972, part III, p. 23). See Chapter 2, pp. 37–39.

Chief Farmer. Title used for appointed or elected farmer leaders. Used for the presidents of the village, district, regional and national representative committees/councils of the UGFC. See Chapter 3, pp. 54–56, and Chapter 6, pp. 147 ff.

CMB *Cocoa Marketing Board* (Gold Coast, Ghana, State). Established in 1947 with responsibility for the marketing of Ghana's cocoa crop, the regulation of producer prices, and the provision of certain services to the producers (crop pest control etc.). See Chapter 2, pp. 40 ff.

CPC *Cocoa Purchasing Company*. A subsidiary of the CMB, established in 1952 to buy cocoa and grant loans to cocoa farmers. Liquidated in 1957 after a public inquiry: the Jibowu Commission. See Chapter 3, pp. 59 ff.

The Colony. The southern provinces (Eastern and Western) of the Gold Coast dependency (later the Central, Eastern and Western regions). See map p. 10.

CPP *Convention People's Party*. Political party founded by Kwame Nkrumah in 1949. Banned and dissolved in 1966.

DeGraft-Johnson Committee/Report. Committee appointed in 1966 by the National Liberation Council to enquire into the local purchasing of cocoa (to propose a system to replace the UGFCC monopoly). Reported to the NLC in July 1966. The Report was published in 1967. See Chapter 5, p. 110.

DC *District Commissioner*. Local administrative officer, usually with a party background, appointed by the CPP government.

Gammalin. Insecticide used for spraying against capsids.

GCMA *Ghana (Gold Coast) Cooperative Marketing Association*. The "old" cooperatives, the largest Ghanaian buying organisation in the pre-independence period. "Merged" with the UGFCC in 1961. See Chapter 4, pp. 91 ff.

The Gold Coast. The colonial name of Ghana. At one point it referred only to the coastal provinces.

Jibowu Commission/Report. Commission appointed in 1956 to enquire into the affairs of the Cocoa Purchasing Company. See Chapter 3, pp. 59 ff.

LBA *Licensed Buying Agent*. Firm or organisation licensed by the Cocoa Marketing Board to buy cocoa. See Chapter 2, pp. 40 ff.

Load. Standard measure used in the local cocoa trade. One load equals 60 lbs. One ton equals 37 loads and 20 lbs. Compare "bag", which is a measure used in the transport of the beans to the ports and for shipments. Standard weight of a bag varies but it roughly equals $2^1/_2$ load.

National Delegates' Conference. The annual "congress" of the UGFC.

NLC *National Liberation Council*. The military and police government established after the February 1966 coup. It was replaced by an elected government in 1969. This was again overthrown by the military in 1972 (the National Redemption Council).

NLM *National Liberation Movement.* Ashanti-based political party formed in 1954. It merged in 1957 with other opposition parties into the United Party. See Chapter 3, pp. 65 ff.

Non-capitalist/development. Term used primarily in Communist Marxist literature referring to political-economic systems or measures in third world countries which point in a socialist direction. See Chapter 1, pp. 23–24, 32.

Petty-bourgeoisie. Term used to designate a stratum of small-scale private entrepreneurs within a framework of a commodity producing (market-oriented) and labour-hiring economy. See Chapter 1, pp. 30–31.

Producer price. In this context the fixed price paid to the farmers at stipulated buying points, determined by the government on the recommendation of the Cocoa Marketing Board. Usually fixed on a seasonal basis. See Chapter 2, pp. 40 ff.

Secretary-Receiver. Clerk in charge of the local (village) buying operations of the cooperative societies and the UGFC local branches.

Society. The local unit of a cooperative organisation. Used by the UGFC to designate its local branches.

State capitalism. See Chapter 1, pp. 31–32.

Stool. Symbol of chieftaincy in southern Ghana (compare "skin" in the North). To *enstool* or *destool* a chief mean therefore to elect or depose a chief.

Supreme Council. The governing body of the UGFC. See Chapter 6, pp. 147 f.

Swollen shoot. Virus disease attacking cocoa. Rampant in the late 30's and the 40's, especially in the Eastern Region. A cutting-out campaign was launched in the late 40's.

UAC *The United Africa Company.* A general trading, plantation, and later, industrial subsidary of Unilever with a dominant position in the West African colonial economy. See Chapter 2, pp. 43–44, Chapter 3, pp. 81 ff.

UGFC/ UGFCC *United Ghana Farmers' Council/Cooperatives/Cooperative Council.* Established in 1953 as a general farmers' organisation supporting the CPP. The Council created a cocoa buying organisation (the Ghana Farmers' Marketing Association) in 1957. In 1961, the UGFCC was made the sole licensed agent of the CMB. The organisation was banned and dissolved in 1966.

UGCC *United Gold Coast Convention*. Political organisation formed in 1947 under the leadership of Dr. J. B. Danquah and others. Dissolved in 1952.

UP *United Party*. Political party formed in 1957 as the result of a merger between the NLM and other opposition groups.